Contents

Part IV. Growing Animals

Fifty Years of Biodynamic Farming:
Essays From the Field

Macrocosm

Seedtime and
harvest shall never cease
— *Genesis 8, 22*

Fire Microcosm Air

ETHERIC FORCES

Oekonomeia:
Stewardship Imitating Nature
—*Aristotle*

HEALTH

Spiritual
Physical
Financial
Ecosystem
Community
Animal
Plant
Soil

Quinta
Essentia

Healing the Earth
through Agriculture
—*Steiner*

ASTRAL FORCES

Earth Water

The Economics of Sacred
Agriculture

A BIODYNAMIC
S&S HOMESTEAD FARM
WHOLE FARM ORGANISM

S&S Homestead Farm • Lopez Island, WA 98261 • www.sshomestead.org

Henning K. Sehmsdorf

Printed by Applied Digital Imaging, Inc, Bellingham, WA

Every eternal idea that becomes your ideal, awakens life force in you.
(Rudolf Steiner 1914)

Part VII. World View & Holistic Science

Part VIII. Works Cited

Index

Preface

This book grew from the wish to provide a record of the work done on S&S Homestead, a small, family-owned biodynamic farm on Lopez Island, WA, established in 1970. During the first quarter century, while the owners held full-time jobs at the University of Washington, the farm mostly produced food for the family: meat, eggs, vegetables, and fruit. Farm work was accomplished on long weekends, during summers, and several unpaid leaves taken over the years. By the time the youngest son was eighteen, the owners gave up the security of salaried university positions and moved to Lopez Island. Henning took up full-time farming, while Elizabeth assumed a position at Lopez Island High School. She split her time between teaching at the school and working on the farm, including a designated farm-to-school program which brought students to the farm several times a week to learn about ecological food production.

From the start, the farm sought a role in research and teaching as well as in producing food. The questions foremost in our minds concerned nutritional wholeness, ecological stewardship, island self-sufficiency, and community resilience. The farm adopted biodynamics as the method best suited to accomplish its long-term goals. These were articulated in a fifty-year plan describing the quality of life values, forms of production and future resource base to be implemented over time. Henning made contact with a group of soil scientists and agronomists at Washington State University who were interested in biodynamic management. He was appointed adjunct professor at the Center for Sustaining Agriculture (CSANR), and S&S Homestead was designated a WSU Demonstration Farm. A similar arrangement with Huxley College of the Environment at Western Washington University brought students to the

farm every year. On-farm research projects, workshops, lectures at county extension venues around Puget Sound, and at local and national conferences followed. Henning joined boards of local organizations committed to environmental and social sustainability, such as the San Juan Islands Conservation District, the Lopez Island Land Trust, and currently Transition Lopez Island, a volunteer organization dedicated to the premise that in order to survive climate change, the island has to become "fossil-free by '33." Most recently, he joined the Agricultural Resource Committee, which advises the County government in protecting and restoring agricultural resources.

Early on, S&S Homestead Farm developed an intern- and apprenticeship program under the auspices of the Biodynamic Association. It brought learners from all over the U.S., and from as far away as India and Sweden. As part of the program, interns and apprentices were given the opportunity to carry out on-farm research projects which resulted in reports published on the farm website. Some of these appear in the present book. Other contributors to the book include university researchers, extension agents, and representatives of faith communities dedicated to environmental stewardship and advocacy.

Chapters

This book is organized in seven chapters. The first contains a single essay introducing biodynamics as a personally held, practical idea. The second chapter contains four essays describing what it means to live holistically: real food, self-sufficiency, and teaching. Chapter three explores what is involved in biodynamic plant production: pastoral methods, medicinal plants, marketing to the community through CSA, compost management and nutrient cycling, seed production, and production of grain and bread. Chapter four describes examples of infrastructure development: building a straw bale bunkhouse, two barns, a water catchment system, and a solar powered irrigation system. Chapter five considers aspects of ecological livestock raising: the difference between industrial and holistic methods, biodynamic dairying, and bee management. Chapter six explores the economics of biodynamic management: the viability of small, self-sufficient farms, home food security, the economics of ecological stewardship, marketing a philosophy, and the question of retirement on the commons. The final chapter presents the biodynamic world view and practice from opposing, but complementary, perspectives. The chapter explores qualitative science as modeled in the work of Goethe, and in the philosophical idea of teleology and its practical application to answer the question of how Spirit manifests in matter. The chapter also examines how quantitative science looks at energy use embedded in farm production, and presents two complementary reports on an on-farm research project to assess whether biodynamically managed systems producing their own inputs can be sustainable. Finally, the chapter glances at transformative changes required for U.S. agriculture to meet the challenges of the 21st century.

Contributors

The essays in this volume were authored by the owners of S&S Homestead Farm, as well as by farm trainees, agricultural researchers, and visitors to the farm. They are listed alphabetically here:

Barnett, Tanya, holds a Master's of Divinity from Vanderbilt Divinity School, focused on ecological theology and ethics, and pastoral care. She sat on the boards of Washington Toxics Coalition and the Washington Sustainable Food and Farming Network, and has worked for WSU Extension in Jefferson County. She is also the editor of *Greening Congregations Handbook* (2002) published by Earth Ministry, Seattle, where she was the program

director. She visited our farm in connection with delivering a sermon at the Lutheran church on Lopez Island on engaging religious communities in environmental work.

Bramwell, Stephen, brought a B.A. in international studies and botany from UW. In 2005, he served a 9-month combined internship at the Lopez Community Land Trust and on the farm, dedicated to "Outreach and Development." After his internship, Stephen earned a M.S. in soil science at WSU, with a thesis on "Integrating Livestock and Organic Grain Production in the Inland Northwest," 2008. As of 2009, he taught sustainable agriculture at Evergreen State College for several years before being appointed WSU Extension Director and Agricultural Faculty in Thurston County.

Carpenter-Boggs, Lynn, is a professor in the Department of Soil & Crop Sciences, WSU. In collaboration with Jennifer Reeve, a professor of organic and sustainable agriculture at Utah State U., and John P. Reganold, Professor of Soil Science at WSU, she has published studies on organic and biodynamic farm systems, compost management, and viticulture.

Gigot, Jessica, came to the farm with a B.S. in biology from Middlebury College. During her three month internship she concentrated on biodynamic small-farm systems and food production with emphasis on compost sprays and soil fertility. After completing her internship, Jessica was appointed research assistant while pursuing a master's degree in plant pathology at WSU, with a thesis on "Survival of Phytophthora Infestans on Volunteer Potato Plants." Subsequently she earned a Ph.D. in horticulture, also from WSU, followed by a Master's of Fine Arts earned at Seattle Pacific U. She and her husband grow herbs, lamb and produce on their small farm in Bow, WA, and she offers educational and art workshops through her Art in the Barn series, as well as making a name for herself as a published poet and musician.

Haden, Andrew, arrived on S&S Homestead Farm with a B.S. in sustainable agriculture from The Evergreen State College, and as a Visiting Scholar, CSANR, WSU. During his three month summer internship, 2002, he compared the performance of S&S Homestead Farm as an ecosystem with the structure and functions of mature natural ecosystems, and presented his research at the WSU Northwest Symposium on Organic and Bio-Intensive Farming. He subsequently submitted the results of his study to the Department of Rural Development and Agro-Ecology at the Swedish U. of Agricultural Sciences, and was awarded the M.S. degree. Andrew is today considered one of the nation's foremost experts in biomass systems and technology, and is the founder and president of Wisewood Energy, a design engineering firm in Portland, OR.

Hök, Johanna, came to the farm with an M.S. in pharmacognosy from Karolinska Institute, Sweden, and as visiting scholar from CSANR, WSU. Her three month summer internship in 2002 concentrated on the functions of medicinal herbs in a closed farm system. Johanna returned to Karolinska Institute to earn a Ph.D. in pharmacognosy, and she now teaches and does research at the Institute, including studies in practitioners' use of shared concepts in anthroposophical pain rehabilitation.

Lia, Barry, holds a Ph.D. in neurobiology from the U. of California, and is a researcher with interests in the epistemology of cognitive sciences, and neuroethics. He is also WA State Coordinator for the Biodynamic Association, and the owner of Lia Biodynamic Consulting, Seattle.

Murgatroyd, Lisa, earned a B.A. at Humboldt University. During her eighteen month traineeship, 2007-2008, Lisa, together with Jesse Pizzitola, implemented Lopez Community Farm CSA. She also took on the coordination of the Lopez Island Farm Education (LIFE) program, which established the farm-to-school program as a part of the Lopez Island School curriculum. After her stay on the farm, Lisa completed a master's degree in creative writing. Lisa married Jesse, and their daughter was born in 2019. They left Lopez Island, first to manage a vineyard and then to run their own organic farm in California. Lisa currently works as an organization development consultant, facilitator, and social entrepreneur. Inspired by inclusive methods that honor emergence and the complexity of today's organizations and communities, she has designed, launched, and managed pioneering social enterprises in education, food and land systems, climate change, and transportation solutions across the U. S. West.

Palmer-McCarty (now Lee), Kelley. After attending our farm-to-school class, Kelley studied at Fairheaven College (WWU). In 2009, she did a summer internship at S&S Homestead Farm, where she learned orchard management and did research on Goethean science and plant morphology. After taking a bachelor's degree in environmental science, ecology, ornithology, and art, Kelley graduated with a degree in nursing science, and is now a public health nurse in communicable disease for the Whatcom County Health Department.

Pizzitola, Jesse, brought a bachelor's degree from Humboldt University, as well as completion of a three-year apprenticeship in biodynamic farming at Live Power Farm in California, before he (together with Lisa Murgatroyd) established Lopez Community Farm CSA during his eighteen month stay on S&S Homestead Farm, 2007-2008. Called back to California to manage his mother's vineyard, he later ran his own organic farm near Petaluma.

Prime, Katy, came to the farm with a B.A. in anthropology and focused her six month internship in 2005 mostly on the relationship between nutrition and human health. After the internship, Katy attended the U. of Idaho and earned a M.A. in sociology. She married Stephen Bramwell, and their daughter Mae was born in

2013.

Questad, Jenelle (now Kvistad), in 2001 came to the farm with a B.S. in Nutrition from Bastyr University to pursue a three month summer internship concentrated on sustainable food production and nutrition. She later married farm apprentice Brian Huntington (now Kvistad), and together with him developed the organic food store Blossom Foods on Lopez Island. They are the parents of two children.

Reeve, Jennifer, is Associate Professor of Organic and Sustainable Agriculture at Utah State U. In collaboration with Lynn Carpenter-Boggs and John P. Reganold, both professors of crop and soil science at WSU, she published studies on organic and biodynamic farm systems, compost management, and viticulture.

Reganold, John P., is Professor of crop & soil science, WSU. In collaboration with Lynn Carpenter-Boggs and Jennifer Reeve, he published studies on organic and biodynamic farm systems.

Research Council of the National Academies: Committee on Sustainable Agriculture Systems in the 21st Century (J. P. Reganold, WSU; D. Jackson-Smith, Utah State U.; S. S. Battie, Michigan State U.; R. R. Harwood, Michigan State U.; J. L. Kornegay, N. Carolina State U.; D. Bucks, Bucks Nat. Resources Mgt.; C. B. Flora, Iowa State U.; J. C. Hanson, U. of Maryland; W. A. Jury, U. of California; D. Meyer, U. of California; A. Schumacher, SJH & Co; H. Sehmsdorf, S&S Homestead Farm; C. Shannon, U. of California; L.A. Thrupp, Fetzer Vineyards; P. Willis, Niman Pork Ranch Co.

Rural Roots and U. of Idaho Research Team: Colette DePhelps, Executive Director, Rural Roots; Cinda Williams, Extension Support Scientist, Plant, Soil and Entomological Sciences, U. of Idaho; John Foltz, Associate Dean, College of Agricultural and Life Sciences, Professor of Agricultural Economics U. of Idaho; John Potter, U. of Idaho Graduate Research Assistant, Agricultural Economics and Rural Sociology; Ariel S. Agenbroad & Karen Faunce (Professional Writing Team).

Scilipoti, Jan. Henning met Jan during a seed saving meeting in 1998. They reconnected while helping an island neighbor build a straw bale house. Jan, who holds an M.A. in interior design, was then developing the enterprise In-Shelter to make straw bale housing more widely available on Lopez Island. We agreed to host an on-farm workshop in straw bale building during the summer of 2001, with the project goal of building a low-cost bunkhouse.

Sehmsdorf, Henning, served a three-year apprenticeship and graduated from business school in Germany before earning a B.S. in philosophy of science from the U. of Rochester, NY, and a Ph.D. in the humanities from the U. of Chicago. Henning manages S&S Homestead Farm and is the Director of S&S Center For Sustainable Agriculture. He is married to Elizabeth Simpson, and together with her published *Eating Locally & Seasonally: A Community Food Book for Lopez Island (and All Those Who Want to Eat Well)*, 2021.

Simpson, Elizabeth, Ph.D., UW, is co-manager of S&S Homestead Farm, and is centrally involved in the farm's educational outreach. She is married to Henning Sehmsdorf, and together with him published *Eating Locally & Seasonally*.

Wilson, Tasha. After participating in our farm-to-school class in 2003-2004, Tasha, then a senior at Lopez Island High School, served a three month summer internship during which she participated in heirloom bean trials in collaboration with Dr. Carol Miles, plant systems specialist, WSU. Her report, "Ecological Food Production: Lopez Island Farm-to-School Project," was published in the Tilth Producers Quarterly (2004). Her senior thesis "Seeds of Life" (2005) describes her work with Dr. Miles.

Acknowledgements

I wish to thank the institutions, organizations and individuals who have encouraged me to undertake this book. First and foremost, I thank our apprentices, interns and the students from colleges, high schools and Waldorf schools who came here over the years, and spurred our learning about sustainable farming and biodynamic management.

I thank the Biodynamic Association for entrusting me with the task of mentorship. In particular, I thank Professor Gigi Berardi, recent director of the Institute for Global and Community Resilience at Huxley College, and herself a homestead farmer and passionate cheese maker, who brought her students to the farm year after year and always encouraged our educational programs.

I also thank Dr. Steven Jones, WSU grain breeder and now director of the Bread Lab in Burlington, WA who helped us develop our own wheat and rye strains grown on the farm. I thank WSU soil scientists Lynn Carpenter-Boggs, Jennifer Reeve, and crop scientist Steve Fransen for collaborating with me in a case study of farm sustainability. I thank Rural Roots and the U. of Idaho for including an assessment of the farm in their study of marketing strategies of selected farms in Idaho, Oregon, and Washington. I thank the National Research Council for including me in a committee tasked with developing strategic recommendations for transforming American agriculture to meet the challenges of rapid changes and growing ecological crises in the global economy.

I thank CSANR, County Extension, and the San Juan Islands Conservation District for years of support of on-farm workshops and research, and help in securing federal grants, and Lopez Island Schools for supporting the class "Ecological Food Production" taught on S&S Homestead for many years.

I want to thank island neighbors who have supported us in innumerable ways over the years, giving advice, buying our products, and helping in emergencies. I want to thank the Lopez Community Land Trust for envisioning holding this farm in trust for the island community in the future.

I also want to thank the families of Mary Coffey and Rafael Velasquez, who settled on the farm when I became an octogenarian and are now doing the heavy lifting, so that I can turn my remaining energies, for instance, to this book project.

Finally, I want to thank my wife and companion ("One who breaks bread with another"), Elizabeth Simpson, without whose love and patience none of my life's projects would have seen fruition.

Part I. Biodynamics

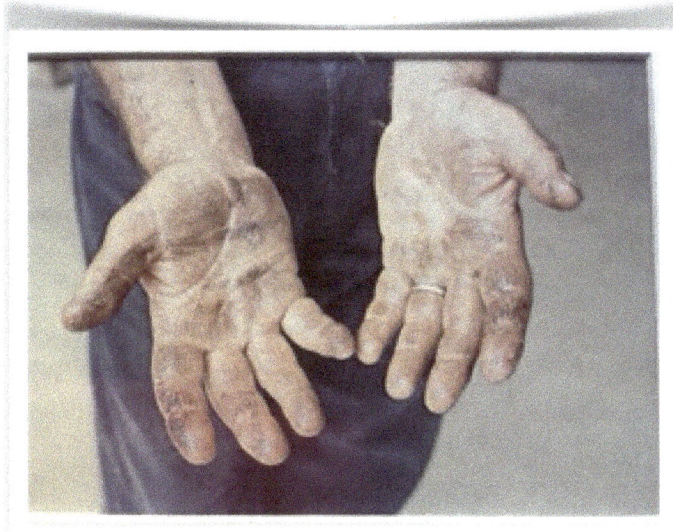

Biodynamics: Personal History of an Applied Idea

Henning Sehmsdorf, 2021

My heart leaps up when I behold
A rainbow in the sky:
So was it when my life began;
So is it now I am a man;
So be it when I shall grow old,
Or let me die!
The Child is father of the Man;
And I could wish my days to be
Bound each to each by natural piety.[1]

Intimations

Born in 1937, I grew up in economically distressed post-war Germany, living in a town near Dresden we saw flattened by firebombing on two nights in February, 1945. The bombing killed tens of thousands of civilians. My father returned home uninjured in body but broken in spirit after six years of fighting in France and Russia. My mother struggled to keep nine children alive by growing food in a large home garden and gleaning the fields of nearby farms. My youngest brother, weakened by malnutrition, died from tuberculosis. The family was forced to abandon our home and flee to the West. The three oldest brothers, including myself, were sent to a boarding school for refugee children, newly established by the Lutheran Church, where we stayed for seven years. The rest of the children were parceled out among grandparents, aunts and uncles throughout Germany.

Nevertheless, in spite of tragedy and relentless trauma, my childhood and early adolescence were seeded with experiences and impressions from which grew a deep sense of the irreducible goodness of life, and of the earth from which grow plants, animals, and nourishment, as sacred. After the bombing, my mother carved a woodcut showing a man with his hand on the shoulder of his son by a tree they just had planted. Surrounding the woodcut my mother had inscribed Martin Luther's words: "And if the world were to perish tomorrow, I would still plant my little apple tree today!"

My grandmother, who baked our bread, instilled in us the sense that this and all food is sacred. The farmer, whose fields we gleaned as children, would gather his household at the edge of the field before planting or harvesting, and say a prayer. The farmer's wife would come out to the field midday with a basket of food, unhitch the cow to milk her and give us the warm, life-giving substance to drink. Later, when I was a teenager, the school sent us to nearby farms to help with the harvest, and a portion of the harvest was sent back to feed us. In school we sang hymns that celebrated nature as the garment of the indwelling divine. One Pentecost, I witnessed a bull garlanded with spring flowers being led to a great iron ring on the outside wall of the village church where inside the sacred communion wine was stored in a shrine. Cows were tied to the ring, and the Pentecostal bull covered them, leaving me with the deep impression that sexual conception and new life were sacred too.

Those early experiences became intimations of the kind of farming I intuitively adopted as appropriate for growing my own food, which I later recognized as biodynamics. Eventually reading Rudolf Steiner, I discovered that his world view encompassed the relation of Spirit and matter in farming. In the foreword to Steiner's *Agriculture Course*, we find the pronouncement that a central product of biodynamic farming is the producer. The right farming practices will lead the farmer's Spiritual development from the merely personal to the cosmic.[2]

Rudolf Steiner

Rudolf Steiner (1861-1925) was born in a region of Hungary, then part of the Austrian Empire, and now located in Croatia. He received his scientific education at the Institute of Technology in Vienna, and earned a doctorate in philosophy from the University of Rostock, a German port city on the Baltic. His dissertation on the metaphysics of Johann Gottlieb Fichte (1762-1814) showed Steiner to be an epistemological idealist. Fichte was a founding member of the German school of thought that developed in response to the Cartesian dictum "I think, therefore I am,"

and to Kant's establishing "the faculty of reason (as the foundation) of knowledge independent of experience."[3] Fichte raised subjective consciousness to a principle of pure knowledge, arguing that the Absolute comes into actuality in thinking about sense experience. In his discussion of Fichte, Albert Schweitzer found this idea akin to the "nature-philosophies of the Brahmans, Buddha, Lao-tse, Chwang-tse, Spinoza, and the mystics of all ages."[4] Fichte confirmed for Steiner what he had experienced already in childhood, that the "unseen" world of Spirit lived within him much like the world of spatial geometry,[5] and that the "mind itself (is) the organ of perception" of both worlds.[6]

In 1880, Steiner, then nineteen years old, was given the task of writing an exposition of Goethe's natural-scientific ideas for Kürschner's *Deutsche National-Literatur* (German National Literature).[7] He quickly realized his affinity with the phenomenological approach of the great poet-scientist, and in several books on epistemology, he emphasized Goethe's transition from an experimental method (as in his work on Newton's color theory)[8] to an holistic, intuitive approach (as in his work discovering the morphology of biological archetypes — the primordial plant (*Urpflanze*).[9] For Steiner, as for Goethe, Spirit manifests in the phenomena of the natural world.[10] Many years later, in his autobiography, Steiner acknowledged that he owed "the development of his own knowledge of Spirit" largely to his work with Goethe's science.[11]

Between the beginning of the 20th century and his death twenty-five years later, Steiner produced such an enormous body of theoretical and applied work, that it is fair to say that his early death in 1925 was brought on by exhaustion, his life force spent at age sixty-four. He left behind three hundred and sixty volumes of writings, including some fifty books and more than six thousand lectures exploring Spiritual processes in human life and in the cosmos. During the same period, he applied his holistic approach to lasting innovations in many fields, from integrative medicine and science, to education, social and economic reform, architecture and the performing and visual arts.

Between 1904 and WWI, Steiner led an esoteric school in Berlin dedicated to individual Spiritual training. Around 1907, he developed a new art form he called eurythmy, a healing system of harmonious body movement to the rhythm of spoken words. In 1912, he founded the Anthroposophic Society,[12] not as a codified system of religious faith and worship, but as an "association of people whose will it is to nurture the life of the soul, both in the individual and in human society, on the basis of a true knowledge of the Spiritual world."[13] In 1913, he designed the first Goetheanum, a massive domed structure in Dornach, Switzerland. When the wooden building burned down on New Year's Eve in 1922, he redesigned it in concrete to symbolize the metamorphosis of matter through Spirit by the building's atectonic forms. The Goetheanum became the site for the Free University of Spiritual Science,[14] which eventually incorporated sections in mathematics, astronomy, medicine, natural science, social science, agriculture, pedagogy, visual arts, performing arts, literary arts and humanities, as well as a youth section. It also became the site where annually the four mystery plays written by Steiner are performed, as well as the entirety of Goethe's *Faust*.

Together with physician Ita Wegman, Steiner created a holistic, integrative form of medicine based on anthroposophical principles, and in 1921 established the first clinic, in Arlesheim, Switzerland. In 1919, with Emil Molt, the owner of the Waldorf-Astoria Cigarette Factory in Stuttgart, he founded the first Waldorf school to provide higher education for the

children of the factory workers. Following Steiner's guidance, pediatrician Karl König developed the Camphill movement to offer education for people with special needs, and in 1939 he built the first Camphill center in England.

In 1922, Steiner gave a series of lectures on social reform and "associative economics," so called because its goal was an economic system managed by associations of business, finance and consumers for the common good, instead of by the "invisible hand" of capitalist markets, or by governments as in socialist economies. In 1936, this led to the development in America of Rudolf Steiner Foundation Social Finance (RSF), "a financial services organization that offers catalytic investing and giving options, and connects social entrepreneurs with capital."[15]

In 1924, the year before his death, Steiner gave the eight extensive lectures on the renewal of agriculture now known as the Agriculture Course. As the foundation of biodynamics, the Course marks the origin of organic agriculture and its many derivatives, such as regenerative agriculture, eco-agriculture, permaculture, and other ecologically based systems.

Ecologically Based Farming Systems: Organic & Biodynamic

When I came to America as a nineteen-year old, I took a day job in a meat plant to fund attending community college at night. I was

11

horrified by what I saw at the plant: profit-driven cruelty to animals on the kill-floor, unsafe and unsanitary working conditions, and low meat product quality. It contradicted everything I had learned during my childhood and adolescence about the sacredness of food and of the earth from which it comes. I resolved that if I were to eat meat in the future, I would grow it ecologically myself. It took ten years of schooling and the security of a tenure-track position at a university before I could realize my dream of becoming a farmer, and it took several more years before I realized that the farming practices I had adopted from cultural memory were biodynamic.

The National Research Council defines ecologically based farming systems as follows:
~ *"Organic farming systems* emphasize the use of renewable resources and the conservation of soil and water to enhance environmental quality for future generations. They typically rely on crop rotations, green manures, composts, naturally derived fertilizers and pesticides, biological pest controls, mechanical cultivation, and modern technology. Organic meat, poultry, eggs, and dairy products come from animals that are not given any antibiotics or growth hormones. Organic food is produced without the use of most conventional pesticides, fertilizers made with synthetic ingredients or sewage sludge, bioengineering, or ionizing radiation. Before a product can be labeled "organic" in the United States, a government-approved certifier inspects the farm where the food is grown to make sure the farmer is following all the rules necessary to meet USDA organic standards.
~ *Biodynamic farming systems* typically use the full range of organic production practices, but also use a series of eight soil, crop, and compost amendments, called preparations, made from cow manure, silica, and various plant substances. Biodynamic farming also places greater emphasis on 1. the integration of

animals to create a closed nutrient cycle, 2. using an astronomical calendar to enhance the timing of subtle energy flows from the sun, planets, and stars at planting, cultivating, and harvesting times, and 3. an awareness of Spiritual forces in nature. Biodynamic farmers view the soil and the whole farm as an integrated, living organism and self-contained individuality. More than a production system, biodynamic agriculture is a practice of living and relating to nature in a way that focuses on the health of the bioregion, landscape, soil, plant, and human life, and it promotes the inner development of each practitioner. The Demeter Association has certification programs for food and feed produced by strict biodynamic farming methods in different countries."[16]

Outcomes of Applied Biodynamics
For most of the fifty years since its inception in 1970, S&S Homestead Farm has been managed as an integrated, self-sufficient, living organism. After some false starts using rototillers in our vegetable gardens and field plots, and observing that the soil by the end of the growing season had turned to dust and lost its vitality, I instinctively turned to the methods I remembered from my childhood. I recalled working in the family garden and gleaning farm fields, and later working on farms as a high school student. Over the years, these practices were informed by reading about biodynamics, by exchanges with other farmers, and long-term practice. The farm stresses sustainable and ecologically responsible methods. By producing most of the inputs required for growing food and fiber, the farm is able to forgo outside resources. Biodynamic management addresses the environmental, financial and social aspects of the farm, emphasizing close grower-consumer relationships, biodiversity, and a healthy, non-polluting environment.

Lessons Learned

~ The biodynamic farm functions optimally as a self-contained, individual entity.

~ Bounded eco-systems are self-organizing, self-healing, and self-correcting.

~ The farm evolves by improving the health of its soil.

~ Healthy soil protects the environment and produces high-quality crops, better feed for livestock, and better food for human beings.

~ Selective breeding of animals and plant seeds on the farm provides natural pest and disease control through site-specific plant and animal immunities.

~ Balance of plant and animal production ensures farm biodiversity.

~ Animal wastes recycled as compost nourish the plants that feed animals and humans.

~ Interplanting crops and flowers encourages pollinators and repels pests.

~ On-farm production of high protein grass and grains feeds the animals.

~ Farm-produced compost and biodynamic preparations provide sustainable fertility.

~ Use of the astronomical calendar develops awareness of seasonal rhythms, and of the beneficial role of biodynamic preparations in soil and compost management.

~ Use of biodynamic preparations in soil and compost management enhances soil metabolism and plant and animal health.

~ Ecological animal husbandry ensures animal welfare.

~ Nutritional wholeness strengthens human health.

~ Biodynamic practice builds natural and social capital before financial capital.

~ Practice of associative economics ensures long-term economic viability.

~ Biodynamic practice supports Spiritual development.

Life Force

I have always thought that *Live Power Farm,* the Decaters' biodynamic farm in California, was aptly titled because the name not only reflects that they farm with horses, but is a direct translation of Bio-Dynamics, which means "Life force." Biodynamics goes beyond organic by positing a metaphysical principle that sustains and shapes all life. By contrast, the USDA defines organics by requiring that substances used to fertilize crops or fight disease must be organic rather than synthetic.

Biodynamics, on the other hand, considers substances the carriers of vital life force, comparable to what in Polynesian tradition is known as *mana,* in India as *prana,* and in China as *chi,* and what some contemporary Christian theologians regard as the Holy Spirit, "the energy that sustains the whole universe."[17] This life force is both physical and metaphysical, material and Spiritual. To Rudolf Steiner this meant that farmers and gardeners produce not only food and fiber, but take on responsibilities as stewards of the whole earth, and for the bodily and Spiritual health of plants, animals and humans.

It follows that a biodynamic farmer thinks of soil as a living organism whose life force must be nourished and strengthened, rather than as a physical medium holding up the plants grown in it. Biodynamic management involves intuitive and meditative practices as much as scientific observation and technical applications. In contrast to conventionally produced and processed foods, biodynamic foods are rich in vital energy beyond chemically identifiable nutrients. Steiner emphasized that our diet not only determines our physical well being, but also strengthens or stunts our Spiritual development.[18] He predicted ecological disaster and world-wide ill health, unless agriculture returned to the ancient perception of the earth as sacred. Today these sentiments are widely echoed in the writings of such ecologists as Rachel Carson, and farmer-poet Wendell Berry, and indeed in the findings of the U.N. Panel of Climate Change, and of scientists throughout the world.

Steiner was a self-declared clairvoyant possessed of a mystic temperament, even though he grew up in a non-religious family, and during his childhood and youth had no formal relationship to the Church. At age thirty-eight, he had a visionary experience he described as "meeting Christ." The experience "culminated in my standing in the Spiritual presence of the Mystery of Golgotha in a most profound and solemn festival of knowledge."[19]

Analytical psychologist C.G. Jung (1875-1961), saw in Christ the archetype of the human Self. He would have characterized what befell Steiner as a numinous experience, the "seizure of the individual by the Holy Spirit,"[20] an experience commonly described in Christian tradition as "being saved." Jung held that the experience amounts to the precipitation of the unconscious archetype of the Self into consciousness. It is activated by personal or historical circumstances, but its final cause is the unconscious image of wholeness prefigured in every human being.

Characteristically for Steiner, however, he looked for the promised Second Coming of Christ not as a physical presence in the historical world, but as an "etheric" manifestation always standing behind all matter.

The ether as manifestation of indwelling Spirit is an ancient idea on which Aristotle's cosmology is based. It found its equivalent in Spinoza's concept of God's coming into being in ongoing natural processes (*natura naturans),* and in Goethe's concept of Spiritual teleology shaping organic development in plants and animals (*Urpflanze).* Steiner's thinking about the ether aligned with Renaissance scholars Giordono Bruno and Agrippa von Nettesheim. They posited that God created the world through a series of emanations, proceeding from pure Spirit to the world soul to the etheric forces crystallizing in the physical world of matter, organic and human life.[21] For Steiner, Christ was present in the here and now of the world, a perception shared by many biodynamic farmers. This is how I experience the natural world, an experience shaping daily farming practice in countless ways. More than anything else, biodynamics is a practical idea, and its truth is verified in ecological sustainability, economic stability, personal and community health, and in the well-being of domestic and wild plants and animals on the farm.[22]

Notes

_____

[1] William Wordsworth 1802. "The Rainbow."

[2] Steiner, Rudolf 1993. *Spiritual Foundations for the Renewal of Agriculture (The Agriculture Course* held at Koberwitz, Silesia, June 7-16, 1924). Junction City, OR, *ix*.

[3] https://en.wikipedia.org/wiki/Immanuel_Kant#Theory_of_perception. Retrieved November 10, 2021.

[4] Schweitzer, Albert 1960. *The Philosophy of Civilization*. New York, NY, 206.

[5] Steiner, Rudolf 1925. *Story of My Life*, chapter I. https://wn.rsarchive.org/Books/GA028/English/APC1928/GA028_c11.html. Retrieved November 10, 2021.

[6] Ibid, chapter XI.

[7] Volume 33, 1884, I – XIV. Later published as: Steiner, Rudolf 2000. *Nature's Open Secrets : Introductions to Goethe's Scientific Writings,* London, UK.

[8] Cp. Lia, Barry & Henning Sehmsdorf 2008 (2021), "Goethe's Color Theory on the Farm," below.

[9] See Kelley Palmer-McCarty 2000, "Biodynamic Farming & Goethe's Concept of Plant Morphology," below.

[10] For further discussion of the place of Steiner's thinking in the history of philosophy, see Sehmsdorf, Henning 2016, "The Spirituality of the Soil: The Idea of Teleology from Aristotle to Rudolf Steiner," below.

[11] Steiner 1925, op. cit. chapter XII.

[12] Anthroposophy means "Wisdom of Man."

[13] Statutes of the General Anthroposophical Society, Article 4. Quoted in Johannes Hemleben 1975. *Rudolf Steiner: A Documentary Biography*. Rye, UK, 142-48.

[14] The term "Spiritual Science" is a translation of German "Geisteswissenschaften," which strictly speaking means the humanities, i.e. the disciplines studying the different aspects of culture, intellectual and Spiritual life, including the esoteric. https://sshomestead.org/wp-content/uploads/What-does-Rudolf-Steiner-mean-when-he-employs-the-term-Geisteswissenschaft.pdf. Retrieved November 10, 2021.

[15] https://rsfsocialfinance.org/. Retrieved November 10, 2021. For a discussion of associative economics on the farm, see Henning Sehmsdorf 2021. "Farming for Health: The Economics of Stewardship," below.

[16] National Research Council of the National Academies 2010. *Toward Sustainable Agricultural Systems in the 21st Century*, Washington, D.C., 21.

[17] Rohr, Richard 2019. *The Universal Christ: How a Forgotten Reality Can Change Everything We See, Hope For, and Believe*. New York, NY, 100.

[18] Steiner, Rudolf 2009. *Nutrition: Food, Health and Spiritual Development*. London, UK.

[19] Steiner, Rudolf 1923. *What Is Anthroposophy? 3 Lectures*. Dornach, Switzerland, 5.

[20] Jung, C. G. 1958. *Psychology & Religion: West and East*. Princeton, NJ, 149.

[21] Storl, Wolf D. 2013. *Culture & Horticulture*. Berkeley, CA, 59ff.

[22] For further discussion, see Sehmsdorf, Henning 2016. "Spirituality of the Soil: The Idea of Teleology from Aristotle to Steiner," below.

Part II. Living the Holistic High-Life

Real Food On The Farm

A Workshop Presenting Nutritionally Vital Foods That Promote Health and Healing

Katrina Prime & Stephen Bramwell, 2005

The farm in May is bustling with the vigor of spring; warm soil, vegetable seedlings, fruit and flower buds abound. At S&S Homestead on Lopez Island, we thought it would be the perfect time to host a workshop. We wanted to celebrate all of the food available on the farm in May, and to identify basic principles of eating well, such as eating seasonally, locally and traditionally. We think of this as eating "real food," a process that allows deeper connections between quality of food and quality of life. This workshop emerged from the fact that, increasingly, it is difficult to overlook the poor state of human health in the United States. But while a health crisis seems apparent, nutritional advice has become so varied and sporadic that one can hardly make heads-or-tails of it. Our workshop asserted that small-scale food production and the right methods of food preparation and preservation offer alternatives to conventional agriculture and conventional medicine, neither of which adequately promotes health in this country.

Held on Saturday, May 14, the workshop began with two presentations. Physician Roy Ozanne discussed the fact that today serious degenerative diseases such as obesity, childhood diabetes, cardiovascular disease, cancer, ADD and depression are becoming commonplace. Dr. Ozanne chalked these up as the unintended by-products of industrialized agriculture, which uses poisons, synthetic fertilizers and intensive refining in lieu of nutrient-rich soil and traditional food processing.

Dr. Ozanne then discussed the benefits of traditional foods as eaten by our ancestors. "If you give the body what it needs, it will build a beautiful temple for us for a lifetime," he said. "If you give the body an inadequate diet, you'll develop all kinds of problems." He identified what the human body needs: chemical-free whole foods, raw dairy, saturated fat found in animal products, and even small amounts of raw meat!

The second speaker, Lynn Parr, a nutritional consultant, focused on the mechanics of digestion and the value of enzymes. Parr identified the three causes of weakened immune systems as "empty" food, toxins, and poor digestion. Her solution was to "re-mineralize, re-enzymize and re-bacterialize" our farms, our foods and our bodies.

For the producer, this translates into non-chemical agricultural practices; for the consumer, fresh, local, nutritionally whole and fermented foods. We welcomed Parr's distaste for the current anti-bacterial craze and noted a very real, if comical, correlation between the sterility of many modern kitchens and sterile soil.

Elizabeth Simpson and Katrina Prime followed Parr's introduction to the beneficial role of bacteria and enzymes with a presentation on lacto-fermentation of dairy products and vegetables on the farm. They demonstrated how to make butter, yogurt, raw milk cheese, and a variety of fermented vegetables, explaining that pre-industrial, traditional cultures relied on these processes in the absence of refrigerators and modern canning methods.

Many of the foods described and consumed at this workshop were foods eaten for centuries by our ancestors. We learned that their value is tried and true, earning the designation "real." They are healthy not because science has told us so, but because humanity has survived on them in good health for countless generations. They are not a fad diet; they are as real now as they have always been. Simpson repeatedly quoted Sally Fallon, who popularized Weston A. Price's research on the benefits of traditional diets in her book *Nourishing Traditions*. Fallon argues that human culture has long been tied to bacterial culture. If you think otherwise, try to forego cheese, yogurt, wine, beer, and many other culinary pleasures. In

other words, fermentation not only keeps us healthy; it is an essential part of human culture. That we are losing, i.e. forgetting, these processes is matter for grave concern.

With this in mind, Simpson and Prime went on to say that bacteria introduced through fermentation enable the body to absorb nutrients just as soil bacteria enable plants to absorb nutrients. Fermented foods improve digestion, have higher nutrient contents and accelerate nutrient absorption. Most modern diets are sorely lacking in live cultures and enzymes. In a sense, we are no longer getting an essential life force through our foods. To demonstrate live foods, Simpson and Prime whipped a bowl of fermented cream into rich and flavorful butter, harvested quark curds drained in cheesecloth, and prepared a salad of raw rutabaga and carrot fermented in the whey left over from making the quark.

Now everyone was ready for a feast of traditional foods typically eaten on S&S Homestead in May. The meal was served outside in the sunshine, in a pasture bordering the woods. And in the memorable words of one attendee, "the most interesting banquet I have ever been to," was about to begin.

We served the meal in five courses:
~Appetizers: Clear broth with dropped egg, herbed French bread and butter; *Gravlaks* (salmon fermented Norwegian-style), green salad, whole wheat bread, and *Siegerrebe* (a German grape grown at a Lopez vineyard).
~Main dishes: *Boudin blanc* (white sausage made from chicken breast, pork fat and spices) with sauerkraut, potato salad, and organic beer (from a Lopez brewery); *steak tartare* (raw minced steak topped with chopped onion, a raw egg, salt, pepper, capers served on sourdough rye bread; *Fenalår* (leg of lamb fermented Norwegian-style), pork liver paté, and beef tongue aspic, together with fermented carrot and rutabaga salad.

~Desserts: milk and egg custard, strawberry-rhubarb pie, pear tart.

As each course was served, Henning Sehmsdorf provided a thorough description of the foods, explaining how they were grown on the farm, and the preparation and health benefits of each. Participants sated themselves on food, drink and the bucolic environment. Everyone was provided with a thirty-page description of fifty foods available in May on the farm, including items not on the menu, written by farm intern Katrina Prime.

Some of the foods people found unusual included the fermented salmon and leg of lamb. Many cultures eat raw or fermented fish and meat. Traditionally, the salmon is prepared by filleting and spicing the fish, and burying it in the ground for two to three days. In the modern version, a refrigerator works nicely. Cured leg of lamb is another traditional north European food. After a two-week salt lather, a two-week brine bath, and minimally three-month of dry curing, it is thinly sliced and served with dark bread, cheese and beer. Curing raw meat saves energy, enhances nutritional value by preserving enzymes, and puts us in touch with valuable cultural traditions. They are the epitome of self-sufficiency.

Aspic, another traditional food, is the gelatinous product of boiled pig's head and pig's feet encasing cooked sliced meat, such as beef tongue, hard-boiled eggs and vegetables. For people wanting to waste nothing, aspic is the ultimate embodiment of using the whole animal snout to tail. Unlike most cooked foods, gelatins attract water and ease digestion.

Steak tartare is an example of the ubiquity of raw animal protein found in every traditional food culture the world over. Of course, eating raw meat makes it imperative to know where the meat is from and how the animal was raised, a precaution that is valuable for any food.

Fermented vegetables include everything from cabbages to root crops. Cultures around the world have harnessed microbial action to preserve foods through lacto-fermentation. As with cured meats, fermenting raw vegetables enhances nutritional content as opposed to preparation methods that damage it. Live fermented vegetables are largely missing from American diets, depriving the digestive tract of enzymes that break down foods.

The day was wrapped up nicely — digestion in action -— as Henning led a strolling tour of the farm. S&S Homestead Farm raises food in accordance with biodynamic principles that bring vital life force to the soil and everything that grows in it. Addressing unresolved tensions between conventional, organic and biodynamic, Henning argued that biodynamically grown foods are rich in vital energy and not just in chemically identifiable nutrients. By way of demonstrating this, Henning introduced the relationship between human health and farm health. He started by showing the baby chicks and the steaming compost piles, appropriate places to begin, as these represent the beginning and cyclic renewal of life on the farm.

His discussion went on to blend philosophy and practice in demonstrating cyclical versus linear approaches to farm management. For example, Henning talked about water cycles, a process nature accomplishes through oceans, atmosphere and ecosystems. He demonstrated how S&S Homestead Farm harvests rainwater off rooftops, channels it to a pond for storage, and then re-circulates it for use when needed.

Nutrient management is also cyclical and starts with animal husbandry. The farm employs summer rotational grazing, winter strip grazing on a sacrifice field, and builds compost piles from manure from feeding stalls, fruit and vegetable matter, chipped brush and grass clippings. Henning noted that a cyclical farm system depends on coordination and balance rather than linear inputs and outputs. For example, participants saw how the farm builds self-reliance by growing barley on the winter sacrifice field after the cattle are removed in spring. This practice utilizes a "waste" product –- manure -– and supplies an otherwise purchased input -– fertility. Animal and plant systems complement one another.

In the end, we felt the workshop was a much appreciated and yes, uncommon "banquet." Participants left understanding that eating *Real Food* begins with a few simple principles such as seasonality, traditional preparation, and healthy and holistic farming. Happy stomachs and inspired minds indicated that we had eaten well.

S&S Homestead Farm

Self-Sufficiency on a Small Family Farm
Henning Sehmsdorf, 2005[1]

The following is an edited version of a keynote address delivered at Rural Roots Small Farm Conference: "Making The Local Connection" held at the University of Idaho on March 19, 2005. The conference organizers asked me to speak about "Living the Holistic High-Life: Self-Sufficiency on a Small Family Farm." At first, the suggested title seemed rather ambitious, even hyperbolic; but actually, it is well-chosen. The term holistic refers to the idea that reality is made up of organic or unified wholes that are greater than the sum of their parts. A self-sufficient, biodynamic farm is an excellent example of such an organic whole. And what is "high life?" To me it means, first of all, health in its broadest sense: healthy people, healthy animals, healthy soils, a healthy environment and a healthy community. "High life" also stands for meaningful work that is its own reward because it is interesting, complex and challenging, all of which a small, integrated farm offers. And, finally, "high life" means that you can pay your bills, i.e. that your self-sufficient farm is economically viable and can support your family.

S&S Homestead Farm on Lopez Island off the northwest coast of Washington State, is a fifty-acre family farm where we raise our own food and food for others in the island community, in accordance with biodynamic processes. We believe that: ~ everyone should be able to eat healthy foods produced on local farms; ~our environment can be strengthened by sound agricultural practices: ~ people should be aware of how and where the food they put in their bodies is raised; and ~ young people should learn to produce food and live sustainably.

Our farm satisfies most of its own needs. Sheep, cows, pigs and chickens produce the manure that fertilizes the gardens and orchards; in turn, crops from gardens and fields are fed back to the animals. We use rotational grazing, which means that our cows and sheep are eating high protein and enzyme-rich grasses and legumes from fields they fertilize as they graze. We move the animals often from pasture to pasture, so that they do not suffer from parasites. Living on the same farm all their lives, they have developed natural immunities and don't need medication.

In nature, different kinds of plants grow in the same space, to their mutual benefit. Following that model, we interplant fruit, berries, vegetables, herbs and flowers so that pollinators come and pests stay away. We keep gardens and fields fertile with composts and cover crops, to make the micro-life of the soil as lively and varied as the plants that grow in it. Our vegetables, fruits and animal products are nutritionally whole and delicious. We do not use any herbicides or pesticides, and our fertilizers are produced on the farm. We think that our farm: ~ is sustainable: it will be able to produce food forever; ~ is self-sufficient: we produce beef, lamb, pork, chicken, eggs, and dairy products as well as fruits and vegetables, using fertilizers and other inputs produced on the farm; ~ is economically viable: sales of meat, eggs, dairy products and vegetables bring in enough income to provide for farm expenses, and infrastructure, and ~ is ecologically responsible: the sun is the source of energy harvested through plants, animals, and an intricate solar powered rainwater irrigation system.

Let us explore two related topics — holistic health and holistic economics, and then end by suggesting some ideas on a holistic future for S&S Homestead Farm and farming in America in general.

Holistic health

Everything my wife, Elizabeth, and I produce on our farm is grown organically and, in the last few years, biodynamically. Needless to say, organic farming was around long before it came to be called that. In fact, until the 1860s when German chemist, Baron Justus von Liebig, laid the groundwork for artificial fertilizers by identifying the importance of nitrogen, phosphorus and potassium (N, P, K) to plant growth – until then, all of agriculture was organic and mostly depended on animal manures for soil fertility. In some parts of the world, as in Japan and China, mulching and recycling animal and human wastes had kept the same fields fertile for over forty centuries. What Liebig and his successors did not understand, of course, was that, while the application of chemical fertilizers produces bumper crops in the short run, in the long term water-soluble chemicals kill the micro-life in the soil, thereby destroying its life-giving vitality. By the 1920s, about the same time when chemical agriculture became widespread in the U.S., soils

in some parts of Europe were already showing signs of accelerating erosion and falling productivity due to the application of chemical fertilizers. Farmers in eastern Germany turned to scientist and philosopher Rudolf Steiner, who created the biodynamic method of gardening and farming to restore soil health by improving the humus content. Biodynamics is thus the oldest form of what is now known as organic farming, and in Europe it is still the dominant mode of organic food production.

Biodynamics goes beyond organic, however, by involving an etheric element that cannot be measured by mechanistic science but is essential to the vitality of soils and therefore to human health. The USDA today defines organics mostly by requiring that substances used to fertilize crops or fight plant disease must be of organic rather than synthetic origin. Biodynamics, on the other hand, considers soil substances agents of life force or energy comparable to what in Polynesian tradition is known as *mana*, in India as *prana*, and in China as *chi*. In the eyes of Rudolf Steiner, the presence of this all-important life force means that farmers and gardeners are not merely producers of food and fiber but have responsibilities for the health of the earth, plants, animals and humans. Biodynamic farming does not consider the soil and everything that grows in it exploitable resources, but rather living organisms whose life force must be nourished and strengthened. In contrast to conventionally or even organically produced and processed foods, biodynamic foods are rich in vital energy and not just in chemically identifiable nutrients.

The ideal unit in which to practice biodynamics is a garden or small farm, where plant and animal organisms above and below ground support each other in a self-renewing and self-regulating whole. Such a unit most likely produces more than it needs to sustain itself and so is able to feed the local community.

In describing biodynamics, Steiner combined non-Cartesian metaphysics with traditional European knowledge. Here I want to tell you about a personal experience that helped me understand biodynamic thought and practice long after the event. I was a child during WWII in Germany, and in 1945 my mother's parents came to live with us, escaping the bombing in nearby Dresden. My grandfather, a physician, was a kindly man who always had time for the nine children in the house, but he had a peculiar dietary habit that fascinated me as a child (I was eight years old then), but that I didn't understand until much later. At mealtime, *Opi* would carefully nurse a residual morsel consisting of all the best on the plate, say meat, potatoes and vegetables. This he called the *Machtbissen* (power bite), suggesting that the nourishing energy of the whole meal was somehow vested in that last bite.

Many years later, when I was a student at the University of Oslo, learning about pre-industrial belief systems and material culture in northern Europe, I was surprised to encounter the same notion of the "power bite." It was even called by the same term in Norwegian, *maktbiten, makt* meaning "force of nature." In oral traditions and customs shaping work practices on the farm and at sea, this force was sometimes represented as energies of human thought or feeling (*hug*). At other times, the "powers" (*maktene)* were imagined as various nature beings upon which a farm's success or failure, health or disease, and the general prosperity of individuals and the whole community, depended.

Looking back on my childhood experience I came to realize that my grandfather's mealtime custom reflected an ancient holistic perception of the vital, life-giving energies of food as represented by that last "powerful" bite (*pars-pro-toto*). This concept survives today in ethnomedicine throughout the world, and in the practice of homeopathy. It also underlies

the biodynamic practice of applying minute doses of fermented herbal preparations to the soil to aid in the humus-forming process. I do not know, of course, whether my grandfather's food custom was seriously meant or merely intended to tease out the child's imagination. It certainly worked on my imagination in thinking about the biodynamic concept of the life force coursing through the sun, earth, plants, animals and humans, and how that energy is carried in feed for animals and food for people.

I have been fascinated to read about how medical scientists and nutritionists look at the vital energy in foods that are particularly rich in enzymes. Dr. Edward Howell, based on pioneer research he did in this field from the 1930s–1980s, came to the conclusion that enzymes are not merely protein catalysts without which no metabolic process in the body would be possible, but carriers of vital energy: "Enzymes are much more than catalysts. . . (Rather, they) are charged with energy factors just as a battery consists of metallic plates charged with electrical energy."[2] The term "energy" comes from Greek *energeia* and simply means "work." In physics, energy is the capacity to do work or produce change. Heat, light, sound, electricity and chemical processes are all forms of energy. But let's think about this for a moment in terms of dynamic processes on the farm. When I pour diesel into the tank of my tractor and turn the ignition, the heat of exploding carbon molecules drives the pistons that turn the wheels and pull the plow. Or, when the sun hits the solar panels on my rainwater irrigation pump, excited electrons in the panels and the wires drive the motor that moves the water out of the pond to the field. That's the energy of the sun at work in the diesel fuel and the electricity.

But when the sun hits the blades of grass and leaves of clover, solar energy is converted in the chlorophyll into sugars that nourish the forage plant and the soil organisms living at the plant roots. The same sugars feed the dairy cow, who ingests the grass and clover and turns the energy into delicious milk rich in protein, vitamins and enzymes to nourish her calf and the lucky people who drink her milk. When the sun shines on the cow, her skin converts the solar energy into essential vitamin D. In these instances solar energy is converted to plant, animal and human use through biochemical processes. But in all cases we are talking about energies of the sun.

The point worth making is that normally we understand the life-giving energy of the sun solely in terms of physical and chemical, i.e. material, manifestations, which is the subject matter of science. What we usually don't think about is solar energy in metaphysical terms, i.e. beyond its myriad physical manifestations. That is, I believe, what the Chinese mean when they talk about the life-giving energy flows they call the *chi*. It is what Rudolf Steiner refers to when he talks about the ethereal energies flowing behind and through the entire cosmos, the sun, the earth, plants, animals and humans.

I am aware that most people are made uncomfortable when a farmer talks about metaphysics or Spirituality in relation to human health, food production, soil management or animal husbandry. For the last 2,000 years Judeo-Christian tradition has drawn a separation between matter and Spirit. In the seventeenth century Sir Isaac Newton and René Descartes established the concept of the physical world in purely mechanistic terms without which Western science and technology would be unthinkable. This division, however, has never existed in Eastern tradition, or, for that matter, in pre-scientific folk cultures of the West. This leads to the puzzling situation where, for example, *tai chi* and acupuncture are widely practiced in Western countries — and even endorsed by health care plans — in spite of the fact that Western science cannot explain

what *chi* is, or describe the meridians by which this mysterious energy flows through the body.

In Rudolf Steiner's holistic perception health is fundamentally based in Spirit, which to him is equivalent to cosmic energy. American biologist and writer, Barbara Kingsolver, makes the same connection while acknowledging that it is mostly ignored in contemporary society: "Modern American culture is fairly empty of any suggestion that one's relationship to the land, to consumption and food, is a religious matter. But it's true: the decision to attend to the health of one's habitat and food chain is a Spiritual choice. It is also a political choice, and scientific one, a personal and convivial one."[3].

Of course, it finally doesn't matter whether a farmer accepts Dr. Steiner's philosophy, but the practice of the holistic principles of agriculture espoused in biodynamics clearly leads to healthier soils, animals, foods, and people. So what does this mean in terms of holistic health and, quite specifically, how we grow and process our food, and care for plants, animals, soils, and the environment on our farm?

Everybody would probably agree that there is something fundamentally wrong with our current industrialized food system dominated by a handful of international corporations. The newspapers are full of reports about the obesity epidemic and about skyrocketing rates of cancer, cardiovascular disease, childhood diabetes, ADA and ADHD in school children, and on and on.

The other day I was talking to the Director of Quality Assurance and Food Safety for the Seattle office of Food Services of America. To him, food safety meant sanitary standards in food production and distribution, making it possible to ship foodstuffs globally so that folks could enjoy Kobe beef from Japan, strawberries from Chile, and lettuce from Arizona in any season. Quality meant that the apples from New Zealand had no spots on them, the tomatoes from Israel were deep red in color, and the pork chops from Iowa were all of the same size. In other words, "safety" is defined in terms of pathogen control, and "quality" as cosmetic perfection. However, for all the commitment to cleanliness and apparent perfection, foodstuffs offered by the industrialized food system cannot nourish us because they are lacking in life force. Nutritional wholeness is sacrificed on the altar of a low price, convenience and choice.

A holistic view of health clearly requires a different approach, starting with the soil. At our biodynamic farm: ~ we treat the soil with minute homeopathic preparations made from fermented herbs grown on the farm to sustain the soil microorganisms which enable the transfer of soil nutrients to the plants; ~ we compost all farm wastes and treat composts with the same biodynamic preparations to fix nutrients and prevent their loss to leaching; ~ we minimize tilling the soil because a soil already enlivened with microorganisms, bacteria, fungi, earthworms and other soil inhabitants is naturally friable and capable of holding air, water and nutrients; ~ we control pests and plant disease by keeping the soils healthy instead of suppressing symptoms through synthetic chemicals; ~ we keep farm animals healthy and productive by feeding them nutritionally whole forages, grains and vegetables grown on the farm; ~ we feed ruminant animals — sheep, beef and dairy cattle — as nature intended, i.e. by maximizing green forages and minimizing grains to produce dairy and meat proteins that are high in omega-3 acids, cancer-fighting conjugated linoleic acid, and vitamin D, and to reduce the levels of E-coli that bedevil meat and dairy animals raised on grain-based diets in feedlots; ~ we eat the vegetables, fruits, grains, meats and dairy products as close to their natural state as possible, preserving the full complement of vitamins, minerals, proteins, phyto-chemicals,

and enzymes; ~ we do not pasteurize our milk to preserve its enzymes, vitamins, and healthful bacteria, nor do we homogenize it since homogenization interferes with absorption of milk calcium by the body; ~ we learn from the food wisdom of traditional cultures to enhance the enzyme content of food by fermenting: milk into butter, cheese, yogurt and whey; ~ we use the whey to ferment vegetables and pickles, and salt to cure meats and fish to increase their enzyme content and enhance their flavor. ~ Last, but not least, we supply neighbors and customers with as many of these foods as we can. As the term implies, holistic health means the well-being of the whole community in which we live, and not just the health of our own families and the immediate farm environment.

Holistic Economics

My wife and I are often asked two questions: "Can you make a living running a small, highly diversified and self-sufficient farm like yours?" and "Can your model be replicated by others?" I believe that the answer to both of these questions is yes. At a recent seminar on Agricultural Systems and Nutrition held at UW Medical Center, an agricultural consultant for the food industry made the blank assertion that American agriculture will survive only as long as it is profitable. Of course, everyone agrees, but what does profitability mean? For food giants like Phillip Morris, Walmart or Archer Daniel Midlands, profitability no doubt implies seven to ten figure salaries for CEOs and commensurate returns for shareholders, but it surely does not mean equal returns for farmers, farm workers, or for the communities from which the profits are extracted. For a small, self-sufficient farm, on the other hand, profitability means something entirely different.

By USDA standards S&S Homestead Farm is not even a commercial farm because the total economic value of our annual production is less than $50,000. In a typical year we produce about $15,000 in fresh and processed vegetables and fruit, a portion of which is marketed through a CSA; $12,000 in beef, pork and lamb sold on a custom basis; and $3,500 in dairy marketed through cow shares, for a total of a little more than $30,000. About sixty percent of total production is sold, the remaining forty percent is consumed by the farm household, including three or four interns. After deducting fixed production costs (such as depreciation for buildings, machinery, fences and water systems) and variable or direct production costs such as seeds, machine hire, fuel, supplies, utilities, taxes, insurance, and so on, and after factoring in the cost of our internship program, we are left with twenty-four percent in net profit, about $7,500.

By industry standards, the profit percentage is high, but the cash profit is low, but so are our living costs. Because S&S Homestead Farm produces its own food and sells more than half of it, the household nets a surplus of more than $15,000 in the food category. Because all members of the household live where they work, we have minimal transportation costs, about one-seventh of the national average. Similarly, because we live on the farm, and used our own labor to build our house long ago (and have stayed in place), our housing costs are a fraction of the national average. In fact, average price increases in housing since we built our home in 1970 net us a substantial annual gain in equity.

The comparison of health costs is particularly instructive. Just about a year ago the *New York Times* (January, 2004) reported that for the first time in history Americans are spending more on health care than on food. Forty-five percent is paid by public spending, the rest by personal insurance or out-of-pocket, about $5,984 per household annually. By contrast, our household spends minimally for medical and dental check-ups and minor consultations per year, and nothing for drugs, medicines, food supplements or other health aids, largely due (I

am sure) to the quality of the food we eat and the exercise we get in producing it. Even our entertainment budget for which the typical American household spends about five-and-a-half percent of income, is only about one-tenth of the national average.

Of course, to a degree this difference reflects personal choice. I have often wondered why we feel less in need of entertainment. Is it not possible that life on a small farm is so interesting, challenging and inherently satisfying that less entertainment (or "getting away from it all") is needed? What could be more entertaining than watching spring lambs perform their games as the sun sets, or more compelling than the drama of a calf being born, or of new life in the seed breaking forth into tender shoots in the garden or on the apple tree?

But back to numbers: In sum, what do these statistics tell us? After providing for food, housing, transportation, health care and entertainment, which amounts to seventy-five percent of average household spending, S&S Homestead Farm shows a surplus of $7,500, which is almost exactly the same amount calculated by the U.S. Department of Labor to pay for the remaining twenty-five percent of typical household spending: utilities, household supplies, clothing, personal care products, education, charity, tobacco, insurance, and pension. In other words, by these standards, the farmers on S&S Homestead do make a living. I am the first to admit that most agricultural economists would not accept the above calculations as proof of economic viability. They would argue that my wife and I are exploiting ourselves and our investments in the farm, and in a sense they are right. It is common practice in calculating profit to provide for an opportunity cost on your capital. Opportunity cost means the monetary return you could earn with the equity currently tied up in an asset if invested "for best use" (i.e.

highest possible return). Our equity in the farm is now at least $500,000. The usual formula to calculate opportunity cost is equity × two × T-Bond rate, which means that the farm would have to return at least $50,000 above cost to be considered "profitable." Furthermore, my wife and I are not paying ourselves wages, which if calculated at $20/hr at 1,800 hrs/year would amount to $36,000 (for the farmer), and another $18,000 (for the farmer's wife), for a total of $54,000 in labor costs. In sum, our farm income would have to exceed $100,000 to be considered "profitable."

So, is the farm economically viable? Obviously, that depends on how you look at it. What if my wife and I had kept our university jobs and were now making $100,000 per year or so? We could have saved our money until retirement and then bought ourselves a farm to make just enough to live on and stay out of debt, which is actually what we did except that we didn't wait until retirement.

The whole argument reminds me of passages in *Walden* where Thoreau defines "the cost of a thing" as "the amount of what I will call life which is required to be exchanged for it, immediately or in the long run." Thoreau observes that "spending. . . the best part of one's life earning money in order to enjoy questionable liberty during the least valuable part of it [meaning during retirement], reminds [him] of the Englishman who went to India to make a fortune first, in order that he might return to England and live the life of a poet. He should have gone up [to the] garret at once."[4]

Unlike the Englishman who went to make his fortune first, we started our farm venture at the very outset of our academic careers, enjoying the immediate health and economic benefits of producing enough food for the family, and gradually building the enterprise as resources and time allowed, while always following a fifty-year farm plan integrating economic viability with quality of life.

This brings me to the second question, how our model of home-based food self-sufficiency and economic viability can be replicated. There are many ways of doing it, not necessarily the way we have done it. For many years, my wife and I had two professional salaries to help pay for the purchase of land and development of the farm infrastructure. This is not unusual. I don't think I know many (if any) farmers on Lopez Island who did not bring outside income, savings or earnings from a previous enterprise, or an inheritance, to establish themselves on the land. And this was true of most agriculture in the U.S. until the growth of large-scale and government-subsidized industrial farming after WWII. Except in places where people homesteaded in isolation, traditional family farms relied on outside income to provide cash for shoes, clothing, a kitchen stove, fencing materials, while the land, livestock and buildings were mostly inherited from parents who lived and died on the farm (much as the Amish still do today). A farmer would sell butter and eggs for ready cash. His daughter would teach school in town, his son would take a winter job in a local sawmill. But the main support of the family came from the farm. In our case, for twenty-five years our outside jobs supported the farm that produced much of our food. For the last eleven years the farm has supported itself. To feed itself, a two-person household on Lopez Island would not need a complex farm like ours, nor nearly as much land as we work now.

In 1970, I started growing fruit and vegetables, chickens and rabbits on about one-quarter acre of the ten-acre piece I had bought on Lopez Island, which cost me one year's salary before taxes. (By comparison, today the same piece, if you could find it, would cost my wife about four to five years of her current half-time teaching salary at the local school). My wife and

I held full-time jobs at UW until 1994, but for almost a quarter-century our quarter-acre garden provided the family with a bounty of fresh and frozen vegetables, fruit, eggs, chicken and rabbit meat. A couple of years after I started growing our food on the farm I made a deal (you might call it an associative contract) with my neighbor. I would help him with the haymaking for his cattle in exchange for running a cow of our own with his herd. Whenever he brought in a bull, we would have a calf to slaughter a couple of summers later. All of this part-time food production not only provided us with vital and flavorful food year round, but it made a huge difference in our household budget. In 1994, after my youngest graduated from high school, my wife and I left our positions at the university, I to become a full-time farmer and she to take a half-time position at the local school. With our savings we purchased an additional five acres, leased more land from two neighbors, built a barn and other farm buildings, bought our own bull and slowly increased our beef herd, added more chickens, sheep, pigs, and finally a dairy cow, increased our vegetable production and started to grow enough to sell substantial quantities to our neighbors, and make a profit.

We follow a few simple rules to keep our farm economically sustainable: ~ we follow a fifty-year farm plan to integrate holistic quality-of-life values with economic viability; ~ we keep it small. My wife and I manage the animals, vegetables, greenhouse and orchard by ourselves; ~ we feed ourselves first, and sell the excess to the community; ~ we incur no debt; we save money for water systems or outbuildings before building them; ~ we maintain a self-sustaining system; ~ we breed and raise our animals on the farm, feeding them what the farm produces, so that they gain natural immunities from living in one place, just as we do; and ~ we minimize purchased inputs, fuels, fertilizers, or medications, thus maximizing the economic opportunities of self-sufficiency.

This brings to mind another anecdote. Last year, during a three-day workshop on biodynamics we held on the farm, a woman participant approached Elizabeth and said that the most important thing she had learned was that two people could run a self-sufficient, sustainable farm. She said, "My husband and I want to do this, but have met with only discouragement from our friends and family who say we will go broke and work ourselves to death in the process. But here I see healthy animals, beautiful and productive gardens, and that you and your husband work without haste or stress and yet make this place run like a well-oiled machine. It gives me hope that we can do the same."

In conclusion, let me suggest some ideas about a holistic future for S&S Homestead Farm, and about farming and the food system in America. First a few remarks about Holistic Planning and Management as a methodology. Quite a few years ago, I got involved in a two-year training course offered by Alan Savory's Center for Holistic Resource Management. Savory's model for testing and managing ideas during planning and implementation was developed when he farmed in Africa, but later grew into a methodology that could be applied to any field or resource base.

Savory's ideas and methods seemed strangely familiar to me, because I had thought along similar lines for a long time. But they supplied me with a new and precise language to organize my goals for the farm. Considering the "whole" under management (people, resource base and money) by relating it to an overarching goal involving quality of life values, forms of production and a vision of what the resource base – the farm – should look like in the future, seemed intuitively right and helpful. Holistic Resource Management taught me to identify the building blocks of the farm ecosystem, the water and mineral cycles and energy flows on and through the farm, as well as the social dynamics of the surrounding island community. I learned to think of human creativity, money, and labor as tools to achieve larger holistic goals. I learned methods of making, testing and re-testing plans, looking for weak links, cause and effect, and impacts both on the biological and the social environments. The benefit of this method for any farmer is not only to help maximize efficiency of resource use and increased profitability. It helps answer the ultimate question of why the farmer is farming at all and to what purpose, and to let the answer to that question shape everything that happens on the farm on a daily basis.

For me, the answer to that question has to do with the future of agriculture in America, with public health and environmental stewardship, and with the survival of a democratic society. All of these concerns are intimately connected and have direct consequences for how we imagine the future of this farm. America is now importing more than fifty percent of all foodstuffs and, if current trends continue, in the near future will import most of the food consumed in this country. We hear from agricultural economists that large-scale American agriculture soon will no longer be able to compete in the world market for labor, land, water and other resources and likely will be phased out within a generation or so, except for the production of certain subsidized commodities such as soy, corn and wheat.

The consequences of the trend for large-scale agriculture to focus on commodity markets far away rather than feeding folks at home, manifest right now in Washington, one of the largest food-producing states in America. More than 100,000 children are going to bed hungry every night in our state, and many families can feed themselves only with the help of food banks. Between 2001 and 2003 an astonishing two million hungry families relied on food

banks in King County alone, in addition to 586,000 children who came on their own.[5] The declining nutritional quality of foodstuffs provided by global markets causes more than sixty percent of the population of my home state to be overweight, if not clinically obese. Looming threats of bio-terrorism make large-scale production and distribution systems acutely vulnerable. All of this leads to the inescapable conclusion that our current food system is fundamentally and chronically insecure.

The remedy to these problems is to strengthen small-scale, local agriculture and community food systems. However, the remedy can be implemented only if we think and plan holistically beyond the profit motive to embrace nutritional, environmental and community health as farm goals. Small, local farms must become resource self-sufficient and community based. Farmers must think beyond their own lifetimes and make the training of the next generation of farmers an integral part of their own production goals.

Over the years, S&S Homestead Farm has tried to meet this challenge through the development of an educational outreach and research program organized through a self-supporting, state-registered non-profit organization, S&S Center for Sustainable Agriculture (SSCSA). Educational programs include classes in ecological food production for high school students offered in collaboration with Lopez Island Schools, a curriculum-based internship program offered in collaboration with WSU's Center for Sustaining Agriculture and Natural Resources (CSANR), where I hold an adjunct faculty position. We host on-farm workshops, farm tours, and public presentations organized through conservation districts and WSU Cooperative Extension in San Juan County and on the near mainland. On-farm research has focused on demonstrating the technical and economic feasibility of small-scale production

31

methods that are environmentally sustainable, enhance farm self-sufficiency and support local food security.

We have been able to attract grants to support these efforts. In 2001–2002 we received a small grant from SARE (Sustainable Agriculture Research and Education, a program funded under the Farm Bill) to grow barley on small acreage using appropriately scaled equipment. The successful experiment demonstrated how to prevent nutrient run-off in winter sacrifice areas, while providing the farm with animal feed and meeting the need for local organic grain.

Another grant we received from SARE linked the ecological food production class with a farm-to-school project that supplies the local school cafeteria with fresh greens for their lunch menu. We hope that the success of this project will lead to a permanent school curriculum in environmental and nutritional health.

A third SARE grant supports ongoing replicated field trials comparing farm-produced biodynamic soil stimulants with lime applications to balance soil pH, increase availability of N, P, K, micronutrients and soil organic matter in small-scale forage and hay production.

A fourth project supported by a grant from EQUIP (Environmental Quality Incentive Program offered by the Natural Resource Conservation Service, NRCS) allowed us to research and develop a solar-powered irrigation system that collects rainwater from the barn roofs, stores the water in a 750,000 gallon pond, from where it is returned to irrigate the orchard and vegetable production sites during the typical summer drought. This minimizes demand on limited groundwater and benefits plant health through irrigation with soft rainwater instead of hard groundwater. While

these grants are typically small, they benefit both the production side of the farm and its educational outreach by focusing our energies on finding solutions to specific problems and by bringing research expertise from the land grant university to our small farm.

During the last few years we have benefited enormously from collaboration with university and extension agents and researchers bringing their know-how in engineering, soil science, microbiology, plant and forage systems, and agricultural economics. We have also been able to obtain modest support for our interns through these grants, so that students pursuing graduate degrees in various fields have opportunities to integrate their research interests with on-farm training. This year, for example, we will be hosting three interns pursuing M.S. degrees in soil science, nutritional science and agricultural economics at WSU and Bastyr U. in Seattle. One of these interns will be dividing his time between work and study on the farm and developing our Future Farm Project through the Lopez Community Land Trust (LCLT).

My wife and I feel strongly that the production capacity, cumulative experience and research-based knowledge accruing over the years on a holistically managed, small farm should not be allowed to vanish once the current owners get too old to carry on the work. The Future Farm Project envisions collaboration between the land trust which would own the farm, the Lopez Public Schools, and WSU's CSANR which would develop a region-wide training program on the farm.

Together with CSANR, we have applied for an implementation grant from SARE and are waiting to hear this spring whether the application has been successful. My wife and I hope that the project will go forward. We share Thomas Jefferson's view expressed in a letter to John Jay in 1785, that the "cultivators of the earth" were the surest guarantee of a free society, because a citizenry whose livelihood was independent, were "tied to their country and wedded to its liberty and interests by the most lasting bonds."[6] Updated to the urgent concerns of the twenty-first century, a self-sufficient small family farm also offers other, equally important, solutions to the problems of nutritional and environmental health, local food security and protection against bio-terrorism, as well as economic viability.

I want to end with another compelling piece of data gleaned from the seminar on Agricultural Systems and Nutrition I mentioned before. Adam Drewnowski, Professor of Medicine and Epidemiology who organized the seminar, presented what he called epidemiological maps of New York City and Seattle to show the correlation between zip codes and obesity rates. The connection seems absurd because we prefer to believe that obesity reflects personal choice in foods, but the maps provided overwhelming evidence that obesity is directly related to affordability. The households clustered around the perimeter of Central Park commanded median incomes of $180,000, while the households just north of the park had incomes of less than $10,000. Obesity rates among the well-to-do Central Park residents were between 4–7%, while those of the working poor just north ranged between 23.5–28.3%.[7]

These data show that the claim that American households on average spend 13.2% of their income on food is misleading. Surely a family that commands a six-figure income has access to the very best, i.e. nutritionally vital, fresh and flavorful food, while a family with an average income just over five percent of the Central Park incomes, will have to choose the cheapest and nutritionally deficient foods. This does not mean, of course, that the well-to-do always make wise food choices, nor that the working poor necessarily have to be obese. Rather, it

seems that in this age of advertising, the nation as a whole is addicted to much-traveled and nutritionally depleted foods whose principal virtue is that they are abundant and cheap. For example, former president Bill Clinton, surely a man of large appetites, was known for his love of the Big Mac, a love that probably contributed to his massive cardiovascular problems that forced him into radical heart surgery to save his life.

On the other hand, both in New York City and Seattle, Drewnowski found community neighborhoods where median incomes were low, but so were obesity rates. Interestingly, these neighborhoods are populated by recent immigrants who grow substantial amounts of food in urban gardens. Their native food traditions probably help them choose and prepare foods in healthful ways. In sum, culture, health and economics go hand in hand. It would seem that everyone could enjoy the holistic high-life if we strengthened community-based, local food production. For the sake of a healthy environment, healthy communities, and healthy people, food growers and food consumers need to re-establish the local connection.

Notes

[1] A preliminary version of this essay was published in *Biodynamics* nos. 253-254 (2005).

[2] Howell, Edward 1994. *Food Enzymes. Health and Longevity*, 2nd ed., Twin Lakes, WI, 17.

[3] Kingsolver, Barbara 2003. "A Good Farmer," *The Nation* (November 3).

[4] Thoreau, Henry David 2004. *Walden,* Boston, Mass & London, UK, 24 and 44.

[5] "Building a Sustainable Community Food System in Seattle and King County: Concept for Developing a Local Food Policy Council," Washington State University, September 27, 2005. (http://king.wsu.edu/foodandfarms/documents/ SeattleKingFPCconcept.pdf). Retrieved July 9, 2021.

[6] Jefferson, Thomas 1785. "Letter to John Jay." See https://www.goodreads.com/quotes/7571241-cultivators-of-the-earth-are-the-most-valuable-citizens. Retrieved July 9, 2021.

[7] Drewnowski, Adam 2007. "Disparities in Obesity Rates: Analysis by ZIP Code Area," *Social Science & Medicine*, Dec; 65(12): 2458-63.

S&S Homestead Farm

Some Thoughts on Eating
Jennelle Kvistad, 2009

In this age of nutrition information, we are both blessed and cursed. On one hand, we now understand more about the biochemical processes in our bodies that make us tick, or not. There is a wealth of information to choose from how best to eat. On the other hand, many have lost the ability to think for themselves and listen to the needs of their bodies. Despite all the research on diets and the marvels of modern food technologies, our nation is dealing with alarming levels of heart disease, obesity, diabetes and other major nutrition-related diseases. We are eating nutrient-deficient foods that make our bodies crave more. We then eat more processed non-food. And so the cycle continues. We are over-eating, yet we remain undernourished. We grow up being told that eating right is important, vegetables are good for us, and will make us strong like Popeye the Sailor Man. But what does eating "right" mean? What is so "good" about vegetables? What does it really mean to be "strong?"

Eating "Right"
In many ways, this is the most difficult aspect of eating to define. Factors include: ~ how environmentally responsibly the food was produced, processed, and transported; ~ conditions experienced by the food workers involved; ~ the genetic affinity we have for ethnic foods; ~ health conditions and diseases that require us to eat a lot of, or stay away

from, particular foods; ~ and Spiritual or ethical beliefs that speak to the appropriateness of what we put in our mouths. As you can see, eating "right" in this day and age is not a simple task! Many of us ignore this task and go about eating whatever is convenient and tastes acceptable.

This casual approach to eating causes more damage than we realize. Some of us go to the other end of the spectrum, taking eating so seriously that we label ourselves vegetarian, vegan, raw foodist, lacto-ovo vegetarian, macrobiotic, fruitarian — the list goes on. The most important aspect of eating is that we eat with a conscience. Get educated about all parts of your food system, global, national, and local. Then make decisions based on what feels right to you. You have the power to make serious political and ethical statements just by the food you put in your mouth!

The "Goodness" of Whole Foods
When determining if a food you are about to eat is a whole food, there are a few things to consider. How close is it to how it is found in nature? How many ingredients does it have? A carrot that has been shredded, combined with processed grains, preservatives and other ingredients, then vacuum-packed in plastic and sold as a "healthy" bran muffin, is not a whole food. Vegetable oils and fruit juices are not

whole foods; they are only a part of what was once a whole food. These partial foods can still be healthful, as long as your diet consists mostly of whole foods.

This is not to say that you can't be creative in working with the whole foods you get. As Evelyn Roehl writes, whole food "encompasses a whole cuisine — a diet consisting of whole foods, and the ways of preparing them to gain the most nutritional and culinary benefits. Basically, whole foods are found in as close to their whole, natural state as possible."[1] Vegetables and other whole foods contain the components essential to life.

These components include macronutrients: fats, carbohydrates, and protein, and micronutrients: micro-and macro-minerals, and vitamins.

Fats

It is unfortunate that we are so afraid of fat in our culture. We need fat! Without it, none of the cells in our bodies could function because fats are essential components in membrane structure. Fats transport the lipid-soluble vitamins A, D, E, and K through our bodies. They are vital components in our cell membranes and provide padding and insulation around our vital organs, protecting them from trauma and temperature extremes. It's the type and quantity of fat we eat that determine appropriate intake. Our bodies need three essential fatty acids: linoleic acid (LA), linolenic acid (LNA), and arachidonic acid. We get these only from our diet. Concentrating on getting these essential fatty acids in our diets is just as important as limiting saturated and hydrogenated fats. The liver makes cholesterol from saturated fats; therefore, saturated fats and hydrogenated vegetable oils (which contain high amounts of saturated fats in place of their once polyunsaturated ones) raise serum cholesterol. Fats also move flavors around in a prepared dish. So don't be shy! Learn how to use fats and oils properly and your body and

your tastebuds will thank you.

Carbohydrates

Carbohydrates are our main source of energy and should constitute the bulk of our diet. Unfortunately and unfairly, carbohydrates are right behind fats as the most feared dietary component. It cannot be argued that our brain prefers glucose, as fuel and glucose come from the breakdown of carbohydrates. The real villain is not carbohydrate itself, but the shift in our culture from complex carbohydrates to refined and simple sugars. Complex carbohydrates are starch and fiber in foods such as fruits, vegetables, and whole grains. These provide stable blood sugar levels as opposed to the rise-and-fall levels in simple sugar and refined flour products. It makes much more sense to lean away from refined sugar and flour products that promote disease and obesity rather than cut out all carbohydrates, including complex fibers.

Proteins

Chemically speaking, proteins are complex molecules variously combined from twenty-two naturally occurring amino acids. Proteins perform many functions. We need them for building all the tissues in our body, for balancing bodily fluids and salt, for acid-alkaline balance, and for energy. They also help build important substances such as enzymes, hemoglobin (the iron-bearing protein in the red blood cell responsible for oxygenating our bodies), hormones, and antibodies. Amino acids are grouped in categories of essential, semi-essential, and nonessential. The eight essential amino acids are: isoleucine, leucine, lysine, methionine, phenylalanine, threonine, tryptophan, and valine. These we must get from our diet.

There are lots of conflicting opinions about adequate protein intake for humans. It has been said that vegetarians must combine their foods correctly in order to get all the essential amino

Absorption Increased By:

~ Body needs during growth, pregnancy and lactation
~ Hydrochloric acid
~ Vitamin C
~ Blood loss or iron deficiency
~ Meats (heme iron)
~ Protein foods
~ Citrus foods and vegetables
~ Iron cookware
~ Copper, cobalt, and manganese

Absorption decreased by:

~ Low hydrochloric acid
~ Antacids
~ Low copper
~ Phosphates in meats and soft drinks
~ Calcium

acids in one meal, hence the popular "beans and rice is nice." New information is coming out to say that we do not need to worry about eating all the eight essentials in one meal; as long as we get a varied diet, we will get adequate amounts. One thing is for sure, we do not need the gluttonous amounts of protein that most North Americans consume on average. Symptoms of excess protein consumption include acidic blood, calcium deficiency, and a tendency to carcinogenic and other degenerative diseases. It is tragic that, while we are seeing these signs of protein overload in our country, there are still hundreds of millions of people in other countries dying from kwashiorkor and marasmus, two serious protein-deficiency diseases.

Micronutrients

We are more likely to be deficient in essential minerals than in vitamins. This is due to the fact that commercial farming erodes topsoil and removes minerals from the soil, usually replacing only nitrogen, phosphorus, and potassium in the form of chemical fertilizers. The food we grow is only as good as the soil in which it was grown. Commercial agriculture employs practices that literally "kill" the soil, thereby creating a "dead" product. Industrialized food is likely to be fortified mostly with vitamins along with some minerals, like calcium. Furthermore, if the mineral or vitamin is not in a form that is bio-available to the body, then it does no good to dump a bunch of it on a product, because it won't be absorbed by the body. Iron is one of the most finicky micronutrients when it comes to bio-availability. This mineral is a good example of how important it is to eat a varied diet. If we lack iron, we will not get adequate oxygen to our cells due to decreases in hemoglobin. Iron deficiency anemia is a common problem, especially in childhood, youth, and for women during their child-bearing years (due to loss of blood through menstruation and increased needs during pregnancy). Anemia can result in functional impairment of the body, reduced work capacity, behavioral and intellectual impairment, reduced resistance to infection, and impaired temperature regulation. There can be problems with too much iron intake as well. It occurs mainly in men and post-menopausal women because they do not shed iron in menstrual blood. Excess iron in the blood may increase the risk of atherosclerosis and heart disease. The mechanism for this is not yet known, but is thought to be associated with increased oxidation and free-radical formation.

There are many factors affecting the absorption of iron. Average iron absorption is eight to ten percent of intake. Iron from flesh foods (heme foods) is more readily absorbed than iron from non-flesh foods, being as high as ten to thirty percent. On the other hand, phytates in whole grains and oxalates in certain vegetables, like spinach, bind with iron and make it unusable by the body. This is why some vegetarians have trouble meeting their iron needs from diet alone. The absorption of iron is a slow process, so a high-fiber diet can decrease absorption due to fast intestinal transit time. This is not to say that a person should not eat fiber! Nor is it being suggested that no one adopt a vegetarian diet. The point is that it pays to have some knowledge of how your body relates to the food you eat. Each nutrient has its own absorption rate, and getting to know all of them can seem an unrealistic and daunting challenge. Be at ease! There is an answer. Getting a variety of fresh, locally grown, organic food is a huge part of staying healthy. Eating a variety of colors means you are getting a variety of vitamins. Having some green chives and parsley on your eggs in the morning, some red cabbage and carrots on your salad for lunch, and cauliflower as your dinner vegetable is a good example of eating from the rainbow. Elizabeth calls this "health insurance."

Strong Like Popeye the Sailor Man

Ever wonder why Popeye would open a can of spinach when he found himself in a bind? Well, spinach contains iron, which is needed for biochemical processes that make us strong. Popeye should be downing a fresh spinach salad, with locally grown hazelnuts, farm-fresh eggs, and spinach just picked from the dirt, but the point is that food contains all the components essential to life. So when our kids say "Blech! I don't like spinach!" tell them it will make them able to play longer and run faster.

Water

Water covers more than seventy percent of the earth's surface. The human body is two thirds water. Most fruits and vegetables are over ninety percent water. Water is essential to all forms of life. We are constantly expelling water through urination, perspiration, and respiration. We must take care to replenish our bodies of this invaluable nutrient. I immediately notice when I become dehydrated: my throat becomes scratchy, I develop a headache, and a bit of an attitude. I used to work at a health food store selling vitamins. Many people approached me with the question, "What can I take to improve the way I feel? I seem to be a bit on the lethargic side." More often than not, I found that they were not taking in adequate amounts of water. I would tell them to drink water often and in between meals, then come back if they were still feeling the same and talk vitamins. Never saw a one again. People with diets high in animal products and salt require more water intake. Those who eat mostly vegetables, fruit, and sprouted foods require less. Climate is also a factor. Those living in dry, hot, and windy areas require more water than those in cold and damp climates. Nutritionally, it is best not to drink large amounts of water with your meals because stomach acid and digestive enzymes will be diluted and your food will not be properly digested. Drink water one half hour before or one hour after you eat your food.

Handling Fresh Produce

Micronutrients can be enhanced or destroyed by how we grow, harvest, process, and store the foods we eat. The food grown in mineral-deficient soil will be deficient in minerals. A head of lettuce may be fresh and full of nutrients the day it is picked, but if it has to travel from California to Lopez Island, it will likely lose many of the vitamins that were once available. Broccoli is an amazing vegetable full of vitamins, but if you boil it for ten minutes in water, most of them will leach out or be destroyed by the heat. Having some basic knowledge about how to properly handle fresh produce is essential to reaping the benefits of the food. In general, cook your vegetables just long enough to make them tender without reducing them to mush. Cutting your vegetables too small creates greater surface area for nutrients to leach out (which isn't an issue if you are using the water, as in a soup); cutting them too big subjects heat-sensitive nutrients to longer cooking times. Aim for medium size chunks. Steaming and sautéing are generally better than boiling because there is no water for the nutrients to leach in to.

The Raw and The Cooked

To cook or not to cook? Variety should apply to food preparation as well as to types of food. Raw foods contain more nutrients and enzymes than in their cooked form. Some foods, however, such as broccoli, carrots, and potatoes, are difficult to digest raw, and cooking makes them easier on our system. Grains, nuts, and seeds can be soaked or fermented rather than cooked, which helps them retain enzymes and heat sensitive nutrients. But, again, eating a variety of cooked and raw foods will ensure that you are getting a wide range of nutrients in absorbable form. I enjoy slow-cooked soup and fresh bread too much to ever become a raw foodist. But I acknowledge the benefits of fresh, unadulterated fruits and vegetables.

Table Manners

For goodness sake, sit down when you eat! Too many people eat standing at the counter! They crack open a can of this and a box of that, many times not even bothering to put it on a plate. They are robbing themselves of the opportunity to be fully nourished in the moment. Part of the problem is that our families have become fragmented and we no longer respect the ritual of eating. We look at

food as fuel and nothing else. When we eat, we must realize that we are not only renewing our body cells, but also our Spirit. Eat slowly and acknowledge the flavors. Eating can be a sensuous experience. Sharing meals with family and friends is another important aspect of sustaining those relationships and further enriching the eating experience.

Cynthia Lair writes: "Deep in our cells we know that eating whole, fresh, natural foods is the best nourishment for body and soul. Eating whole foods can help feed the desire for wholeness within ourselves. This Spiritual benefit is magnified when the entire family partakes of nature's bounty together. Not only are the individuals of the family enriched and nourished, the family is strengthened as well."[2]

On S&S Homestead Farm, Henning and Elizabeth put strong emphasis on the importance of a well-prepared meal. They take great care to plan meals and prepare the food that is grown right on the farm. There is always a dinner presented that far surpasses anything offered even at the finest restaurants.

Notes

[1] Roehl, Evelyn 1996. *Whole Food Facts: The Complete Reference Guide*. Fairfield, CT. https://www.google.com/searchtbm=bks&hl=en&q=%E2%80%9Cencompasses+a+whole+cuisine%E2%80%93+a+diet+consisting+of+whole+foods%2C+and+the+ways+of+preparing+them+to+gain+the+most+nutritional+and+culinary+benefits.%E2%80%9D. Retrieved July 6, 2021.

[2] Lair, Cynthia 1996. *Feeding the Whole Family*. Seattle, WA, 8.

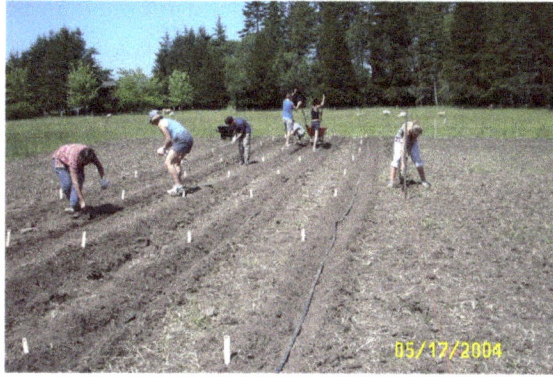

Teach Your Children Well
(On The Farm & In The Community)
Elizabeth Simpson, 2009

Our farm has always been a favorite place for Lopez folks to bring their guests, especially visiting grandchildren. Those visits are an education for them, and for us, too. Some of our experiences with younger people showed just how de-natured their lives are, and how far most of them are from understanding the sources of their food. There were the two teenage boys who refused to get out of the car because they were afraid of getting their new white sneakers dirty. There was the young man who pointed to a cabbage growing in the garden, and asked, somewhat disdainfully, "What's that?" There was the little boy who tasted Loveday's fresh milk, made a face, and said, "I only like store milk."

Those responses range from annoying to alarming. Children who don't want to get dirty. Children who have never seen vegetables growing. Children who have no idea how pure, unprocessed food tastes. It was, in part, our experiences with these visitors that prompted us to create the Ecological Food Production class that we have offered to high school students for the past five years. The class is different every semester, depending on what is going on at the farm, so many students have

taken it up to three consecutive semesters. Students come to the farm on Monday and Thursday afternoons, and work with Henning, or me, or both of us.

Henning teaches them about: ~ rotational grazing, pasture management, and animal husbandry; ~sustainable water systems that make use of rainfall catchment and solar power to irrigate gardens and orchards; ~ orchard care – pruning, propagation of berries, and care of fruit trees; gardening – bed preparation, composting, the preparation and use of biosprays to enhance the immune systems of plants and the micro-biological life in the soil; ~ planting, direct seeding, and care of plants and beds; ~ sustainable use of farm resources, such as cutting and peeling logs for on-farm construction; ~ the inter-connectedness of people, plants, animals, insects, and soil life.

I teach them how to: ~ seed and transplant in the greenhouse; cook and bake from scratch; ~ process raw milk into many products, including yogurt, butter and cheese; preserve foods, including freezing, drying, canning, and fermenting; look first to the garden, second to your shelves, third to your freezer, before going to the store; ~ read labels to see what ingredients are in the processed foods available at the store. (Anything they can't pronounce they probably shouldn't eat.)

Both of us teach them to: ~ know where their food comes from; pay attention to health, political, environmental, and social issues relating to food; ~ take joy in growing, processing, preparing, and celebrating food as part of family and community wholeness.

As a result, students have: ~ started home gardens; ~ changed the purchasing, cooking, and eating habits of their families; ~ made sustainable living the focus of their studies, publishing papers, creating senior projects, and choosing majors related to sustainability.

Early on, Henning and I aimed at changing the school cafeteria menu. Dana Cotton, the school chef, was happy to include our garden's greens in lunches. At the urging of our students, she soon went beyond requesting standard lettuces, and ventured into root crops, brassicas, and leafy vegetables. On Mondays, students would gather foods that they had helped to plant and grow, such as beets, kale, chard, lettuce, spinach, carrots, sorrel, cucumber, and mustards. During the week, they would go into the kitchen on their own time and help to wash and prepare the vegetables.

One of our students started a "Beef for Kids" campaign in the community, putting coffee cans in local businesses so that people could donate money towards the school's purchase of locally raised, grass fed beef, to replace the commodity beef supplied by USDA. In the years since this student project began, the school has been able to obtain nearly all of its meat from local sources, and always notes on the monthly school menu that the beef in the hamburgers, tacos or spaghetti sauce is local and organic. Students don't want anything else now.

Dana attended an institute in Berkeley, California on providing healthy foods in cafeterias, and what she learned there has changed the school's food culture entirely, from the yogurt offered during nutrition break to the organic milk offered by the glass (not the carton), to the whole-grain breads, pastas and rice they now serve. Students love the food, and the effects on their behavior and academic performance (soon to be measured by a study sponsored by Johns Hopkins University) are already noticed by staff, teachers and administrators.

The change in the school menu from commodity foods to locally grown, organic foods was a long time coming, and not without difficulties. For the first four years we offered it, Henning and I struggled to have our course supported (or even acknowledged) by the school. We taught without compensation. We worked with administrators who gave our efforts lip service, but little meaningful support.

Finally, a new principal and new superintendent took our ideas seriously, but could not squeeze funding from a tight budget for an ideal that most folks saw as quixotic. We needed an angel. She arrived as Michele Heller, a local person who is concerned about global problems and determined to solve them locally. Her first move was to offer the district a high tunnel hoop house so that students could grow vegetables for the cafeteria. The school board responded with understandable concerns about funding, maintenance of the hoop house, and the impact on an overstrained curriculum. Henning and I offered to have the hoop house placed on our property, to provide water and fencing, and to locate a person who would help maintain it, and work in the district to create and implement a farm-to-school, farm-to-cafeteria program.

Henning and I wanted to turn our south field into a CSA, following our belief that this island can and should feed itself. We searched for a young couple to develop and run it, hoping that one of them would have the skills to bring farm, cafeteria and elementary school curriculum together. Jesse Pizzitola and Lisa Murgatroyd responded to our outreach, arrived at the farm a few days after school started, and plunged into the community and school projects with some farming experience and a lot of faith in what we were trying to do. Michele funded Lisa's position at the school. Rhea Miller, from the Lopez Island Land Trust, built the beds in the hoop house, brought in compost, and planted seedlings. Lisa took over from there. Her work in the elementary school and cafeteria were so successful that the LIFE (Lopez Island Farm Education) program became a model for school districts all over the country.

The school board, now convinced of the viability and usefulness of the program, gave staff time and school resources to advance the project. Laurie Parker, a film producer, made a pilot video on Lisa's and our work with students. Michele, Laurie, Lisa, Jennelle Kvistad, Jean Perry and I produced two well-attended community events on food issues.

In the winter of 2008, Michele instituted monthly community dinners. Two local chefs, Kim Bast and Jean Perry, plan menus of seasonal foods, purchase produce from island farmers, and create simple and delicious soups, salads, breads, vegetable dishes, and desserts from those foods. The cost is $5.00 for adults, $3.00 for children, $15.00 for a family. On those evenings, the cafeteria is transformed: tablecloths and simple, handsome plates and bowls, table decorations, elegantly printed menus listing the foods, providing recipes, and identifying the farms that provided the produce. There is the hum of hundreds of folks talking, laughing, and eating. Usually, over 300 people are served, and the food is abundant.

The goal of these dinners goes beyond creating an affordable community feast: Jean, Kim, and Michele are downright subversive. They believe that when people taste kale sautéed with garlic, chocolate beet cake, pumpkin bread, leek and potato soup, they will learn how delicious simple, healthy foods are, and how easily prepared; perhaps this experience will change the way they eat.

Such a change would impact personal health, community economics, and global politics. Wendell Berry writes, "Eating is an agricultural act." And also, "How we eat determines, to a considerable extent, how the world is used."

Teaching students through the LIFE program, providing forums for community education and discussion, and feeding families at the community dinners, brings home in concrete ways, how people can become healthier, more self-sufficient, and less dependent on a wasteful and irresponsible global food system.

Part III. Growing Plants

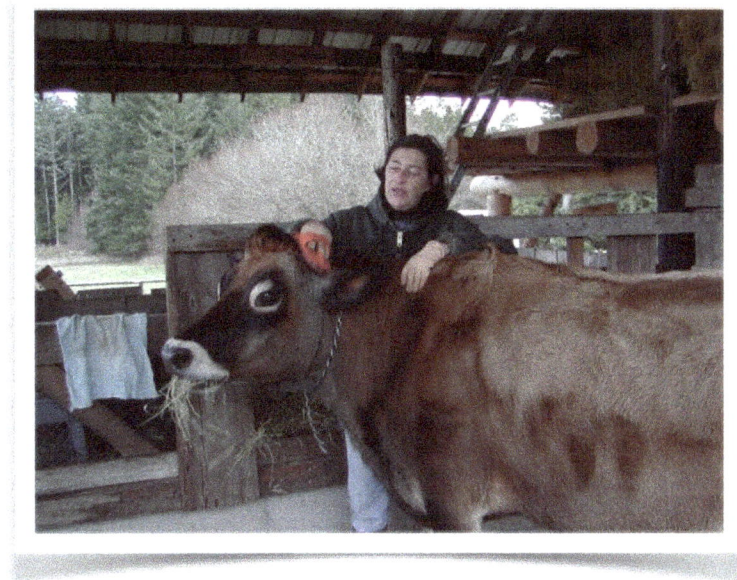

Pastoral versus Industrial Food Production
Elizabeth Simpson, 2007[1]

It is an honor to be with you all today at this Lopez Island Community Meeting on Food. This is a very familiar scene. Every celebration I have ever attended on Lopez has had food as its centerpiece. Weddings, memorial services, graduation parties, and community events — all feature foods, and mostly they are potluck. Usually the dishes people bring are local — often from their own gardens. So when Rhea Miller[2] asked me to speak about the importance of eating locally, I realized that I had the chance to participate in an ongoing, and lived, community conversation. Food is the most intimate act we perform. Three times a day, or more, we put food into our bodies. The choices we make about what foods we ingest have many ramifications.

To me, the term "local" means a number of things. The first of these is freshness. The first strawberry, warm from the bush, the first peas from the pod, or the first ripe tomato, provide a delight to the senses we can't get from imported produce. But freshness also means nutritional wholeness. According to a Pennsylvania State University study, spinach can lose 90% of its vitamin content within twenty-four hours of harvest.[3] What you get with well-traveled spinach is fiber and water. The news about health problems today is bad news: over the last twenty years, there has been a dramatic rise in cardio-vascular disease, cancer, obesity, diabetes, digestive problems, ADD, ADHD, asthma, allergies — the list goes on and on. And these illnesses are striking younger and younger people. Our island physician, Dr. Bob Wilson, says that he is seeing children in elementary school who have early-onset diabetes and are at risk for heart disease. All of these illnesses are related to a diet of processed and de-natured foods, most of which come from far, far away. The issue of shipping foods long distances goes beyond the illnesses they contribute to. I have never wanted my food to be better traveled than I am. Most foods available in grocery stores have been transported 1,400 miles from field to plate, changing hands at least six times on the way. These crops have been chosen for their ability to withstand long journeys and many days in the produce section and your refrigerator, so qualities of nutritional value, crop variety, taste, and scent, have been neglected in order to privilege staying power.

Food miles — the number of miles it takes for food to get to the marketplace — are a real consideration, especially as we see the world's oil supply dwindling. One kilo of asparagus from California uses four kilos of aviation fuel to get to Maine. The use of fuels in the production, distribution, and transportation of food is a primary source of the greenhouse gases that are responsible for climate change. Crop variety is seriously undermined by foods that are raised for shipping. Since fruits and vegetables are selected for staying power, large growers generally do not raise heirloom varieties that may be delicious and nutritious, but will turn to mush during transport. These varieties are disappearing, and with them, the necessary genetic diversity that protects a species from disappearing when threatened by pests or disease.

Fifty years ago, there were 5,000 different seeds available from seed catalogs. Today, there are 500. This is because large seed companies are buying up smaller ones. If you are a gardener, you probably get ten different seed catalogs per season. The names are different, but the same company may own them all. That parent company gets the cheapest seed it can from a single source — which may be in Europe, or India, or China — and its first task is to dump seeds that are not useful to commercial growers. These are usually heirloom varieties, because they must be grown out every year — the seed does not store well. So the biodiversity we depend on for our food security is best sustained by growing our own, or buying locally. Small farms that sell to local folks can grow heirloom cultivars for their flavor, nutrition, and variety of color and size. Not only seed production, but food production, is being consolidated into fewer and fewer hands. Five corporations control the world's grain supply.

Ten companies — Nestlé, Pepsico, Coca-Cola, Unilever, Danone, General Mills, Kelloggs, Mars, Associated British Foods, and Mondelez — control almost every food and beverage brand in the world.[4] Just as with seed catalogs, the names and labels are misleading. A hundred different brand names can be owned and controlled by a single company — Phillip Morris, tobacco manufacturer, among them. Labels on seed catalogs mislead us about who controls and distributes the seed. Labels on processed foods also mislead us in the same way. I grew up with TV ads for the Jolly Green Giant and believed that, all by himself, he

48

grew and distributed the vegetables always pictured behind him. However naïve that was on my part, at least there was a real corporation that controlled the fields and foods that came from them. But today, Green Giant grows, produces, and distributes nothing. Its farms and factories were sold off years ago, and only the brand name is marketed.

More troubling is what labels don't tell us. Companies that produce and market genetically modified foods spend billions of dollars to keep labeling laws from being passed. They don't want you to know about the fish gene in your tomatoes or the Bt corn in your tacos. You might choose not to buy the product. And, if you react to the foreign gene, or the bacterium or antibiotic or protein used to insert the gene into the food, the cause of your allergic reaction or illness can't be traced back to them.

Environmental concern is another issue very much at the heart of eating locally. Wendell Berry says, "Eating is an agricultural act." I understand this statement to mean that, when we choose the foods we put in our bodies, we are supporting either agribusiness or local production. Agribusinesses are clustered in several spots in the country, but I would like to focus on the Midwest as an example. Pig farms and corn. Setting aside all other issues for the moment, let's look just at pollution. Effluent from pig farms and chemical runoff from cornfields in the Midwest flow from contributory streams into the Mississippi River, destroying life in the river all the way up the food chain. From there, the poisons flow into the Gulf of Mexico, where they have created a dead sea the size of New Jersey.

Environmental concerns include not only what is present, but what is lost, when small family farms are gobbled up by agribusiness. A case in point is my husband's brother-in-law, Friedrich Beckmann, who is a farmer in northern Germany. When his father ran the farm, there were fields with varied crops. Five families worked and lived there, managing dairy, meat, poultry, vegetable, and fruit production. Storks nested on the roof of the main house. The frogs, fish, and snakes that fed the storks lived in the ditches and streams that crossed the fields. When Friedrich took over the farm, he tore out the crops that his father had raised for local sale. He sent the five farm worker families packing, and replaced the labor of their hands with gigantic machines. He filled in the ditches and streams, because the machines he uses need level surfaces. The frogs, fish, and snakes died, and the storks that had fed on them stopped nesting there. He began to mono-crop, raising whatever foodstuffs the government would subsidize — acres of rapeseed, wheat, or apples, for example. When the price of apples was undercut by cheaper apples from Spain, the orchards were knocked down. When Spanish apples suffered blight and stopped producing, Friedrich replanted apple trees. Friedrich is running a business, not a farm. He works on machines, not with the soil. The wildlife now on his land consists only of animals he likes to hunt. He and his wife maintain a minimal garden, a few poultry, and exist mostly on processed foods from the grocery store.

I use this example because I have visited there, and seen the changes. But envision a farm in Kansas before WWII. There was a farmhouse, surrounded by trees, a picnic table under them. There were barns for hay, a corral for horses that pulled the plows and harrows, chickens and pigs and a few cows for eggs and meat and dairy production. Now those family farms are gone, and the towns that served them are dying. The farms have been replaced with agribusinesses, supported by subsidies, growing one crop — mostly corn. The soil and water are polluted with pesticides and herbicides. Birds and wild animals are gone, because there is no habitat to support them. There are only acres of corn or soybeans as far as the eye can see.

What happened to the animals? They have been moved into CAFOs — Confined Animal Feeding Operations. "Factory farms" is a more common term, though the word "farm" is misused here. "Factory" is appropriate, since animals in these operations are treated as machines — "production units" — and are unable to live anything like the lives they would have lived on that Kansas farm. Laying hens are crammed five or six to a cage. The stress caused by overcrowding and the frustration of every natural instinct a chicken has causes them to rub their breasts against cage walls until they are featherless and bleeding. It also causes them to cannibalize their cage mates, so they are "debeaked" with a hot knife. As their production begins to fall off, they are starved of food and water and light for several days. The assumption is, that as the hen begins to die, she tries to produce life, and will give a few

more eggs before the end. Broiler chickens are not kept in cages, not out of humanitarian kindness, but because they might scar the breast meat if they rub off their feathers. Unless organically raised, they are fed antibiotics, hormones, and even arsenic (which increases their appetites) in order to hasten their growth. They must be slaughtered at seven weeks of age. Their growth is so rapid that often their legs break and their hearts stop before the slaughter date.

Henning and I once tried to raise some Cornish cross hens. Because our chickens run around outside all day, they are muscular; they make excellent broth, but their meat is difficult to chew. Hungry for fried chicken, we got a dozen Cornish cross chicks, and tried to raise them naturally. Because these chickens are overbred — bred strictly to grow fast and produce tender meat, they have no immunities. So we built them their own little pen inside the chicken run, and fed them the same home cooked meals our laying hens get — ground barley, fresh vegetables, cooked potatoes, and table scraps. These chicks were the sweetest little birds I'd ever raised. They had trusting, gentle natures, and funny, croaky little voices. But they did not act like chickens. They did not forage for food. They did not run around. They did not establish little hierarchies among themselves. They did not take dust baths or sun baths, or have much to say to each other. And, sadly, some of them could not defecate. I will spare you the description of how I helped them. But it was awful to watch their beaks and combs grow pale, and see how listless they became. We were not giving them antibiotics, nor high-powered commercial feed, and they were not bred to survive in the natural world. We managed to take six through to maturity. I have no memory of eating them. The whole experience was too sad.

The overall picture of CAFOs is an ugly one. Thousands of pigs in a single building, each in a cage so narrow it cannot turn around. Because they are bred to produce lean meat (a big selling point in fat-conscious America), their natural hardiness is bred out of them. When a stranger enters the building he or she is met by the screams of terrified, overstressed animals. They may be reacting to the hazmat suit that the visitor is required to don before entering the pig building. These animals have no immunity to disease.

CAFO cows, just like factory-farmed pigs and Cornish cross hens, are bred to be producers of profitable foods, far from the way nature created them. Cows should eat grass. They have four stomachs that allow them to produce protein from cellulose. In natural circumstances, cows graze, ruminate, and fertilize their pastures with their manure. CAFO cows are bred to process unnatural feeds — cardboard, stale pastry, grains, and feed concentrates. They stand crowded in feedlots up to their knees in their own waste, do not move in pastures, raise their young, or in any other way, live out their lives as cows should. They too are pumped full of antibiotics to prevent the spread of disease. Indeed, half of the antibiotics produced in America are administered to dairy cows, and a good percentage more are administered to

chickens and pigs. You get them in the milk, eggs, and meat. As a result, the number of antibiotics effective against infections in humans is diminishing rapidly.

There are other health concerns relating to CAFO animals. The lethal strain of e-coli bacteria that has appeared in apple juice, beef, and most recently, spinach, was unknown before 1982 — evidently it evolved in the guts of feedlot cattle. Grass-fed cattle are not susceptible to that strain. The Centers for Disease Control and Prevention estimate that our food supply now sickens seventy-six million Americans every year, putting more than 300,000 in the hospital, and killing 5,000.

In Japan, every animal that goes through a slaughterhouse is carefully inspected. In our own county, a USDA inspector travels with the Mobile Unit that goes from farm to farm and slaughters animals on site. The inspector (who loves coming to Lopez Island, by the way, because the animals are so healthy here), evaluates the condition of every animal in the field, and inspects each carefully after it is slaughtered. So local folks buying local meat know that it is safe.

In commercial slaughterhouses, the focus is not safety, but speed. Beef slaughterhouses process five thousand head of cattle per day. USDA inspectors, if present at all, are not allowed to stop the line. If an inspector sees a visibly diseased animal go through the line, he or she can tag the animal, and hope someone on the other end will take note, but that is all.

The people employed in these slaughter

houses have the most dangerous jobs in America. Because of the speed with which they must work, and the repetitive motions they use, they suffer lacerations, loss of limbs, broken bones, torn muscles, slipped discs, pinched nerves. Every year one quarter of the meatpacking workers in this country — about forty thousand men and women — suffer a serious work injury or illness. Thousands more go unrecorded. Workers are discouraged from reporting injuries, and many avoid doing so for fear of being fired. Only about one-third belong to a union that could represent their interests. Most of the non-union workers are recent immigrants, many are undocumented, and many are hired "at will," meaning they can be fired without warning, for any reason. Government oversight of plant safety is so minimal that a single plant can expect a visit from an Occupational Safety and Health Administration representative once every eighty years.

The issue of social justice extends throughout the industrialized food system. When you look at cucumbers raised on commercial farms, think of the people who harvested them. Chances are, they came from Mexico or Central America, leaving hometowns and families, possibly crossing the border illegally, at great cost and danger. They work for less than minimum wage. They take in insecticides and pesticides through their skin. Usually, in the camps they live in, there is no hot water to wash those off, and they and their children get sick from eating food from the fields.

Understand one thing: we bring them here.

NAFTA, the North American Free Trade Agreement, created a treaty through which all protections for local producers were eliminated. United States agribusinesses, subsidized by our tax dollars through the Farm Bill, dump cheap corn into Mexican markets, at a price with which local farmers cannot compete. Since the inception of NAFTA, ten million farmers in Mexico alone have lost their land, and must come to the United States and pick our crops in order to support their families.

The health, environmental, ethical, and social costs of the industrialized, corporation-run food system are high. But the bright side of this dark picture is that each of us has a choice. Not only can we buy local foods, but we can also change the system. Consumer demand puts organic milk and meat and fruits and vegetables on the shelves of grocery stores, and on our plates in restaurants. We have plentiful examples right here on Lopez. When you buy local foods:
~ The food travels from the farmer's hand to yours, not thousands of miles;
~ You have a personal relationship with the farmer, and you can see how the food is grown and how the animals are raised;
~ You strengthen the food security of your community;
~ You help to preserve the open spaces and complex ecology that farms create;
~ You avoid pesticides, herbicides and genetically modified foods that may have unforeseen consequences for human and environmental health;
~ You help preserve a local farming economy;
~ You cultivate a local food culture.

This last point is especially interesting. In his latest book, *The Omnivore's Dilemma*, Michael Pollan points out that Americans who are not recent immigrants to this country have no particular food culture. There is no tradition to tell us what to eat and how to eat it. People who grow up in a strong food culture, such as the French, know, for example, to buy fresh and local foods, and to eat them at a carefully set table, slowly, spiced with lots of conversation. The practice of "grabbing" a muffin and a latte and consuming them in the car on the way to work is surely a habit that inhibits healthy choices, the pleasure we should take in food, and a healthy digestive system.

Mainstream America's lack of a clear food culture numbs us to issues of health, the environment, ethics regarding the treatment of animals, and makes us prey to fad diets, fast foods, and the devaluating of eating as a communal and celebratory act.

Fortunately, all across America, small farms are providing wonderful fresh foods to people through farmers' markets, farm stands, CSAs, local restaurants, and grocery stores. These are, interestingly, under the "radar" — the census does not even list farmer as a possible occupation any longer. But they are there. And they are here.

Look around. San Juan County is the only county in Washington where the number of farms has increased in the past twenty years. And Lopez, the most rural of the large islands, has a growing population of people who move here to farm, and young families who take up farming, because there is

support for them here. We all have a choice about what to eat, and on Lopez Island we have a food culture and a community tradition of making local foods the centerpiece of our celebrations.

There is more that we can do. We can cultivate a garden, or a gardener. We can choose local greens rather than greens bagged in plastic from California. We can take our children to a local farm and show them where the meat, and milk, fruit, and vegetables on their dinner plates come from.

I want to thank the organizers of this event for bringing us together, and making us aware of the power of our choices.

Notes

[1] Portions of this presentation were included in the introductory chapter to Simpson, Elizabeth & Henning Sehmsdorf 2021. *Eating Locally and Seasonally: A Community Food Book for Lopez Island (and All Those Who Want to Eat Well)*, Lopez Island, WA, 5-14.

[2] Assistant Director, Lopez Community Land Trust, Lopez Island, WA.

[3] Burrows, Sara 2018. "Vegetables Lose Half Their Nutritional Value by the Time They Get to the Store: Another Reason to Grow Your Own," *Return to Now.* Net/2018/09/23. Retrieved 7/13/2021.

[4] Taylor, Kate 2017. "These 10 Companies Control Everything You Buy," https://www.businessinsider.com/10-companies-control-food-industry-2017-3. Retrieved 7/13/2021.

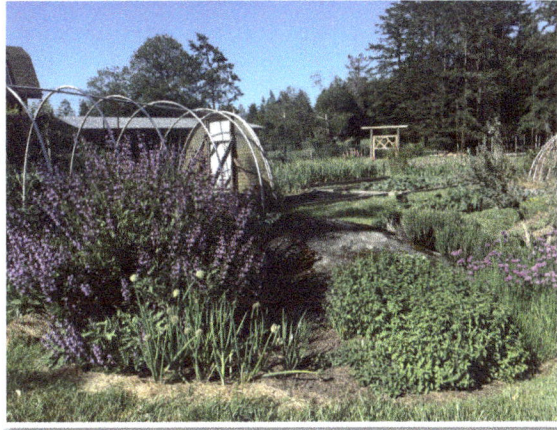

Medicinal Plants on S&S Homestead
Johanna Hök, 2002 & Henning Sehmsdorf, 2021[1]

Background

Since the dawn of humankind, plants have been used to heal and prevent disease. Herbal medicines surround us everywhere in nature. Despite urbanization and the industrialization of pharmaceutical production, the use of herbal medicines has not waned. However, most herbal medicines sold in western countries today are in the form of shelf-stable pills and tablets, that is, in industrialized form. Most people can no longer identify the medicinal properties of plants growing in their own backyards. Used the right way, herbs are an important resource for curing minor ailments and diseases, and for toning the body. Herbs increase the available energy of organs or body systems, and prevent disease. As a local source of medicine, the use of herbs is inexpensive and sustainable, and can reduce health care costs.

Medicinal plants can serve multiple purposes on self-sufficient farms like S&S Homestead. We will discuss the qualities of herbs as valuable medicines and tonics for humans and animals. We will show their use as companion plants in the garden, their role in repelling pests, their culinary and nutritional qualities, and their function in the farm organism in the form of biodynamic preparations.

In the herb garden and around the farm, some two dozen different perennial and annual medicinal herbs are cultivated. There are also many useful wild herbs growing on the farm and nearby. In the right climate and soil, most herbs will grow vigorously. In fact, most common weeds are valuable medicinal plants, among them dandelion, mustard, chamomile, stinging nettle and comfrey.

As with any medicines, herbal medicines have to be used with respect and caution. While some mild medicinal herbs can be consumed as regular food, others should be taken only under the guidance of an herbalist. Medicinal herbs should be taken for the right reason, in the correct dosage and for an appropriate length of time.

Practical Work With Herbs

A few generations ago most people knew how to collect, prepare and use medicinal plants. This knowledge has diminished in a culture where dietary supplements and pharmaceuticals are in common use. One way to revitalize knowledge of herbal medicines is to bring them back to organic gardens and farms to supply growers with many of the medicines they need. Instead of using a chemical vermifuge for sheep, for instance, herbal treatments can be found. One of the main benefits of an herbal garden is that the medicines are available whenever you need them.

However, it might be convenient to store some of your favorite herbal medicines as preparations. Preparations store better than fresh herbs and, made in large batches, are available whenever needed. To dry herbs or to make infusions, tinctures and syrups, the herbs should be harvested in the morning just after the dew has disappeared. This is when volatile oils, containing most of the active ingredients, are at peak level, and the herb contains minimal water.

The medicinal quality of herbs varies throughout their life cycles, and therefore should be harvested at certain, specific times of the year. Stinging nettle, for example, must be harvested before going to bloom in spring or fall, since old plants develop compounds that may be toxic to the liver.

During the dry summer, herbs hung in a bag upside down will desiccate fully within two weeks. A quicker method is to use a dehydrator. Infusions are made by steeping twenty grams of herbal material in 500 ml water for five to fifteen minutes, depending on desired concentration. The standard dosage for infusions from mild and tonic herbs, such as chamomile and mints, is three to four cups a day (500 ml). For highest medicinal value, infusions should be used within twenty-four hours.

In the case of roots, barks or twigs, infusions are not enough to extract their medicinal properties, and a decoction is a more effective. Twenty grams of dried material or forty grams of fresh are mixed in 750 ml cold water, brought to a boil and simmered for thirty to forty minutes, strained and the liquid stored in a cool place up to forty-eight hours. The standard dosage for decoctions is three to four

doses (500 ml) per day.

In making herbal preparations, it is crucial to keep all equipment and the environment scrupulously clean. All jars used should be sterilized.

A syrup is a blend of an herbal infusion and a sugar base. Mix one part of concentrated infusion (steeped for thirty minutes) with one part honey while the infusion is still warm. Stored in a cool, dark place, syrups will keep for six months. Syrups are especially suitable for children as cold remedies because the honey is soothing for sore throats.

Alcohol tinctures preserve the medicinal qualities of an herb for more than six months. Three parts forty-per cent vodka are mixed with one part of herb (200 g dried plant material or 300 g fresh to one liter of alcohol).

Chop the herbal material finely and put it into a sterilized jar. Pour the alcohol over the herbs to prevent oxidation. Shake the jar for one to two minutes, label with name, date and strength, and store in a cool dark place for ten to fourteen days. Shake the jar for one to two minutes every day. When tinctures are ready, use a press or strainer to extract the alcohol from the plant material. The tincture can then be stored up to two years in a cool dark place. Tinctures are the strongest kind of herbal preparations, and it is important to check appropriate dosage for each preparation.

Phytochemicals

Practical knowledge about herbs is essential for making medicines. It is also useful to know their chemistry. Most of the medicinal and common properties of herbs and plants, their flavors and colors, derive from phytochemicals. These secondary metabolites are not part of the essential plant metabolism, but help plants survive, and have important medical uses. For instance, some phytochemicals act as insect repellent and others are toxic to predators.

Unlike many vitamins, most phytochemicals are not destroyed by cooking or other processing.[2] Medicinal plants can be grouped by the pharmacological actions of the secondary metabolites they contain. Phenols, for example, are compounds that reduce inflammation when taken internally. Salicylic acid found in white willow (*salix alba*) and thymol found in garden thyme (*thymus vulgaris*) are two examples of phenols.

Tannins are compounds present in most plants. The harsh, astringent taste of tannins discourage animals from eating them. Tannins contract the tissues of the body and have a positive effect on the immune system.

Coumarins are found in a wide variety of plants and are responsible for many different biological actions. Bergapten, a coumarin in lovage and celery, can be used as a sunscreen.

Flavonoids are found in many plants and generally have anti-inflammatory properties and maintain healthy blood circulation. They are thought to keep cancer-causing hormones from latching on to cells.

Anthocyanins are pigments that give flowers and fruits blue, purple or red color. They help keep the blood vessels healthy.

Glucosinolates are found exclusively in the mustard family (*cruciferae*). They irritate the skin, but applied as poultices to painful joints, they increase blood flow and remove waste products. Glucosilinates are present in mustard and radishes.

Volatile Oils

Volatile oils are extracted from herbs to produce essential oils. This is a wide-ranging and versatile group of medicinally active substances, often containing tens and hundreds of active compounds. The volatile oil of

German chamomile (*chamomilla recutita*), for instance, contains the anti-inflammatory sesquiterpene azulene. Triterpenoid saponins are strong expectorants and may aid in the absorption of nutrients. Many steroidal saponins have a hormonal function. Certain saponins may prevent cancer cells from multiplying.

Cardiac glycosides can be found in various herbs, for instance, foxglove (*digitalis purpurea*). A farmer needs to be aware of which plants contains cardiac glycosides because they are toxic. As implied by the name, cyanogenic glycosides contain cyanide. However, the cyanogenic glycosides have some valuable medicinal properties, for instance, suppressing coughs. They can be found in the flowers of elderberry (*sambucus nigra*).

Bitters are a wide group of compounds linked by their pronounced bitter taste. Generally, bitters stimulate the salivary glands and digestive system, improving appetite and strengthening the digestive system. Increased digestive "power" in turn enhances the absorption of nutrients.

Alkaloids have three common characteristics: they are found in plants, their atomic structure forms organic rings containing at least one nitrogen atom, and they are pharmacologically active, having anti-asthma, anti-cancer, anti-arrhythmic, anti-bacterial, analgesic, and other such properties.

The Herb Garden at S&S Homestead Farm

Below are the domestic herbs most important to the farm in terms of their medicinal and culinary value. By herbs we mean non-woody plants we routinely consume as condiments to vegetable and meat dishes. We list them in approximate order of preferred use: alliums (garlic, onions, chives, shallots, leeks), parsley, thyme, rosemary, oregano, marjoram, sage, cilantro (coriander), peppermint, lovage, lemon balm, hyssop, borage, echinacea, fennel, catnip, raspberry, marshmallow, vervain, comfrey, and the flowers of calendula, nasturtium, and lavender. Among wild herbs (or weeds) common on this farm, the stinging nettle, horsetail, yarrow, dandelion, valerian, and chamomile, the latter two being both wild and domesticated, are highlighted both for their medicinal value and their use as biodynamic preparations to "enliven" the soil.

Allium (allium sativum) is the generic name (derived from the Latin word for garlic) for a mono-cotyledon, flower-bearing plant species cultivated the world over that includes garlic, onions, shallots, scallions, chives, and leeks. We grow all of these on the farm and they are part of our daily diet as condiments to vegetables and meats.

Alliums are energetic power houses containing high levels of phytochemical compounds that avert illness. Organic sulphur is a major characteristics of alliums, making up between one and five percent of the dry weight of the bulbs. Other vital compounds include carbohydrates, flavonoids and saponins.

Garlic (allium sativum). Widely known for its use in the kitchen, garlic is powerful medicine. The phytochemicals it contains include volatile oils (alliin, alliinase, allicin), selenium, and vitamins A, B, C and E. Garlic fights ear, throat, nose and chest infections, and lowers blood sugar levels. It reduces cholesterol, prevents strokes and lowers blood pressure. And it is safe to use in large doses!

We also use garlic externally for its antiseptic and wound healing properties. The raw juice is applied directly to infected wounds, bites, and stings. To protect trees from borers, we plant garlic (or nasturtiums) around young trees. As any allium, garlic will repel rabbits. However, cows and sheep intuitively know that garlic is health food. It is an important component in

vermifuges. If garlic is planted in the pastures, both cows and sheep will eat it eagerly! We blend it with salt or other feed to get the animals used to the taste.

Experiments with garlic on the farm confirm its use as a repellant for aphids and other insects on various plants. For garlic spray as an insect repellent, add three to four ounces of chopped garlic to two tablespoons of mineral oil. Refrigerate for twenty-four hours. Add one pint of water and one tablespoon of fish emulsion, and stir well. Strain and store in a glass container (no metal). Before using, dilute with water to a concentration of 1:20. Spray on infested plants. The anti-insect spray for the cow was prepared as an infusion and sprayed directly on her skin for fly control.

Onions (allium cepa) are herbaceous biennial plants grown for their edible bulbs and green scapes. They contain antioxidants and compounds that reduce cholesterol levels, fight inflammation, and decrease triglycerides. In winter and spring, we grow scallions (*a.cepa var. cepa*), and bunch onions (*allium fistulosum),* which are chopped whole and added to dishes, and in the summer yellow and red storage or bulb onions, the most widely cultivated species of the genus *allium*. An onion variety prized for its subtle flavor is the shallot (*a. cepa aggregatum*), which we mainly use for sauces and salads. A green onion grown all year are chives (*allium schoenoprasum*), a mildly onion-flavored herb we routinely add to salads, soups, vegetables and sauces. Rich in flavonoid antioxidants, regular doses of chives help fight inflammation, even cancer, and improve heart and skin health, while easing the digestive process. Another member of the onion family are leeks (*allium porrum)* native to the Mediterranean and Middle East. The edible portion of the plant includes the bulb and the tightly bound leaf sheaths. Leeks have high levels of flavonoids, vitamins A, E, and K, and fibers.

Alphabetical List of Medicinal Plants on S&S Homestead:

 Borage

 Calendula

 Catnip

 Chives

 Coriander

 Comfrey

 Echinacea

 Fennel

 Garlic

 Hyssop

 Lavender

 Lemon Balm

 Lovage

 Marjoram

 Marshmallow

 Nasturtium

 Oregano

 Parsely

 Peppermint

 Raspberry

 Rosemary

 Thyme

 Sage

 Vervain

Plants for Making Biodynamic Preparations:

 Chamomile

 Dandelion

 Horsetail

 Nettle

 Oak Bark

 Valerian

 Yarrow

Parsley (*petroselinum crispumis*), is native to Europe and the Middle East. Widely used in cuisine, it is rich in phytochemicals such as vitamins (A, C, and K), antioxidants, folate, and minerals (potassium, calcium, magnesium). A single tablespoon of fresh chopped parsley provides more than seventy per cent of the recommended daily intake of vitamin K, strongly supporting bone health. The plentiful supply of vitamin A in parsley helps protect the surfaces of the eye, and the antioxidants lutein and zeaxanthin guard against macular degeneration. Parsley is a natural diuretic and helps reducing blood pressure.

Added to soups, sauces, salads, or sprinkled over dishes, it is the most commonly used herbal seasoning on the farm. We also make it into a flavorful pesto.

Garden Thyme (*thymus vulgaris*) is a variety of wild thyme (*thymus serpyllum*), native to southern Europe.[3] Garden thyme is cultivated world wide. The aerial parts are used for their excellent culinary qualities as well as their medicinal properties. The green leaves contain vitamins B-complex, C and D. Other phytochemicals include volatile oils (thymol, borneol, methylchavicol, cineole), flavonoids (apigenin, luteolin), tannins, chromium, fluorine, iron, silicon, and thiamine. Research has validated the traditional use of thyme as an antiseptic.

Garden thyme has been traditionally used for mitigating gas and reducing headaches and mucus. Herbalists prescribe thyme for asthma and hay fever. On the farm, we make thyme syrup and tincture. Thyme is an excellent expectorant. The syrup is mild and tastes good in tea or hot water. The recommended dosage is twenty ml (four teaspoons) three times a day. Thyme tincture is stronger and the recommended dosage is five ml (one teaspoon) two to three times a day. An infusion of thyme should be taken for the same complaints in a

dosage of fifty ml, three times a day.[4] Rodale suggests planting thyme here and there in the garden. It deters cabbage worms and whiteflies.[5] The literature also indicates that thyme has insecticidal properties.[6] Thyme is also thought to treat all kinds of digestive complaints in animals. One handful, brewed and finely cut, is recommended mixed with food. Thyme is also known as a worm expellant, anti-septic, and insecticide. The literature suggests using thyme infusion as an insecticidal spray.

Rosemary (*rosmarinus officinalis*) grows in sandy, rocky places, mountains, hills, or on sea cliffs, which is why it was named "Rose Marinus" (Dew of the Sea). Rosemary is a valuable herb that "raises the Spirit." It has bitter and astringent properties. Teas, tinctures and oils made from rosemary leaves have been used traditionally as tonics. Rosemary is believed to strengthen memory and soothe the nervous system. Rosmaricine, one of the key components in the leaves, has been shown to be a stimulant and a mild analgesic. Phytochemicals include volatile oils (borneol, camphene, camphor, cineole), rosmaricine and rosmarinic acid, diterpenes tannins, and flavonoids (apigenin, diosmin).

Applications include treatment of headaches and migraine, poor circulation, and poor function of the adrenal glands. Rosemary is a good companion plant for cabbage, beans, carrots, and sage.[7] This perennial deters cabbage moths, bean beetles and carrot flies. A strong brew of rosemary or the infused oil, functions as an insecticide. For animals, a hot strong brew of Rosemary is used externally for body and limb massage to treat rheumatic or arthritic problems. The herb is also used traditionally for problems involving the heart, convulsions, epilepsy, chorea, paralysis, impure blood, gastritis, diarrhea, dysentery, obesity, and torpid liver. Rosemary flowers are a major food source for bees.

Oregano (*origanum vulgare*). Although very similar in appearance to sweet marjoram, the two herbs are quite different in their chemical structure. Historically, oregano has been more widely used as a medicine. Medicinally, oregano can be described as a hot and dry herb. It contains the following phytochemicals: tannins, resin, sterols, flavonoids, and volatile oils (carvacrol, thymol). Research has proven the antibacterial and antifungal properties of volatile oils in oregano. The herb is strongly antiseptic and can therefore treat respiratory conditions such as coughs. The diluted oil can be applied for toothache or painful joints. Oregano should not be taken as medicine during pregnancy and the essential oil should not be taken internally. Like marjoram, when planted here and there in the garden, it improves the flavor of vegetables.

Sweet Marjoram (*Origanum majorana syn. majorana hortensis*) has stimulant and antispasmodic properties. The phytochemical content of marjoram includes three per cent volatile oils (sabinene, linalool, carvacrol and other terpenes), and flavonoids. It has been used to treat flatulence, colic and respiratory problems. It makes a good general tonic, helping to relieve anxiety, headaches and insomnia. However, the herb should not be taken as medicine during pregnancy and the essential oil should not be taken internally. Sweet marjoram is good to plant here and there in the garden because it improves the flavor of most vegetables.

Sage (*salvia officinalis*).The Latin name *salvia* comes from *salvare*, meaning "to cure." Sage is an evergreen bush cultivated all over the world. It thrives in sunny conditions. Although best known today for its culinary use, sage is also an important medicinal herb with astringent, antiseptic and relaxing properties.

There are a number of varieties of the herb. The purple sage (*salvia officinalis purpurascens)* is

the one preferred for medicinal use. It it rich in volatile oil (fifty percent thujone), diterpene bitters, flavonoids, phenolic acids, tannins, and esterogeic substances.

Sage is used in gargles for sore throats and as a digestive tonic and stimulant. The fresh leaves can reduce pain from stings and bites.

Containing estrogen-like substances, sage helps with menopause problems, reduces sweating and helps the body adapt to hormonal changes. Sage is sought out by ruminant animals. Given to a lactating cow, sage increases the milk yield and makes the milk refreshing and tonic. Sage is a companion plant to cabbage, carrots and rosemary. Sage deters cabbage moths and carrot flies. Sage is antagonistic to cucumbers.

Cilantro (*coriandrum sativum*), consumed either in leaf form or as seed known as coriander, the herb is credited with reducing the risk of heart disease, diabetes, obesity, and seizures, as well as raising energy levels and healthy hair and skin. It contains immune-boosting antioxidants, vitamins A, C, and K, and the leaves have folate, potassium, and manganese. We use it to flavor meat dishes and sauces.

Peppermint (*mentha piperita*), also known as *mentha balsamea* (wild peppermint) is a wild cross between water- and spearmint indigenous to Europe and the Middle East, but now widely cultivated throughout the world. Added to foods, fresh or dried, or consumed as a tea or in alcoholic beverages, peppermint is known for its cooling properties. Mint is rich in vitamin A (critical for eye health and night vision), manganese, iron and folates, flavonoids, saponins, alkaloids, tannins, steroids, glycosides, volatile oils and is a potent source of antioxidants.

Studies found that applying peppermint oil mixed with gel or water to the nipple relieves pain from breast feeding. On the farm, this application has proven useful in treating the teats of lactating animals. Peppermint tea is helpful in alleviating symptoms of the common cold or flu, and is a welcome herbal tea because of its refreshing flavor. Drinking the tea or chewing mint leaves reduces bacteria causing bad breath.

Peppermint oil contains menthols thought to be an effective remedy for irritable bowel syndrome and to relieve indigestion. Some studies have shown that inhaling the aroma of peppermint oil increases memory and lessens anxiety, fatigue, and frustration.

Lovage (*levisticum officinale*) is a warming and tonic herb for the digestive and respiratory systems. The phytochemistry of lovage includes volatile oils (bergsten, psoralen, umbelliferone), plant acids, phthalides resins, gums, and beta-sitosterol. It treats indigestion, poor appetite, flatulence, colic and bronchitis, promotes menstruation, and relieves period pain. It also improves poor circulation. The phthalides have sedative and anticonvulsant properties. However, lovage should not be used as a medicine during pregnancy. The leaves are harvested as needed for fresh use. In fall, bunch foliage and stems are hung up to dry. Lovage enhances flavor and health of plants if planted here and there.

Lemon Balm (*melissa officinalis*), which belongs to the mint (*labiatae*) family, has a long tradition of use as a tonic that raises the Spirit and comforts the heart. It contains volatile oils (citral, caryophyllenen oxide, linalool and citronellal), flavonoids, triterpenes polyphenols, and tannins. It is a calming herb, relieving insomnia, menstrual cramps and cold sores, and reducing the chance of outbreaks. The aerial parts can be used fresh or dried and made into a delicious evening tea. The juice from the leaves is also used as a first-aid remedy to soothe cuts and insect stings. As other herbs in the mint family, lemon balm

is a good companion to cabbage and tomatoes. improving the health and the flavor of these vegetables. Lemon balm also deters white cabbage moth.

Hyssop (*hyssopus officinale*) is a shrub celebrated in the Bible for cleansing victims of plague, leprosy and chest ailments, and symbolically for cleansing the soul. For the Jewish Passover, hyssop sprigs were used to sprinkle blood on the door to fend off the angel of death, and in Egypt it was used for religious purification. The phytochemicals in hyssop include diosmine, volatile oils, flavonoids marrubin, and tannins.

The efflorescence of this beautiful perennial shrub has been traditionally used for lung problems such as chronic coughs and asthma. It has a dual action, encouraging the production of mucus and acting as an expectorant. The leaves and flowers make an excellent calming and tonic tea. The blue flowers of hyssop are very attractive to bees. Planted in the cabbage bed, hyssop deters cabbage moth.[8] Hyssop has traditionally been used in the Nordic countries as a vermifuge for weak lambs and kids.

Borage (*borago officinalis*) is a hairy annual with intense blue flowers highly attractive to bees. Borage contains mucilage and tannis, as well as pyrrolizidine alkaloids that are toxic when consumed in large quantities; therefore, its medical use is controversial.

However, the small flowers have a cucumber-like taste and are beautiful used in salads. Borage is a good companion plant for tomatoes, squash and strawberries; borage deters tomato worm.

Echinacea (*echinacea angustifolia & e. purpurea*) is native to North America and thrives in the climate of the Pacific Northwest. Research has proven echninecia's ability to raise the body's resistance to bacterial and viral infections by stimulating the immune system. It is also good for allergies and asthma. Echinacea contains volatile oil (humulene), alkamides, caffeic acid, esters, polysaccharides, echinolone betaine potassium, and vitamins A, C and E.

It was used by Native Americans as a remedy for toothaches and sore throats. Medicinal preparations are made from combinations of the roots, flowers and leaves. The flowers are gathered in full bloom and the roots in autumn when the plants are two to four years old.

Fennel (*foeniculum vulgare*) is an aromatic perennial herb with feathery leaves. Indigenous to the shores of the Mediterranean, it has become naturalized in many parts of the world. Fennel stalks can replace celery in soups and stews, and complement roasted chicken, meats, and seafoods. The fronds do well as a garnish, or chopped, used like dill or parsley. An infusion of the leaves enhances digestion and makes an excellent after-meal tea. Fennel contains volatile oils, coumarins, sterols, oleic acid, flavonoids, and vitamins A and C.

Fennel was popular among dairy farmers because it increases the yield of milk and gives it a sweet taste. In the garden, fennel is antagonistic to most plants and should be planted in a separate pot or away from other herbs or crops.[9]

Catnip (*nepeta cataria*), long recognized for its phytochemical and medicinal properties, is a highly aromatic herb renowned for use in cooking and as tea, and is grown to distill essential oils for use in foods, flavors, and personal hygiene.[10] It contains polyphenols, phenolic acids (caffeic and rosmarinic), flavonoids (luteolin, apigenin and glucosyl, and glucuronyl derivatives). Catnip is anti-inflammatory (inhibits nitric oxide).

Well known for the volatile compounds causing euphoria in felines, as an herbal tea, the herb was purported in Native American and Black American traditions to treat inflammation, digestive ailments, infantile colic, toothaches, the common cold, anxiety and as a blood depurative. Europeans and the Chinese used catnip leaves in cooking and herbal infusions. Consumption of polyphenols have been reported to reduce the risk of cancer, cardiovascular diseases, neurodegenerative diseases and inflammation, and could be useful for people suffering from Alzheimers. Anti-oxidant activity by phenolic acids and flavonoids in catnip have been shown to reduce inflammation, and hazardous free radical formations have been confirmed. On S&S Homestead catnip tea is a favorite beverage.

Raspberry (*rubes idaeus*) is a fruiting shrub whose leaves themselves are a nutritionally potent agent. They are high in Vitamins C, E, A and B, and contain significant amounts of magnesium, potassium, calcium, and phosphorus, as well as the essential trace minerals zinc, iron, chromium and manganese. The vitamins and minerals become available in tea made from the dried leaves.[11] They also contain gallotannins, flavonoids, quercetin, anthocyanin, cumaric, ferric and ellagic acids.

The phytochemicals in red raspberry leaves act as antioxidants, protecting against heart disease and mental aging problems, and the ellagic acid may decrease the risk of cancer, inhibit growth of tumors caused by carcinogens, and protect DNA against damage by radiation therapy. The leaves are astringent and stimulating, traditionally used to treat throat infections, diarrhea, and to aid in childbirth. Studies in humans and animals have shown that extracts of raspberry leaves strengthen the uterine muscles, and stimulate milk production.[12]

Marshmallow Root (*althaea officinalis*), a perennial herb native to Europe, Western Asia, and Northern Africa, has been used as a remedy for thousands of years for wound dressing and to treat digestive, respiratory, and skin conditions. Phytochemicals in marshmallow, of which the flowers are particularly rich,[13] include alkaloids, flavonoids, phenolic acids, saponin, tannins, and antioxidants.

Marshmallow Root contains mucilage enzymes which loosen mucous and inhibit bacteria. It eases coughs, skin irritation and wounds, pain, diuretic, digestive and heart problems such as inflammation linked to cardiovascular disease, and serves as an anti-oxidant protecting cells from free radicals. It strengthens the immune system. As a tea, it acts as an analgesic. A powerful herb, it should not be ingested for more than four weeks at a time. We rotate it as a stimulating tea, but also use it as a tincture.

Vervain (*verbena officinalis*) was a favorite of Hippokrates and the Druids. who used it on their farms as medicine for people and animals. It is harvested in the summer when the flower is in bloom.Vervain is now widely recognized for its health promoting properties, such as "anti tumor effects, nerve cell protection, anxiety- and convulsion reducing properties, and antimicrobial activities."[14] Vervain contains volatile oils, bitter iridoids, alkaloids, mucilage, and tannins.

Comfrey (s*ymphytum officinale*), a shrub native to Europe, Asia, and North America, grows up to five feet tall, producing clusters of purple, blue, and white flowers, long, slender leaves and black-skinned roots. Traditionally, comfrey ointments have been used to heal bruises, pulled muscles and ligaments, fractures, sprains, strains, and osteoarthritis. But since the roots and leaves contain the alkaloid allantoin, a potentially toxic substance, in 2001 the USDA banned the sale of oral comfrey products because the toxins in the plant can cause liver damage, cancer, genetic mutation, or death.[15]

Farmers, however, prize comfrey as a nutritious animal feed rich with fifteen to thirty per cent dry-weight protein content, rivaling some legumes. As pig fodder, comfrey can comprise up to eighty to ninety per cent of the diet, and for poultry up to fifty per cent, and improve egg quality. Cows like to eat it, and mastitis is reduced in cows fed comfrey.[16] Presumably, potential liver damage from long-term use of comfrey is of little concern in short-lived slaughter animals.

Comfrey also provides ecological benefits by attracting pollinators with its showy flowers and habitat for beneficial insects, thereby helping to keep the garden pest-free. Comfrey leaves make a nutrient-rich mulch, and fermented compost tea made from comfrey is a potent fertilizer.

Calendula (*calendula officinalis*). The flower petals of this orange-yellow herb have anti-inflammatory properties and are used as a skin soother. External application of calendula helps prevent the spread of infection and speed healing. Calendula contains carotenes, flavonoids, glycosides, triterpenes, volatile oils, lycopine, saponin, resin, and sterols. It is antiseptic and antifungal.

Calendula should not be confused with marigold (*tagetes*). In *Rodale's Herbal Garden*, calendula is called the workhorse of pest deterrents. The plant repels Mexican beetles, nematodes, and other insects. It is ideal to plant throughout the garden and under trees and bushes that might be targets for these pests. For a colorful winter garden, it can be planted in fall.

Nasturtium (*tropaeoleum majus*) This colorful plant offers spicy, palatable leaves and flowers. Nasturtiums have more value in the garden than in the apothecary. However, all parts of the plant appear to have antibiotic actions, and the juice from the leaves and flowers makes an antiseptic wash for external use, both for humans and animals. It contains the phytochemicals glucocyanates, spilanthol myrosin, oxalic acid, and high levels of vitamin C. Nasturtiums deter aphids, squash bugs, striped pumpkin beetles and are good companions plants for radishes and cabbage. They should also be planted under fruit trees.

Lavender (*lavendula officinals syn. l. angusitfolia*) is an important relaxing herb, but is more widely known for its sweet aroma. Phytochemically, lavender contains cardiac glycosides, terpenes, steroids, coumarins, flavonoids, sponins, tannins, anthocyanates, and essential oils.

The flowers of lavender are picked in high summer and often dried or distilled to produce essential oil. The essential oil has very low toxicity and is used externally for headaches, insect bites and relieving pain. Massage a few drops of oil into the area of pain (massaging the temples with the oil eases headaches). The oil stimulates blood flow and relieves muscle spasms. For insomnia, take one half to one teaspoon of lavender tincture with water. A multitude of phytochemicals in lavender have anti-fungal and anti-bacterial properties and protect against diseases of the central nervous system.[17]

Wild Herbs & Biodynamic Preparations

There is increasing evidence that wild edible herbs, fruits and vegetables contain higher quantities of phytochemicals than their domesticated counterparts. Commercial cultivars selected for size, appearance, shipping and storage-ability, showed decreased range of phytochemicals found in wild herbs, vegetables and fruits, and lessened flavor and nutritional wholeness.

The modern strawberry is a telling example. When strawberries were first mentioned in ancient Roman literature for their medicinal use, the reference was to the wild forest berry (*fragaria vesca*) used to treat depression. As early as the 14th century, the French were cultivating strawberries in their gardens, and in order to narrow genetic variability, the plants were propagated asexually by cutting off the runners.

In the 17th century, the cross-breeding of wild strawberries from North America (*fragaria virginiana*) with wild strawberries from Chile (*fragaria chiloensis*), gave rise to the modern strawberry prized for its size and juiciness. However, the deep, spicy flavors and phytochemical diversity of the wild strawberries were lost.

Inversely, wild edible plants provide potential for crop improvement in genetic variability and propagation in a large range of habitats. It has been shown that domestication of wild plants is least deleterious if cultivated in the open field rather than in non-soil media — as in hydroponics where fertilizers are suspended in water instead of in soil — or under cover, as in a greenhouse.[18] None of the results of these botanical studies are surprising to the biodynamic grower who knows that embedding the plant in the soil open to the sky is essential to the influx of "vitalizing" energies from the cosmos.

Biodynamic preparations are used to restore the plant vitality lost by domestication. The preparations consist of various substances applied in homeopathic doses to plants and soil. The preparations — fermented chamomile, dandelion, nettle, yarrow, valerian, horsetail, and oak bark — enhance plant and soil metabolism.

German chamomile (*matricaria chamomilla syn. chamomilla recutita*). The hooded flower heads of German chamomile have been used for digestive problems ever since the 1st century A.D. Unlike Roman chamomile (*anthemis nobilis*), German chamomile is an annual that must reseed every year. Used fresh or dried in a tea, it promotes sleep and relieves irritation caused by allergies. It contains a-bisabolol, a-bisabolol oxides A and B, chamazulene, and espatulenol.

German chamomile is useful for calming hay fever and asthma, smoothing the skin, reducing inflammation and fever, and soothing the nervous and digestive systems. Chamomile also induces perspiration to flush out toxins, allergens and infections. Externally it can be applied as a poultice to reduce skin irritations such as eczema.

In older times, bunches of chamomile were put up in barns to deter flies. The powdered plant has been used as a flea repellent. In the garden, chamomile improves the growth and flavor of cabbages and onions.

As one of the six plant substances from which the biodynamic preparations are made, German chamomile has a special role to play on S&S Homestead Farm. Because it does not occur in the wild in our area, we grow it in the garden. The dense and bushy canopy, finely divided leaves and shoots topped by yellow flower heads, correspond to a widely branched root system that breaks up compacted soil.

The dominant element in German chamomile is sulphur (from Latin *sol* = sun, plus *ferre*= to carry) which carries the vitalizing force of the sun into the soil. Sulphur works in conjunction with calcium, giving structure to the soil by the ability to bind. The flower heads are collected just as they begin to open (usually in May), preferably on a sunny morning. The flowers are dried on a screen or on paper, and stored in a paper sack in an airy place until autumn. They are packed into the intestines of a slaughtered cow and buried in the garden or in the open field to ferment and absorb the systolic energies of the winter soil.[19] In spring the finished preparation is unearthed and inserted into compost piles to stimulate and organize the sulphur metabolism of the piles.[20]

Valerian (*valeriana officinalis*) is another "wild" plant we grow in the garden, both for medicinal use and as a biodynamic preparation. A perennial flowering plant native to river meadows and on damp edges of woods in Europe and Asia, the mature plant reaches a height of up to six feet. Its scented pink or white flowers harboring sticky, sweet nectar attract bees and other pollinators. Valerian was used as a medicinal herb in ancient Greece and Rome. In the 5th century B.C. it was described by physician Hippocrates, and in the 2nd century A.D. Galen recommended valerian for insomnia. In 16th century Europe, it was used to treat nervousness, heart palpitations and headaches. During WWII, it was used in England to relieve stress due to air raids.

Valerian contains terpenes (valepotriates, esterified iridoid-monoterpenes, valtrate, and acevaltrate), pyridine alkaloids, and organic acids.

Today, extracts of the roots of valerian are widely used for inducing sleep and improving sleep quality without side effects. This remedy is important because insomnia affects approximately one-third of the adult population in the U.S. Insomnia contributes to increased rates of absenteeism, need for health care, and social disability.[21] Traditionally, valerian was also used to treat gastrointestinal pain and spastic colitis. Phytochemically, valerian contains over 150 active components.[22] To this day, scientists do not fully understand the medical effects of Valerian.[23]

Biodynamically, the preparation made from the valerian blossoms stimulates the phosphorus processes related to warmth in the soil. As a diluted spray, it provides frost protection for flowering fruit trees, and promotes the ripening of flowers, fruit, and grains.

In June, opened clusters of flowers are collected in the morning, ground finely and pressed to extract the brownish, redolent juice, which is then fermented in a dark bottle for minimally six weeks. Sprayed on compost heaps, garden beds and fields, it creates a protective enveloping warmth and is beneficial to earthworms.[24]

Dandelion (*taraxacum officinale*) is considered a pesky weed by most people, but in fact provides important medicinal benefits. The plant contains substantial phytochemical substances, such such as flavonoids, phenolic acids, terpenes, carotenoids, coumarins, inositol derivatives, vitamins (A, C, K, E, B), folate, and minerals (iron, calcium, magnesium, potassium).[25] Dandelion has proven effective in the treatment of acne, eczema, high cholesterol, heartburn and gastrointestinal disorders, diabetes, and even cancer.[26] In traditional Chinese, Ayurvedic, and Native American medicine, dandelion root has long been used to treat stomach and liver conditions.[27] Dandelion also has important culinary and nutritional values. The greens can be eaten cooked or raw and serve as an excellent source of vitamins and minerals.

The biodynamic farmer pays attention to the latex-filled roots and stems of dandelion, each supporting a flower bud that opens at sunrise facing east and closes at sunset facing west, showing that the dandelion is a plant gifted with special sense organs for the life-giving energy of the sun. When flowering is over in spring, each floret in the composite flowering head metamorphoses into a delicate, siliceous seed-bearing parachute.

Silica support all sensory functions in plants, animals and humans. In the dandelion, this function arises from the interaction between silica and potassium.

The dandelion preparation is made from the flowers gathered in spring before they are fully opened, then dried and stored until fall, when they are wrapped in the mesentery (the membrane that holds the intestines in place) of a slaughtered cow, and buried to ferment in the ground for the winter. When stirred out in water in spring and applied to fields and compost piles, the dandelion preparation regulates the interaction of silica and potassium in the soil.

Horsetail (*equisetum arvense*), a wild fern native to Northern Europe and America, grows rampantly near our farm pond. Its long, green and densely branched stems and hexagonal, lateral branches consist mostly of silica, a compound comprised of silicon and oxygen. The phytochemical content of the plant includes apigenin, luteolin, quisetumoside A, equisetumoside B and C, nicotine, palustrine and palustrinine, providing anti-bacterial, anti-fungal, anti-oxidant, analgesic, anti-inflammatory, anti-diabetic, anti-tumor, cytotoxic and anti-convulsant benefits, and supporting the health of the skin, hair, nails and bones.

Traditionally, horsetail has been consumed as a tea made from the dried herb and has been in use since the ancient Greeks and Romans the world over.[28] For the biodynamic farmer, horsetail holds a special position. It is not one of the compost preparations, but is sprayed as a tea directly on plants and soil. Because of its siliceous nature, it relates to light and warmth, and as such it is used to control fungal growth under damp and sunless conditions.

Yarrow (*achilles millefoleum*). The yarrow plant does not grow wild on our farm; we could cultivate it from commercial seed, but prefer to harvest the wild yarrow growing profusely on the rocky shores of the island. It has finely divided leaves, and pink or white blossoms; its stems are hard, almost woody. Its phytochemical content features flavonoids, phenolic acids, coumarins, terpenoids (monoterpenes, sesquiterpenes, diterpenes, and triterpenes).

The use of yarrow as medicine is ancient. In Homer's Illiad, Achilles treats the wounds of his injured friend Patroclus with a poultice of the plant; thus the name *achilles millefoleum* (Achilles' Thousand Leaves). Native Americans and early settlers used yarrow for its astringent qualities in wound healing and blood stopping. Traditionally, yarrow has been used to treat fever and cold, dysentery, diarrhea, gastrointestinal problems, and skin problems.

Yarrow is consumed as tea, or by chewing the leaves, and the young leaves and blossoms are added to foods fresh or cooked.[29] Recent phytochemical analysis of yarrow has confirmed that many components of the plant are highly bioactive.[30]

17th century English physician and botanist Nikolas Culpeper described the traditional medicinal uses of yarrow. He added that if fed to cows while the leaves are tender and fresh, yarrow "will double (the cow's milk) production."[31]

The biodynamic farmer looks to yarrow as one of the compost preparations to regulate the potassium and sulphur balances in the soil. The preparation is made by cutting the florets from the dry stalks and drying them. In spring after Easter and by Midsummer at the latest, the moistened flowers are stuffed in the bladder of a stag and hung in a protected place until fall, at which time the preparation is buried in the ground to ferment.[32]

Oak bark (*quercus robur*). The oak may seem out of place in a list of medicinal herbs. Unlike herbs, trees are huge and woody plants. But the oak belongs by virtue of the medical uses of its bark, for instance, as tea for colds, coughs, fever, bronchitis, and diarrhea, and for stimulating appetite and improving digestion. Today some people apply oak bark directly as a compress or add it to bath water to ease inflammation.[33]

Recent research has found that the galls, wood, bark and twigs of the Quercus species contain phytochemical substances that have anti-oxidant, anti-inflammatory, anti-microbial, and anti-cancer functions.[34] The list includes flavonoids, phenolic acids, coumarins, and terpenoids (monoterpenes, sesquiterpenes, diterpenes, triterpenes).

Quercus robur also provides the biodynamic grower with another important compost preparation, based on the calcium content of the bark (seventy-five percent), and tannins (twenty-five per cent). Calcium has anti-fungal properties and the tannic acid acts as an insecticide.

The bark is scraped from the tree, finely ground and stuffed in the cranial cavity of a slaughtered cow and submerged in a barrel filled with leaf sludge over which water flows periodically, for instance, from a gutter. After a year of fermentation, the finished preparation is inserted in compost piles to protect against fungal growth.[35]

Notes

1 Originally composed by intern Johanna Høk in 2002, the essay was edited and amplified by Henning Sehmsdorf, 2021. For the original version, see https://sshomestead.org/wp-content/uploads/Herb%20project.pdf.

2 Balch, James & Phyllis 1997. *Prescription for Nutritional Healing*, 2nd ed. New York, NY.

3 For an excellent recent review of longitudinal studies of thyme, see Snežana Jarić et.al. 2015 "Review of Ethnobotanical, Phytochemical, and Pharmacological Study of *Thymus serpyllum* L," *Evidence-Based Complementary and Alternative Medicine*. https://www.hindawi.com/journals/ecam/2015/101978/. Retrieved May 27, 2021.

4 Chevallier, Andrew 2016. *Encyclopedia of Herbal Medicine: 550 Herbs and Remedies for Common Ailments*. London, UK.

5 Hylton, William H. 1974. *The Rodale Herb Book*, Emmaus, PA.

6 Bairacli, Juliette de 1984. *The Complete Herbal Handbook for Farm and Stable*. London, UK.

7 Jeavons, John 2006. *How to Grow More Vegetables…* Berkeley/Toronto, CA, 169.

8 Riotte, Louise 1998. *Roses Love Garlic: Companion Planting and Other Secrets of Flowers*. Pownal, VT.

9 Jeavons, op.cit, 168.

10 Wu, Qingli. et al. 2018. "Phytochemical Analysis and Anti-Inflammatory Activity of Nepeta cataria Accessions." *Journal of Medicinally Active Plants* 7, (1):19-27. DOI: https://doi.org/10.7275/1mca-ez51 https://scholarworks.umass.edu/jmap/vol7/iss1/4. Retrieved May 27, 2021.

11 Zabel, Steph 2017. "Raspberry: Beyond the Fruit." https://www.cambridgenaturals.com/blog/raspberry-beyond-the-fruit#. Retrieved May 27, 2021.

12 https://www.phytochemicals.info/plants/raspberry.php#:~:text/ Retrieved May 28, 2021.

13 Ewais, E.A. et al. 2019. "Phytochemical Contents of White and Pink Flowers of Marshmallow (*Althaea officinalis* L) Plants and their Androgenesis Potential on Anther Culture in Response to Chemical Elicitors," *Bioscience Research* 16(2): 1276-1289.

14 Lang, Adrian 2020. "What is Vervain?" https://www.healthline.com/nutrition/vervain-verbena. Retrieved June 1, 2020.

15 Reed, William 2002. "FDA Bans Sale of Comfrey Herb." https://www.foodnavigator.com/Article/2001/07/09/FDA-bans-sale-of-comfrey-herb. Retrieved June 1, 2021.

16 "Permaculture Reflections." www.permaculturereflections.com/2009/02/species-of-month-comfrey.html. Retrieved June 21, 2021.

17 Piccaglia, Roberta et al. 1993. "Antibacterial and antioxidant properties of Mediterranean aromatic plants," *Industrial Crops and Products,* 2 (1): 47-50.

18 Ceccanti, Constanza et al. 2020. "Comparison of Three Domestications and Wild-Harvested Plants for Nutraceutical Properties and Sensory Profiles in Five Wild Edible Herbs: Is Domestication Possible?" *Foods* 9(*): 1065.

19 For a detailed discussion of the choice of the various sheaths in which biodynamic preparations are fermented, see Sehmsdorf, Henning, "Biodynamics: Personal History of an Applied Idea," below.

20 Wistinghausen, Christian von, et al. 2000. *The Biodynamic Spray and Compost Preparations Production Methods, Book I.* Stroud, UK, 38-43.

21 Bent, Stephen, et al. 2006. "Valerian for Sleep: A Systematic Review and Meta-Analysis," *American Journal of Medicine,* 119 (12): 1005-12.

22 Patočka, Jiří, et al. 2010. "Biomedically Relevant Chemical Constituents of Valeriana officinalis," *Journal of Applied Biodmedicine, 8: 11-18.*

23 Johnson, John 2020. "What to Know About Valerian Root." https://www.medicalnewstoday.com/articles/valerian-root. Retrieved May 29, 2021.

24 Wistinghausen, op.cit, 67-71.

25 Singh, R. et al."Phytochemical Composition and Functional Properties of Dandelion," *Acta Horticulturea, 1287: 24.*

26 Ovadje P., et al. 2016."Dandelion root extract affects colorectal cancer proliferation and survival through the activation of multiple death signaling pathways." *Oncotarget.* 7(45):73080-100.

27 Wong, Cathy 2020. "Health Benefits of Dandelion Root." https://www.verywellhealth.com/the-benefits-of-dandelion-root-89103. Retrieved May 29, 2021.

[28] Asgarpanah, Jinous 2012. "Phytochemistry and Pharmacological Properties of *Equisetum arvense*," *Journal of Medicinal Plant Research*, 6: 21.

[29] N.a. "Yarrow" 2021. https://www.rxlist.com/yarrow/supplements.htm. Retrieved May 30, 2021.

[30] Saeidnia, S., et al 2011. "A Review of Phytochemistry and Medicinal Properties of the Genus *Achillea*," *DARU Journal of Pharmaceutical Science*, 19 (3): 173-186.

[31] Culpeper, Nikolas 2019. *Culpeper's Complete Herbal: Over 400 Herbs and Their Uses.* London, UK.

[32] Wistinghausen, op.cit, 30-37.

[33] N.a. "Oak Bark." https://www.rxlist.com/oak_bark/supplements.htm. Retrieved May 30, 2021.

[34] Burlacu, Ema et al. 2020. "A Comprehensive Review of Phytochemistry and Biological Activities of Quercus Species," *Forests*, 11: 0904.

[35] Wistinghausen, op. cit, 50-56.

First Year in the Development of Lopez Community Farm C.S.A.

Lisa Murgatroyd and Jesse Pizzitola, 2007

A year and four months ago, we heard about a little place called Lopez Island. We had been touring farms all summer looking for just the right fit. We had started in California, and after visiting about twenty farms we had just arrived on the other side of the country, Massachusetts. It was then we heard that S&S Homestead Farm was offering land, resources, and capital to start a small CSA. We headed back west, making a bee line for an exotic little island in the far corner of the country, drove onto an odd vessel that floated us out into the Puget Sound, and through the magic isles of the San Juans. Sitting on deck, we looked out over the sparkling waters and islands washed in the golden orange of sunset and shared our last beer. It was a moment of almost as much fear as beauty and excitement. We remember feeling raw and on the edge of what might have been panic. Except for a handful of trips to the mainland, we have been on Lopez Island ever since. In that time, with the help of Henning

and Elizabeth, and a great many others from the community, we have developed a plan for a CSA, taken it through its first year, and found that Lopez Island embraced it and wants it to continue. The last year has been filled with calculations, hoping, guessing, mistakes, faith, failures and successes. Jesse had a couple years of experience managing a small CSA before, and Lisa had experience on farms as well, but neither of us had started a business from scratch. It was quite a project that took a lot of work and was ultimately very rewarding. The CSA was a success, forty-five families ate locally for six months because of it, and the work last year paved the way for others to come. This paper will cover what we learned and the techniques we used in developing the CSA.

The Farm

S&S Homestead is a fifty-acre farm. They pasture beef and lamb, chickens and pigs. They grow vegetables and have a tree fruit and berry orchard. They have a milk cow and grow barley and hay for animal feed. What is undertaken on S&S Homestead is not necessarily decided by what will make the most money, but by what will create the most health and harmony among all the parts. All of the enterprises are designed and managed to intertwine with each other and create diversity and health on the farm and to minimize off-farm inputs. Their vision is to put the land into a trust that will give a farming family secure access to farming and living here. The idea of the CSA is to create additional income for a young family to step into the farm. The balance the CSA needs to strike is to earn an income while being integrated into the rest of the homestead. This year, while scheming to make the CSA an entrepreneurial success, we were also striving to keep it consistent with the biodynamic philosophy of S&S Homestead. Luckily, we agreed with this underlying philosophy and are committed to manifest it in the CSA.

Member Management

Building good relationships with our members has been fundamental to our success this year. Our goal was to teach members the value of local food and a food community through enjoyable experiences on the farm and in the kitchen. Our marketing, record keeping, member communications, vegetable pickup, and worker-share design were all thoughtfully undertaken with these goals in mind.

Marketing

Our business goal was to enlist forty-five full-share members for the season. Since soil fertility was still unknown, we decided on a low membership goal, ensuring a bountiful first season, and a good name for the CSA in this small community. CSA pickup began in June, and we set monthly goals to fill CSA membership by May. We created a three-fold brochure on Microsoft Publisher that described the social goals of the CSA, the financial arrangements, schedules, and a bit about us.

Our big marketing debut came on the weekend after Thanksgiving, at the Preschool Bazaar, a holiday craft event, where we publicized the new CSA while selling garlic and sheepskins for the farm. We hung a beautiful cloth sign advertising our new enterprise from the table at the bazaar, and later hung it on the outside wall of the CSA distribution shed the first few weeks of pickup to help members locate the farm. After the event, several people called and signed up, our first members.

At this time, we also set up an email account we checked three times a week and found that ninety percent of our members use email regularly. For marketing's sake, it helped us to make friends who were community organizers and advocates of local food. They passed out brochures and spread the word. Local businesses allowed us to place stacks of brochures on their counters. We posted signs advertising farm memberships with tear-off contact information on the bottom. To encourage sign-up, we advertised an early-bird price, but, being reluctant capitalists, kept the same price level for everyone. We sought out community functions and groups to present what we were offering to anyone willing to listen. Plugs at the Master Gardeners, the Garden Club, a local food charette, as well as Lisa's position at school, provided opportunities for publicity. The Lutheran Church's Food, Faith, and Sustainability group supported us by speaking to the congregation, and several church members joined. We were also able to collect funds from our new member community and from Blossom Grocery to provide two free memberships for a family and a senior in need of fresh food.

Then, unexpected marketing aid suddenly fell from heaven. Debbie Hatch, columnist for the local newspaper, published a New Year's article touting CSA memberships as one of the best ways to "go green." Within a week, we received over twenty new inquiries. But by mid-spring the CSA was still only half subscribed. At this point, we asked our customers to market for us. We wrote the first of weekly harvest newsletters, the Lopez Community Farm Howler, describing what was happening on the farm, and asking members to spread the word about the CSA. Within a month we reached our membership goal. We could now focus on growing the food instead of selling shares.

Membership Record Keeping

For new members, a mail-in registration form collected information about contact, payment plan, and pickup date choices, and informed members of "pickup rules." The information was transferred to a spreadsheet on which payment records were also kept. Paper copies of members' registration forms were filed to guard against bookkeeping mistakes. We gave members three payment options: a one-time up-front payment of $450, a payment plan of four installments, or a worker-member plan of four installments at a discounted price. All members were required to pay a first installment of $150 at time of registration.

Payment records were updated daily as checks were received. Members were called or emailed if payments were late to remind them of their due balances.

Communications

To communicate with members, we used newsletters, emails, and announcements on the pickup board. Without all three methods of communication, we often found that a number of members would not receive the information. We made an effort to keep communications prompt and set a high standard. To create a new community, a CSA farmer bears the burden of developing and communicating the initial relationship structure. Members need to understand the boundaries and opportunities of what they are stepping into.

Vegetable Pickup

New members were sent a welcoming email thanking them, inviting them to visit the farm any time, and reviewing CSA pickup procedure. Closer to the date of the first pickup, all members received another email reviewing the process. Pickup date was either Monday or Friday, 12-7 pm; people were to bring their own bags, call ahead if they weren't going to pick up that day, and any food not picked up would go to Lopez Fresh, the food bank. Pickup began the first week of June and ended the last week of November. We told members to expect that the size of the share would ebb and flow seasonally. We set strict pickup rules, but made exceptions for people with scheduling difficulties, to pick up the next day or switch their harvest day a particular week. The number of special situations proved manageable and made members feel supported in the CSA relationship.

The first week of pickup, we worked near the shed to guide members in the process. This proved helpful in summer when members often sent house sitters or friends in their place. The CSA experience often excited them, and this was great advertising for next year's membership drive. At pickup, we used a large whiteboard to describe what was included in a share each week and how much of each item the member could take home. The vegetable list from top to bottom matched the order of vegetable bins on the counter from left to right. We ordered the list in the way one would layer vegetables in a bag, heaviest on the bottom, delicate on the top. Members were asked to sign in. At pickup each week, our newsletter provided an update on farm happenings, recipes, and information on vegetables in the share. We invited members to contribute and received several great recipes. Three weeks into the season our members were thoroughly trained in how to manage pickup themselves. Members enjoyed the trust and community Spirit involved with pickup and some folks would ramble around for an hour or so to pick out their veggies, chat with their neighbors, and get into the field for u-pick.

The first weeks when u-pick flowers and strawberries became available, we left detailed instructions about where the items could be located in the field, the quantity available to each member, and constructed a marked string path leading to where u-pick items were located. There was some confusion, mostly because people hadn't read the whole announcement. It took some time, but the u-pick was very successful in the sense that it heightened members' feelings of intimacy and familiarity with the farm and saved us time. The u-pick crops were outlined in the brochure before members joined.

A few communication problems arose. One month there was lifting of vegetables going on! Sometimes members forgot to sign in when they picked up. So, it was unclear for several weeks whether folks forgot to sign in or whether vegetables really were being lifted, until one day, when everyone did sign, and one man did not get his vegetables. We knew vegetables had been stolen because we had harvested lots of extra to give the largest share possible. With people taking extra, this really put a kink in our system. Instead of working on another project, we would have to check and re-harvest for what might be missing throughout the day. This was a great inconvenience to us since our daylight hours were so filled with growing food, and these extra tasks were unwelcome.

But in response to the theft, we kept our communications positive instead of accusatory. We wanted to avoid innocent members feeling policed. We reminded people of the importance of signing in, the need for community-mindedness for this venture to survive, and the effect of our actions on other members' ability to receive their fair share. We thanked everyone for doing their part. After some emails, newsletter announcements,

and notes out on the board, the problem stopped suddenly. We were relieved and thankful that by reiterating the social goals of this venture, we were able to appeal to peoples' moral center that put them back on track.

Looking forward, we would do an open-bin pickup scheme again instead of making boxes. It is important to us and to members to feel a sense of trust and community in the CSA so often missing in modern life. Every harvest we expected to spend an extra thirty minutes re-harvesting or tidying the CSA throughout the afternoon, and this gave us an opportunity to talk to members who were around at the time. Overall, this pickup scheme saved us time in comparison to filling member boxes.

Worker Members

Worker member communications will need to be improved in the future. The worker member option was contingent on such members working five hours per month during pickup and receiving a lower member price. We made this very flexible, too flexible, as it turned out. We initially asked all five worker members what they were interested in learning. Most answered that they wanted to learn about growing food but didn't have specific goals. We found worker members to be most helpful during harvest when we gave them specific vegetable harvesting and processing tasks. Generally, they saved us a small amount of labor; some were diligent workers, while others did not nearly fill their promised hours.

To keep accurate records, a sign-in and sign-out clipboard for worker members will be needed. Worker members should call the day before they work, so that appropriate jobs can be prepared for them. Originally, we debated the idea of making worker time a requirement for signing up for the CSA, encouraging the idea of community supported food production. However, we did not require working as a condition for membership this year, because we feared that potential members might be discouraged. After gauging members' motivation and abilities this year, we would not require worker hours as part of the membership contract in the future. As an alternative, we might recruit more worker members, increase their responsibility, and provide more structure for them in working on the farm.

Future Member Involvement

As a final communication for the season, we collected anonymous evaluations about the CSA experience from members. This information, used in coordination with planting and harvesting records, will be valuable in creating a better seeding plan for next year.

As community support for the CSA grows, we envision members taking a larger role in CSA structure. A goal of any CSA is community building: the more we encourage members to have a say in what they are supporting, the more they will feel ownership in the project. We would like to invite all members to a meeting concerning budget, farmer's compensation, price of shares, crop plan, creation of member committees, and outreach. This sort of meeting would build awareness about the actual work and costs of growing foods locally and sustainably, give members input in what and how much is grown. We dream of volunteer committees working on the CSA as a community gathering place, caring for the perennial garden, producing the newsletter, planning seasonal farm celebrations, providing educational outreach, studying food systems and nutrition. Member leadership in these activities would give the farmers more time to grow high quality food for everyone, provide educational experiences for members, and strengthen community on the farm. As the CSA becomes more established, and members grow to feel more empowered through the food, we believe these visions will take form and bear fruit.

Production Plan

In the fall and early spring, while deciding how many members we wanted, how much produce we wanted to grow, and how much to charge, we called on past experience to guide us. The last CSA Jesse worked at had sixty members, gave small shares, and they were cheap. Originally, we planned a sixty-member start-up for Lopez Island Community CSA. For that number of people we wanted to plant five of eight blocks. This would mean thirty thousand square feet for sixty members, or five hundred square feet per member.

Although a small land-to-member ratio, it made sense for the price we charged. By March, we decided to cut down the membership goal to forty-five because we wanted to concentrate the compost we had on hand, and build up the soil with cover crop as much as possible. For the first year it seemed prudent to plant half the field in vegetables, and half in summer cover crop. To cut the vegetable planting down to four blocks, we needed to cut the membership by roughly the same proportion. At forty-five members, with twenty-four thousand square feet, we would have five hundred thirty-three square feet per member.

In a member survey at the end of the season we asked about the quantity and value of the share. The answers generally showed there were not enough vegetables per week for a four-person family, but that the shares were considered a bargain. Next year we would aim at shares to feed a family of four, and wouldn't want too many people thinking that the vegetables were a bargain. We don't want to over-charge for vegetables, but a CSA is not about bargain hunting; the price needs to reflect the value of food.

Splitting shares caused some difficulties. We only offered one-size share this year. We didn't want to deal with the hassle of two sizes. But in the end people split shares anyway, and then complained when amounts didn't divide neatly. To keep members happy, the farmer might offer two share sizes, a half that would be almost as big as this year's share, and a whole that would be twice that size. We would charge on a sliding scale. For the half share the middle of that scale would be somewhere around $450. We would charge a little less than twice that amount for the whole share. Experience with sliding scales shows that in the end the average payment is right down the middle of the range offered. To plan for such a change in production, we would assume that each bed would produce the same as it did this year and change the area proportionately to accommodate the estimated change of production. Although we want to work on increasing production per bed, we wouldn't want to count on it in the plan.

Fertility Management

Thinking about fertility kept us up at night in the beginning stages of the crop plan. Our goal was acreage big enough to provide a living through the CSA, but small enough that the fertility inputs could be produced on the farm. Our major tools for on-farm fertility were compost, cover crops and biodynamic preparations. Early on it became clear that there was not enough on-farm material to fertilize the whole garden the first year. However, it remains our goal to set up a system to provide most, if not all, fertility inputs from the farm.

Compost

The first week on the farm we made two beautiful compost piles with Henning out of sheep manure and spoiled hay. Looking out over the two-acre field, we tried to calculate how much compost we would need for this new garden. Consulting several books, we tried in vain to estimate what a ton of compost looks like in comparison to a cubic yard. Weight is hard to guess and changes as moisture levels do. Even volume is hard to calculate.

A new pile is fluffy and deceptively mountainous; but in a few days it settles. Even if the pile seems metabolically stable, air is mixed in as the pile is turned. The metabolism of the pile wakes up, and the pile shrinks more in the weeks after the turning. By the time you are ready to spread it there is much less volume than originally calculated. Calculation, while important, must be balanced with experience.

Soil tests showed three per cent organic matter content. To raise this level, we collected as much compost as we could from neighboring horse farms and over-applied compost rather than aiming for an exact amount. Our plan was to get the organic matter up to five per cent or so in the first few years. After building up the life in the soil, we could work toward maintenance applications to come mostly from the farm.

We had ten to fifteen yards of compost made from on-farm manures to spread in the spring. To this we added about thirty yards of horse manure from the Meng Farm. We spread compost on the four vegetable blocks, eight to ten yards per block. There was no compost spread on the four cover crop blocks. We spread the compost by loading the trailer with the tractor, towing the trailer out to the fields, and shoveling it on the ground.

In the early fall we found a huge source of horse manure at Casey Buffum's farm. With the tractor, we spent three days loading up our truck and trailer, driving it back to the CSA, and spreading it directly on the fields. We spread about forty yards on the four blocks that were grown in summer cover crop and would be in vegetables next year. The manure was mostly old and mixed with very small wood chips that break down slowly. We figured it would continue to break down in the field in the fall, and be ready to plant in the spring. There would be no need for spreading compost till the next fall.

After spreading the forty yards, we collected another forty and made two windrows. Half of a windrow was layered with spoiled hay to see if it would stimulate break-down, which it did, heating up to 130 degrees, while the rest heated to 110 degrees. We piled spoiled hay around both windrows to minimize leaching over the winter and to keep in the heat. Both piles were prepped biodynamically and covered with a thick layer of spoiled hay. The plan is to let these piles sit and spread them in the fall of 2008, at which time another two piles will be collected for the following year.

Cover Crops

By October 15, 2005, we sowed winter cover crop on the whole field. In spring 2006, we seeded a mix of rye, Austrian field peas, and fava beans. The fall seeding went in too late, and was sown too sparsely. The harsh winter, sparse germination, and small plants made for a very weak cover crop. In fall 2007, we sowed a denser mix of oats, vetch, peas and fava beans in early October and now have a much better looking cover crop. The plants are bigger and closer together, but it would be better to seed even more densely next time.

In spring 2007, we let the sheep eat down the weak winter cover crop. It made great grazing and the sheep droppings fertilized the soil. Next spring, the cover crop will possibly be too big for effective grazing. Unless it is grazed earlier in the spring, it will need to be cut, collected and composted.

To reduce the area to spread compost on, and to increase the area in cover crops, we decided to grow vegetables in only half of the available land. The other half was kept in cover crop all season. Of the eight blocks, every other one would be in summer cover crop each year, and each block would be summer cover-cropped every other year. This way the soil in the blocks will get a break every other year and the organic matter content will be built up.

Of the four summer cover-crop blocks, two were in grain, wheat to harvest, and oats to till in. However, the oats grew so beautifully, we fell in love with them and let them go to flower and then to seed. We ended up harvesting both grain crops. Of the two other summer cover-crop blocks, we planted one in red cow peas and the other in buckwheat. The cow peas did miserably. They germinated, and were healthy, but grew very slowly, perhaps because the summer temperatures were not high enough for cow peas. They let all the weeds grow around them and it became a half-weed, half-pea field. We weeded out some of the worst weeds and tilled the rest in. The buckwheat grew fast, flowered beautifully, and we weed-whacked, and tilled it right into the ground.

Biodynamic Preparations

Biodynamic preparations were not used to their full potential this year, although we had perfect conditions to apply them. All the preparations are made on the farm and there was enough to go around; they were at our fingertips the whole year. We did spray the Barrel Compost in the spring two times, and BD500 once. In the summer we sprayed BD501 once, and in the fall BD500 one more time.

One of the first things to do next year is to create a firm schedule of BD500 and BD501 spraying and set aside time for spraying and not push it off as we did this year. We prepped the two big compost windrows set aside for next year. The composts spread in the spring and in the fall were not prepped. The composts we took from other farms was already partially composted, and we applied them immediately. There was no time for prepping. Now that there is a system to age the manure for a year on the farm before spreading it, prepping the piles will be easier. Biodynamic preparations are very important in establishing a fertility program that relies entirely on on-farm inputs.

Amendments

Tests showed that the soil was basically in good shape but needed some amendments. Besides low organic matter, there was a boron deficiency, as well as need for a general boost in nitrogen, phosphorus, potassium, and lime. I had not used amendments other than compost, lime and rock phosphate before and always assumed that in most cases good compost and biodynamic preps brought in enough minerals to grow good vegetables. One thing that puzzled me was that the soil test treated compost as a source of organic matter and nothing more. It suggested that N, P, K and trace minerals should come from sources other than compost. We knew that compost could provide these nutrients, but didn't have experience with this soil, and no idea how it would react to intensive vegetable gardening. In the end, to make sure that the vegetables would grow satisfactorily, we took the soil lab's advice and spread four hundred pounds each of blood meal, greensand and lime, three hundred pounds of rock phosphate and gypsum, as well as thirty pounds of zinc, and seven pounds of boron over the entire eight blocks (forty-eight thousand square feet), spreading these amendments in the spring a couple of weeks before the soil was dry enough to cultivate.

Tillage Methods

In deciding tillage practices we wanted to choose methods that were gentle to the soil ecosystem, didn't rely on big equipment, and saved us labor. This is a delicate balance, depending on field size, soil texture and structure, and equipment availability. The first year on this field, we were bumping around in the dark to find the right combination of methods appropriate to the field size and soil condition. After one year of tillage management, we still have only a limited idea of what the best practice for this field might be, since we are not yet seeing long term effects. However, in a year's time we saw improvements in the soil structure and feel

fairly certain that the practices we used would work well over the years, both for vegetable production and for the health of the soil.

Opening the Field

In late September of 2006, we opened the field up for vegetable production. It had been mostly in barley for the past three years and before that, pasture. We paid a neighbor to plow the field for the first time in at least thirty years. The ground was dry and very hard. In some areas the plow only sank in three inches. After several days of picking out rocks pulled to the surface, we ripped with a single-tooth ripper borrowed from a neighbor, Eric Hall, at Crowfoot Farm. We first ripped east-west, and then north-south. The crisscrossing broke up the ground maybe one to two inches deeper. A disk, also borrowed from Crowfoot Farm, broke up all the big chunks the plow and ripper had made. Then we planted a winter cover crop and disked it in by the middle of October.

Forking

In Spring, the ground was saturated. It took a long time to dry on its own. We got into the garden by hand-forking the surface. We plunged the digging fork into the ground and popped the soil up. Large air cavities were formed and the soil dried out. Besides drying the soil faster, the forking broke up the hardpan the plow had been unable to cut. It reached deeply into the soil, popped out rocks, and created a surface the rototiller could grab onto.

Forking does a wonderful job. It is very gentle on the soil and does not demand big equipment. However, this method requires a lot of physical work. Of the eight blocks, we forked the four in which vegetables would be grown. Twenty-four thousand square feet was a lot to fork by hand; any more and we wouldn't have been able to finish. We felt great about doing it, but it was hard on the body. Doing it year after year could wear someone out.

However, most likely forking is needed only once. The forking action might have a permanent effect on the structure of the soil, letting in enough air to dry it without compaction.

Or perhaps forking is needed only every other year. If one did want to fork every year, one could find several strong and willing workers to get it done all at once. It would probably take ten good forkers a day or two to fork twenty-four thousand square feet. The forkers could be CSA members paid in vegetables, or others paid cash. One way or the other, we saw very good effects from the forking.

Ripping

The four blocks in successive cover crops were ripped instead of forked. In the spring, the ripper sank into the ground very nicely because of the moisture in the soil. The problem with the ripper available to us was that it only had one tooth. To place the tooth fractures close together, we had to run the tractor so close to the last pass that the wheel ended up running over the last fracture. This creates two potential problems: one is compacting the soil just ripped. The second is to have to make too many passes across the field and thereby further contribute to compaction. However, we found that the tooth broke up the soil deeply and the tractor wheel only smashed down the soil surface, leaving the subsoil fractured. However, smashing down the surface kept the air from penetrating as well and the soil not drying as quickly as the forked soil.

Running the tractor over the field every foot and a half in two directions seemed excessive, but was the only way to get the tooth fractures close enough together. Even then the soil was not broken up nearly as well as in the forked blocks, as we found out. When we had finished forking three of the four blocks, we were so tired we decided to rip the last block. However, as we were preparing beds in that block, the rototiller would not sink in, but skidded over the surface and jumped over chunks. We could not make beds with friable soil, and were forced to go back and fork the whole block before planting. After that the beds shaped beautifully. The ripper unfortunately was not perfect. It broke things up enough in the spring to seed cover crop; however, a disk might have been a better tool for the job. Better yet would have been a subsoiler with multiple teeth. Then we wouldn't have to run over the field as much and the tooth fractures would have been closer together. But Henning pointed out that a bigger tractor would be needed for such an implement, requiring investment beyond what the CSA could afford.

Rototilling

To shape the beds and get good tilth, we used a walk-behind rototiller. It is a small tiller and the tines reach only two feet wide. If we were to buy a new rototiller for this garden, it would be a bigger one. It would save time and energy, but this little one did a good job anyway. The soil is naturally very friable, and after forking, broke up on the first pass. If we forked first, the rototiller left the beds with good tilth ready for planting or seeding.

Irrigation

On Lopez Island, geographically isolated and surrounded by salt water, freshwater is very limited. Out of necessity, island farmers most often irrigate from ponds dug on their property. On S&S Homestead Farm, the CSA water system predated our arrival. A pond was installed at the southeast corner of the farm, draining the whole farm, immediately east of the CSA field. Filling during the winter, it is mainly used for irrigation between the months of June and September, the drought season. The seven hundred-fifty thousand gallon pond was sized to accommodate at least a two-acre intensively cropped vegetable garden. The pond also waters the orchard, the homestead garden, and the greenhouse.

Well Water

There is also a well spigot supplying the CSA field. Though limited for agricultural use by the well association, we found it necessary to use this water for several tasks due to solar-pump system limitations.

These tasks were as follows: watering in transplants (heaviest usage May-July); watering greenhouse seedlings (February-June); watering the perennial garden (April-September); and washing vegetables (June-November). We tried to be conservative, but after Henning received a report from the well association, we realized that our CSA well-water use was far from sustainable. The record showed a fifty per cent increase in well water use between 2006 and 2007, the year we watered in most of the transplants for the growing season, and also watered the perennial garden twice weekly. With its curving bed shapes, the perennial garden was a drip tape nightmare. Thankfully, perennials will be more established next year and will need less water.

For next season, it will be necessary to stop well-water use in the perennial garden. This could be accomplished by watering from the pond using bendable emitter tubing. Another well-water conservation measure that must be implemented next year is watering in transplants and greenhouse seedlings with hoses that attach to the header pipe from the pond irrigation system.

The challenge will be to schedule transplanting within the window of available sunlight. Seedlings must be transplanted as the day cools off in the late afternoon, but must be watered in before the waning sunlight stops the pump. The same challenge applies to watering the greenhouse seedlings, because we prefer to water them first thing in the morning while it is still cool. The plants use water more efficiently in the morning. A simple solution to this problem could be filling watering cans when the sun is shining, and then water any time the next morning. Since we had no meter on the spigot, it was unclear how much well water we were using. Adding a meter and tracking usage will be necessary for next year.

Solar Pump

The CSA irrigation system is powered by two photovoltaic panels supplying a direct current piston pump. With built-in pressure regulation, the system turns on when pressure falls below ten psi and pumps continuously until reaching twenty-five psi. The pump is active during peak solar hours of 9 am-5 pm in summer, and fewer hours in spring and fall. The system is fitted with a sand filter that can be flushed for cleaning. Water is drawn from the pond with a screened hose floating below the water surface in the center of the pond.

The solar-pumped water runs to the main header pipe along the fence on the north side of the field, or can be switched to pump into the underground farm cistern, the catchment tank at the high-tunnel hoop house, or to the orchard. This pumping system had been designed and installed through an EQUIP grant Henning secured from the NRCS (National Conservation and Resource Service) to encourage water conservation in the San Juan Islands.

An initial challenge of this system was that the pump and PV panels were not a perfect electric fit, and the fuse blew at least ten times during the season. Replacing the fuse took only a few minutes, but fuses are expensive and the system worked much better after a larger pump was installed. Since the PV system was fixed at a south-facing angle, the power available for watering varied drastically by hours, days, and seasons. To avoid crop stress during peak summer heat, it is necessary for the irrigation manager to be attentive to water needs, hours of sunlight, and the weather.

Another irrigation challenge was that the PV system made it necessary to water during the hottest time of the day. Any experienced gardener knows that the heat of the afternoon is the least effective time to water plants because of evaporation. During the hottest days of summer, the watering system was pushed close to its limits, but usually was adequate. Another PV panel added to this system would allow for an expanded garden.

Occasionally, we also had problems with the uptake filter in the pond clogging up. The remedy was a swim in the pond to clean the intake filter, a process similar to cleaning a huge shower drain.

Bed Irrigation System

Considering the pumping conditions, we chose a half-inch poly-tubing header to run along the length of the north side of the field. At each vegetable block, sub-headers were attached to the main header. These sub-headers ran across the width of each block. Attached to these were low-flow fifteen ml T-tape drip tapes, with eight-inch emitter spacing, running the length of each hundred-fifty foot garden bed. Each T-tape line was attached to an on/off valve where it met the sub-header line. Garden beds had one-two lines per bed depending on the number of rows per bed.

The Dripworks catalog advertised drip tape needing only four psi to function, allowing us to maximize irrigation hours. During the course of a summer day, the number of lines that could be watered simultaneously peaked at eight lines by 1 pm. During summer, pumping usually began about 8:30 am and ended at 4:30 pm. At the end of the day, only one or two lines could be used simultaneously. There is significant frictional head loss, approximately twenty-five per cent of psi maximum, as water travels across the field westward. The T-tape emitted thirty gallons of water per hundred-fifty feet each hour. There

were approximately fifty-five lines of T-tape in the rotation at peak growing season. Once we became masters of the rotation, we could water everything in two days' time. We estimate that it was possible to water twenty-six lines per day during peak summer hours. That is six lines dripping at a time for eight hours/day, each line dripping sixty to ninety minutes. At this rate we utilized a total of 1,440 gallons of pond water per day in the summer. We probably used half this amount in spring and fall, with the irrigation season extending from May through the second week of October.

Based on these figures, we used 183,600 gallons of pond water for irrigation during this growing season. To our delight, the pond remained at least a third full at the end of the dry season. The pond can support a larger production area.

We heard from experienced farmers that this was an exceptionally wet growing season. Thus, it may not be accurate to assume this year's water use to be average. To create a less labor-intensive system, we recommend putting drip tape in every single row needing watering. This will save thirty minutes per day changing lines in beds. We used this strategy to save money on drip tape, but the little money saved was not worth the work.

Throughout the season, the irrigation manager came to understand that the pond water is full of dissolved nutrients the sand filter does not expel. These nutrients provide fertility for the plants; however, they also cause significant plugging of the drip slits. We noticed this problem particularly in lines with emitter slits pointing up towards the sun. Algal growth caused less water to be emitted from the holes. The plugged lines took much longer to water a row, since standard system pressure was no longer great enough to force water out.

To deal with this problem, we opened the ends

of the tapes and drained them at full pressure. This helped, but in one case we were forced to replace the tape. We began draining all tapes in the field every two months. We have since heard of soaking the T-tape, while coiled, in vinegar to dissolve the growth, but do not know anyone who used this method. Commercial non-organic growers in the Skagit Valley use a dilute bleach solution to alleviate the problem. However, this is not a healthy option for a biodynamic system focused on soil health.

Systems Overview

Though labor intensive, the system has key redeeming qualities. We are off the grid and, thus, self-sufficient regarding water. The CSA could operate without a well, if necessary. Changing water lines frequently took the irrigation manager down to visit all crops up close and personal at least every other day. If irrigation had been on a computer-timer, we might not have paid close attention, and not have noticed plant conditions early enough to take effective action.

Seed Propagation

All direct seeding was done with an Earthway seed drill. For all the small plants we seeded four rows in each bed. The cost of the drill is well worth it and on this scale of growing, it is the most appropriate tool I have used. But we had a lot of irregular plantings because of it. Getting more seed plates might help; we did not have a good selection and often had to use plates not sized quite right for the seeds.

Hoop House

The seeds not direct sown were all started in our thirty foot by ten foot hoop house. The hoop house was constructed of PVC pipes and six millimeter UV resistant clear plastic. It is a thrifty design borrowed from Henning's garden hoop house and has worked very well for us. It keeps rain and wind out, and heats up quite a bit during the day. It has of course, very little insulating value. As the sun goes down, so does the temperature in the hoop house. The propagation tables we built have a gravel top to collect heat and keep it warm a little longer. We worried a bit about our tomatoes in the early spring, but in the end there was no frost damage in the hoop house. The tomatoes and peppers were germinated on a heat pad in Henning and Elizabeth's greenhouse.

After a week or two, we brought the seedlings to the hoop house. Every night we carefully placed two layers of row cover on a rack over the little babies to keep them warm. The tomatoes and peppers did fine. They grew very slowly but became hardy. The other crops did great in the hoop house. We had absolutely no disease and good germination. There was a time in spring when we almost ran out of room. We filled up the three tables and had to bring in pallets to function as tables. There was very little work space then, but it only lasted a week or two before the ground dried and the seedlings were sent out of the womb and into the field. The hoop house was a wonderful place to work during that time. Everywhere else was rainy and cold and mucky, but the hoop house was dry and warm and full of new life.

Potting Soil

We mixed all our own potting soil. The recipe for the mix was not always the same, depending on the materials at hand. We generally followed this recipe: four parts garden soil, four parts finished compost, one part peat moss, one part perlite, two handfuls lime, two handfuls greensand. This is for one full wheel barrow and is really just an estimate. The recipe works fine, but I would play around with it to get it just right. We used cheap plastic trays to grow everything in. Some things we grew in cells, others we grew just in the tray. Both worked all right. The trays are truly pieces of junk. They crack, bend and even melt in the sun. Unless they are doubled up, they will break no matter what. Much better are the wooden trays Henning made from one half by four-inch cedar strips for the frame and one quarter by two-inch bender board for the bottom. The trays he made twenty years ago are still as good as new today.

Pest Management

Our policy for pest management is to tend to the health of the soil. Healthy soil grows healthy plants, and healthy plants are much less susceptible to pests and disease. This does not work all the time, of course, and there were times we had to launch a defense/attack on a particular pest. Throughout the season there were several bugs who enjoyed chewing on leaves here and there, but never did much damage. And then there were a few that went too far.

The Little Ones

Wire worms did the most damage in the garden this year. There was a little in the onions, carrots, and beets, but the real damage happened in the potatoes. Almost every potato had a hole or two, which made the otherwise beautiful potatoes ugly. The good news was the holes rarely went very deep and the potatoes were still usable. Still it was painful to dig them up and see the damage. The wire worms probably came in with the horse manure. Henning has had no problem with wire worms in his garden, and we have heard that the little buggers especially like horse manure. If this is true, it illustrates how important it is to create on-farm fertility for vegetable production. The less we import to the farm the better. As for what to do about wire worms now, we're not sure. We have not researched wire worms enough to make a plan.

We had one semi-damaging aphid infestation. It was on the fall brassicas, just as they were about eight inches tall. The interesting thing is that this crop grew faster than any planting all season long. A few days after we transplanted them, they just skyrocketed. You could almost see them grow. It was exciting until a couple weeks later when they started getting distorted leaves. There had been aphids around all season, but nothing worth worrying about. I think the speed of growth left the brassicas vulnerable to attack, and the aphids took advantage of the plentiful food, and multiplied. We sprayed down the crop with garlic spray, to make things uncomfortable for the aphids. We also sprayed with equisetum tea. The equisetum tea is a biodynamic concoction made from horsetail that stimulates drying forces, pushing the water back down where it belongs, in the ground. With both sprays and the plants getting bigger and stronger, the aphid infestation disappeared. There was no noticeable long-term damage to the broccoli, cauliflower, or cabbage. The Brussels sprouts were somewhat stunted, but still produced enough for the CSA.

The Big Ones

There was a rabbit problem at one point. They got through the fence and chewed on a few vegetables. One or two even tried to burrow in and make the garden home. They didn't do much damage until we transplanted a bed of broccoli, and one night they ate about half of it. Then we did two things: we hunted them, and we got materials to rabbit-proof the fence. The shooting worked well, and we got about ten of them in a month. After that the message went out on the bunny telegraph that that the CSA was not safe. We didn't have any more trouble with them. Rabbit-proofing the fence was slow work, and because we had so much work at that time of the year and the shotgun worked so well, we didn't finish installing the rabbit proofing until late fall. Now the CSA field is rabbit-proof. Hopefully there will be no need to shoot them unless someone is hungry for rabbit.

We also had a small problem with field mice eating seedlings in the spring which we dealt with effectively with mouse traps. In the fall, we had voles in the carrots and beets. They did just enough damage to be bothersome. They were not interested in the mouse traps, but Henning says vole plants work very well in keeping them out.

Last but not least, there were a few rats trying to make their winter home in the tool shed. They poop and chew on vegetables left during the night. We kept it very clean, set traps and hoped for the best.

Deer were a challenge in the fall of 2006. They jumped right through the electric fence, into the field and ate from the vegetable beds. Because they were off the ground as they sailed between the electric wires, they did not complete the circuit and thus did not get shocked. Todd Goldsmith, a new farmer on Lopez and retired electrical engineer and rocket scientist, showed us how to rig the fence so the acrobatic mammals would get a good shock next time even if their feet were off the ground. We grounded every other wire on the fence. This way, as the deer jumped through the fence and its belly touched one wire and its back touched another, one of those wires would be electrified and the other grounded. The deer would complete the circuit and get a nasty shock. We had zero problems with deer in 2007.

Labor

Our farm work peaked and fell with the growing season. Traditionally the farmer rests during winter. However, this first winter was dedicated to planning and building projects which were necessary for the first year. Thus, the late fall and winter labor hours reported here will not be representative of average working hours for an incoming CSA manager.

Unless developing new systems, minimal work is needed between December and late February. This first late fall and winter, the average CSA workload was thirty hours per week. During the peak of summer, the total CSA workload was a hundred and ten hours. During the first year, the labor force consisted of the two of us and some volunteers. Jesse's brother Casey and his

partner Melissa helped with most of the harvest and were invaluable. Henning's summer intern was also a great help, especially at harvesting. She worked with us three mornings a week for the three months she was on the farm.

We expect labor hours to decrease by the end of the first CSA year, as the farmers learn how to be more efficient. For instance, between spring and fall 2007, collecting and spreading compost went from a nine-day to a two-day process after discovering helpful equipment trades in the community, more accessible compost contacts, and learning to spread compost in fall before the field became muddy. Jesse worked on the CSA full-time and Lisa part-time until the school year ended in June. Lisa's part-time job allowed us to stay afloat financially while we got the business up and running. The school job complemented the CSA mission and diversified our income, providing a financial safety net.

We usually worked six days a week, about eight hours a day in fall and winter, and about 12 hours a day in summer. These hours count breaks for meals as part of the work day. We almost always took a full day off on Sunday and usually worked shorter hours on Saturday. Included in that six-day week was one day working on the whole farm in exchange for housing and farm-grown food. Labor needs for the CSA were basically met by just the two of us. During the summer, an S&S Homestead intern provided significant harvest help. Also, we had volunteer help throughout the year, but not enough to reduce our workload significantly. We had help from a high school student who worked four hours per week throughout the season and was paid through the Family Resource Center. His presence on the farm was beneficial for all involved. We enjoyed teaching him about the farm, and he gained skills and perspective.

84

Infrastructure Overview

Elements of CSA infrastructure predating our arrival include the water system, drainage ditches, the electric deer fence, access to the barn for crop storage, tools, and the tractor. Elements of infrastructure added since our arrival in 2006 include a ten by twenty foot vegetable distribution shed with counter and attached tool shed, a thirty-five foot PVC and clear plastic seedling propagation hoop house, three large seeding tables, two of which have gravel surfaces for thermal mass collection, rabbit proofing for the fence, all tools necessary for managing a CSA, and accompanying computer files to be used as templates for record keeping, crop production, and marketing.

Profitability

During the 2007 growing season, with a forty-five member CSA, net income for Lopez Community Farm CSA was approximately $15,000. This level of profit was based on decisions made by the managers concerning size and membership price of the CSA for the 2007 growing season. Certainly, profit for a CSA/whole-farm manager trainee could increase significantly if the CSA were expanded, share price and size were raised, or if members were charged for add-ons to the CSA, such as flowers and berries. Through an end-of-season evaluation, we found that most members thought the CSA share a fair value for its price. Thus, we wouldn't recommend raising the price without increasing the size of the share. Increasing share size and price can lead to significant increase in returns with only a small increase in workload.

Originally, we designed the CSA for sixty members for 2007. This plan would have provided an extra $7,000 income to the farmers. Looking back on the year, we see we could have handled a sixty-member CSA. This expansion would have required more water and fertility inputs, as well as more labor on our part, but would have been doable. If we were

to continue the CSA, we would increase our income by expanding the CSA to sixty members. We would develop a fertility plan to maximize fertility from on-farm sources. We could also begin developing other on-farm enterprises.

Though profit this year from the CSA was lower than most starting salaries, consider the fact that we did not have to pay for rent or food. Our food was of higher quality than we could have purchased at the store. If we estimate housing and utilities at $400/month, and food at $400/month, the income increases by $9,600 per year, for a total of $24,600.

Considering that most small businesses do not see a return on investments for five years, the profitability of this business was huge. We sold out all goods available and realized a profit beyond expenses.

The CSA work during fall and winter of 2006 and early spring of 2007 was light enough that Lisa could work part time at the school to supplement our income. While living on a somewhat frugal budget, after this first year, we were still able to come away with several thousand dollars in savings. In subsequent operational years, since basic infrastructure and systems planning for the CSA are complete, it is possible that the CSA farmers can work other jobs during the off season or develop additional farm enterprises for more income.

Certainly, there is more profit to be made with the CSA and the whole farm than we took advantage of this first growing season. Considering the long term opportunity for the CSA farmers to become whole-farm managers, there are many other potential enterprises that could bring in significant incomes, eventually doubling, possibly tripling, our CSA income for 2007.

The CSA is just the beginning for farm-income

potential. It provides the base for the farmers to get their feet on the ground. Henning and Elizabeth's vision is that the CSA managers spend their first year or two focused on the CSA and then begin envisioning what they would want the whole farm to look like. Pork, goat, chicken, beef, egg, dairy, value-added products, fruit, hot house, grain, educational farm stays, are all options for entrepreneurial expansion on the farm, many of which already exist on a homestead scale.

Henning and Elizabeth are committed to transitioning different systems of the farm, and the income they bring, to new farm managers. For a beginning farmer or couple committed to farming in the islands, this is as perfect an economic situation as one can find, considering it is currently almost impossible to pay for land through farming. This is an opportunity to manage and expand a farm business in a low risk environment, with long term security. The owners encourage new ventures, and want to help a farmer create a lucrative and balanced life on the farm.

All new venture ideas are considered as long as they follow regenerative practices. It is important to Henning and Elizabeth that the farm remains an integrated, balanced whole. The scale of new enterprises must meet these ends. The future-farm project is their labor of love, and their deepest hope is that this land will continue to provide food for the community.

Lopez Island is a challenging place for working people to support themselves. Land and rent prices are high. The opportunity at S&S Homestead is unique because a future whole farm manager can bypass these challenges by accepting the opportunity to farm the land with long term security, without having to buy it.

As most farmers today know, it is very difficult to save enough money on a farming income to buy a home. We believe, however, that through opportunities developing in the community, for instance through the Lopez Community Land Trust, it would be possible for a farm manager and his/her family to acquire a home either adjacent to the farm or nearby, after a few years of saving and investment. After two years of living on the farm, we believe we could have found a way to make housing work for us long-term. The people of Lopez Island are hungry for more local food. They are waiting to receive more options, and the local food market is expanding quickly. If you build it, they will come.

S&S Homestead Farm

Compost Management
Jessica Gigot, 2003; updated Henning Sehmsdorf, 2021

"Soil is the connection to ourselves. From soil we come, and to soil we return. If we are disconnected from it, we are aliens adrift in a synthetic environment. It is soil that helps us to understand the self-limitation of life, its cycles of death and rebirth, and the interdependence of all species. To be at home with the soil is truly the only way to be at home with ourselves, and therefore the only way we can be at peace with the environment and all of the earth species that are part of it. It is, literally, the common ground on which we all stand."[1]

The use of composts and compost teas is a reliable method of applying organic matter to create and maintain fertility in cultivated soils. Sustainable organic farms rely on this soil-enhancing technique to reduce dependence on off-farm inputs and ensure the growth of nutritious crops that are healthy and resistant to pests and diseases. According to WSU soil scientist John Reganold, "Organic matter has a profound impact on the soil; it encourages mineral particles to clump together to form granules, improving the structure of the soil; it increases the amount of water the soil will hold and the supply of nutrients; and the organisms in the soil are more active. All in all, organic matter makes the soil more fertile and productive."[2]

With ninety per cent of US cropland being eroded each year due to conventional agricultural techniques,[3] it is important that conscientious farmers focus on growing soil, through the application of composts, before growing food.

S&S Homestead Farm operates as a small, integrated, biodynamic system, taking all members and components of the farm ecosystem into account when making management decisions. The ultimate goal for the farm is to achieve a closed symbiotic circle of the flora and fauna that make up the whole of the farm. To achieve this symbiosis various approaches are applied. From John Jeavon's technique of double digging to Elaine Ingham's biological approach to microbial soil health through the use of compost teas, S&S Homestead is always adapting new advancements in alternative procedures to place-specific needs on the farm. There is no attempt to subscribe to a single, specific methodology or qualify for organic or bio-dynamic certification.

However, S&S Homestead Farm soil practices are consistently chemical-free and based on a holistic view of natural resource management. A strategic use of compost and compost tea is needed to achieve the farm's goal of self-sufficiency. Previous interns have addressed this need through practical compost-related projects.[4] Building on past intern reports and data, the present Soil Fertility and Fertilizer

Management plan is designed to consolidate existing information with present research into a cohesive outline of responsibilities for future interns. The hope is that this document will be a useful guide in maintaining productivity of farm soils and maximizing the use of on-farm inputs. As an educational enterprise, S&S Homestead Farm always welcomes new ideas and theories, and it is expected that the Soil Fertility and Fertilizer Management plan will be continuously developed to keep pace with the farm's needs and individual intern goals.

Compost

This is a general outline of the system we developed this year for making, curing and storing compost. It involves "active" or "hot" composting techniques, which will be described in one of the following sections. Biodynamic compost preparations are also being integrated into the farm's compost system as S&S Homestead develops its own production of BD preparations in association with the Josephine Porter Institute in Woolvine, VA.

the pile is monitored until it reaches the curing phase. Temperature is an indirect measurement of microbial activity since microorganisms radiate heat as they work[5]. When temperature first falls below ninety degrees, it is time to turn the pile in order to add oxygen and water which will re-stimulate microbial activity and further decomposition. After two to three weeks the pile is ready for the "curing" area underneath a designated coniferous tree, and after another three to four weeks in this spot the compost is done and is moved to a permanent storage area for the winter. As recommended by Eliot Coleman,[6] the storage area was constructed with hay bales as the perimeter and a heavy, rubber horse pad as a foundation. The compost is shredded finely and the pile is then covered with a tarp to prevent excess nutrient leaching throughout the winter.

Joseph Jenkins' handbook on "micro-husbandry" offers an excellent description of the composting process and of suggested annual applications of compost.[7] How much is

Location	Bed Spare	C-Application	Depth	Sub-Total
Main Garden	2000 sq ft	2 times/year	½ inch	288,000 cubic inches
Orchard	1000 sq ft	2 times/year	½ inch	144,000 cubic inches
			TOTAL	432,000 cubic inches
			1 square foot=144 square inches	

Annual Compost Requirement for Garden and Orchard in Cubic Inches

System Description

To build a compost pile, organic waste material is collected in an enclosed cement area. Once enough material has been collected, a pile can be made. It is important that there is enough material before beginning this process because active piles must be assembled at one time. Fresh, green material can lose almost sixty percent of its nitrogen once it has been cut or mowed. After assemblage, the temperature of

enough? Between the orchard, house garden, greenhouse, and CSA field, there are approximately 45,000 square feet (one acre) of bed space in use. Ideally, all of the beds should receive an application of compost in early spring and fall and the orchard perennials, such as the strawberries, need a similar amount sometime before and after fruiting. In addition, compost is used on the farm's flowerbeds, in potting soil mixtures, and in compost teas.

Fresh manures left behind by rotationally grazed animals compost in pastures and hay fields support forage and hay production.

Following are some calculations to assess compost needs for cultivated beds:

In 2002, Johanna Høk set up a good model for converting cubic inches to tons:
~1 bucket of compost = 35 lbs
~1 bucket = 5 gallons = 18.95 liters
~18.95 liters = 18,950 cubic centimeters = 1,155.95 cubic inches (cubic meters x .061 = cubic inches)
~5 gallons = 1,155.95 cubic inches

Therefore the following calculations apply to our compost project:

~1,512,000 cubic inches/1,155.95 = 1,308 buckets
~1,308 buckets of compost x 35 lbs = 45,780 = 22.89 tons.

The calculations show that twenty-three tons is sufficient for garden, orchard and CSA-field bed needs. As Robert Parnes shows, a typical compost application rate is five to ten tons per acre.[8] Since the cultivated bed space on the farm is one acre (45,000 square feet), this figure also reveals that a goal of twenty-two tons will provide the farm's vegetable and fruit production with a secure source of fertility. According to Johanna Høk's projections (2002), a standard compost pile generated one and a half tons of compost in one season, i.e. fifteen piles are needed to provide the required twenty-two tons for vegetable and fruit production. In 2016, S&S Homestead Farm generated an estimated sixty tons of compost, plus forty-five tons of fresh manures deposited directly by twenty animal units (cows and sheep) in the fields to produce an estimated twenty-two tons of composted fertilizer to support green forages and hay.[9]

The piles made in summer 2003 were calculated as follows, based on average initial size:
~Piles 1-5: 36"x 48"x 48" = 82,944 x 5 = 414,720 cubic inches
~Piles 6-11: 48"x 48"x 60" = 138,240 x 6 = 829,440 cubic inches
~Total: 1,244,160 cubic in./1,155.95 cubic in. = 1,076.3 buckets of compost
1,076.3 buckets of compost x 35 lbs = 37,670.8 lbs = 18 tons

This calculation may have been overly generous since most of the piles shrank substantially in the course of the season. The year's compost supply was nearly exhausted by mid-season, a sign of variation from this model. However, we kept track of wheelbarrow loads of finished material (one wheelbarrow = seven buckets = 245 lbs) and this is another way of keeping track of compost in storage. By mid-season, there were twenty-six wheelbarrows x 245 lbs/ 2000 = three tons of finished material left.

Biodynamic Inputs
In July we received the Pfeiffer BD Compost Starter from the Josephine Porter Institute and applied it to four of ten compost piles. The starter material was soaked overnight to allow the bacteria and fungi to re-activate. After twelve to twenty-four hours the starter was potentized by stirring in water in a five-gallon bucket and applied with a whisk broom to each layer of the pile. In the fall, the BD preparations (components of the Pfeiffer mixture) will be added to the compost and left to cure over the winter.

Compost Monitoring & Production
To provide the farm with good, pathogen-free compost and maintain records of compost production, it is necessary to record compost temperature data. This is a useful way to learn about compost and can be incorporated into a research project involving nutrient cycling and the farm's intent to "grow" its own compost by

producing inputs rather than merely cycling farm wastes. A proposal for stages in producing compost from plant materials grown for that purpose follows:

~Stage One (Fall): Barley straw harvested from field, baled and stored;

~Stage Two (Fall): A legume cover crop and companion grass is planted in the barley field and allowed to overwinter;

~Stage Three (Spring): the cover crop is cut, one half baled for hay, the other half built into large piles in the corners of the field. These piles can be turned once during the summer and will be ready as compost for the following spring;

~Stage Four (following Spring/Summer): Barley is replanted and harvested;

~Stage Five (following Fall): Field is left fallow or cover-cropped and the rotation continued on another field. If the cover is not used for hay or compost, it will provide fertility for the next barley crop.

Worm Composting

Worms are great composters! Their castings are five times richer in nitrogen, two times in exchangeable calcium, seven times in available phosphorous, and eleven times in available potassium than the soil they inhabit.[10] Worm castings are a reliable source of fertility for the farm. The castings can be applied directly to the garden or they can be used as a tea. According to research, vermi-composts have significantly higher microbial activity than regular composts.[11] Since most compost is heated, it favors only thermophilic bacteria. Vermi-compost, by contrast, supports a wider range of bacteria and fungi because its temperature remains relatively low. However, the nutrient richness of vermi-compost is always dependent on what goes into the worm bin and since thermophilic temperatures are not reached, it is important to avoid importing pathogenetic elements.

Vermi-Compost System Description

We constructed a worm bin (five by ten feet) using straw bales as a perimeter. The elevated temperature of typical compost piles is too hot for most types of worms, so they need special conditions in order to break down material efficiently. According to the ATTRA (Available Technology Transfers to Rural Areas) web page, optimal conditions for worm composting include temperatures of 60-70 degrees, a pH level of 5-9, and moisture level of eighty to ninety per cent.

There are many types of bins, ranging from small- to mid- to large-scale to accommodate the worms' needs. The structures used to house worm populations can be separated into four major types: stacked bins, windrows, continuous flow reactors, and lateral movement bins. Stacked bins are generally small and used within the home. Windrows have become popular on dairy farms where there is a lot of raw material to process on a regular basis. Continuous flow reactor bins are permanent boxes that have a mesh or grated opening at the bottom to release fresh castings. This allows for the continuous addition of material without having to clean out old material. The lateral movement bins have a variety of forms and often mimic windrows in that they allow worms to migrate to fresh material, thus leaving behind castings for harvest.

The area we created is basically a "framed" windrow that utilizes lateral flow as a management technique. The bin was located in the shade of a coniferous tree. A two-inch layer of greenhouse soil was spread along the base and one bale of straw added to the top as bedding. From piles of rotted dairy manure that had attracted copious numbers of red wigglers (*eisenia foetida*), four wheelbarrows of material containing the worms were added to the worm bin, along with various vegetable scraps. A thin layer of straw was placed over the whole area to deter pests and odor. A framed wire door and tarp cover prevented bird (mostly chicken) predation and maintained a cool, moist microclimate. After a month the amount of food in the box had declined substantially, and there were noticeable increases of worms. Normally, you start with one pound of worms to one pound of food and gauge the addition of food scraps by population growth and productivity. However, since the worms we used were integrated in the rotted manure, we did not weigh them beforehand, which would have been a time-consuming process. Instead we monitored food decreases and encouraged worm migration by adding kitchen wastes and allowing space for harvesting casting. The worm castings were applied to garden beds.

Compost Tea

S&S Homestead is interested in a bio-spray regimen that will replenish essential nutrients to plants throughout the growing season. Although farm-produced compost offers a wealth of fertility, it is often advisable to supplement crops with secondary sources of nutrition. In previous years organic fish emulsion, with an N-P-K ratio of 5-1-1, had been the main nutrient supplement for summer vegetable production, applied to each bed on a two-week rotation. To eliminate this purchased input, I spent this season figuring out how to make compost teas, why they work, and if they can provide sufficient nutrient levels to replace the fish emulsion.

What is compost tea? Creating an active compost tea versus a passive extract (leachate) involves soaking compost in an aerated container of water. Input of oxygen by means of an aerator prevents the teas from becoming anaerobic in the first twenty-four to forty-eight hours, which is the case in most passive methods. Anaerobic conditions can reduce the nutrient level and generate acids, such as butyric, proprionic and acetic acids, which retard plant growth.

In contrast, the active, aerated process extracts macro- and micro-nutrients and stimulates beneficial microbes reproducing in oxygen-rich environments. Research shows that active teas improve soil and plant health through enhanced microbial populations, suppress disease, and improve nutrient cycling and retention.[12] However, there can be three major sources for discrepancies among compost teas: original feedstock, method of extraction, and time interval of extraction,[13] and these discrepancies have been at the root of much debate and confusion over the benefit and effectiveness of compost teas.[14]

On S&S Homestead Farm, we use the Soil Soup Blender™ mounted on the rim of either a five-gallon or twenty-gallon container. Four pounds of compost are suspended in a mesh stocking for varying amounts of time. By this method we create an aerated tea that potentially enhances both microbial life and nutrients, but without lab analysis of the compost and the tea, we are unable to describe our product definitively. Some manufactures of compost tea brewers offer nutrient supplements.[15]

A tea is only as good as the compost it is made of. In a review of current literature about compost tea sources, Richard Merrill and John McKeon distinguish:

~*Fresh manure* rich in macronutrients N, P, K, Ca, Mg, S and micronutrients Fe, Zn, Mn, and Cu;

~*Young or unstable compost*, in which available nutrients have not yet been fixed in microbial biomass, sugars, amino acids, and the chelating agents (humic and fulvic acid) that make them available to plants.

Also produced in young compost are long chain carbon molecules that provide carbon and oxygen for soil microbes, including micorrhiza. The micorrhizal hyphae, in turn, greatly expand the root systems of plants,

Bed Number	Week 1 Date:5/18	Week 2 Date:6/1	Week 3 Date:6/16-6/22	Week 4 Date:7/12-7/14	Week 5 Date: 7/24	Week 6 Date: 7/28	Week7 Date: 8/7	Week 8 Date: 8/10	Week 9 Date:8/16
1				Batch 6/7	Batch 8				Batch 13
2								Batch 12	
3		Batch 2							Batch 13
4			Batch 4						
5	Batch 1					Batch 9			
6	Batch 1								
6a			Batch 3	Batch 6		Batch 9			
6b			Batch 3	Batch 6		Batch 9			
6c									
6d									
6e					Batch 8				
6f					Batch 8				
6g									
7			Fish Em,		Fish Em.				
8			Batch 3				Batch 11		Batch 13
9									
10									
11									
12									
13									
1-1			Batch 3		Batch 8	Batch 9			
2-2			Batch 3		Batch 8	Batch 9			Batch 13
3-3			Batch 3		Batch 8				
4-4			Batch 3		Batch 8				
5-5									
6-6	Batch 1								
7-7					Batch 8				
8-8					Batch 8				
9-9	Batch 1								
10-10	Batch 1								
11-11									
12-12									
13-13				Batch 6					

Compost Application Chart

increasing their nutrient uptake, respiration, and weather tolerance.

In addition to the nutrients provided by fresh manure and immature compost, *well-aged composts* also supply organic chelators, as well as populations of bio-fungicidal microbes that compete with and suppress certain plant pathogens. Teas made from mature composts contain populations of microbes, such as mycoparasites, rhizosphere colonies, hyperparasitic fungi, epiphytic microbes, pseudomonas, azotobacter, trichoderma, and gliocladium. The microbes improve soil structure by excreting organic gums and resins which together with fungal hyphae, bind soil particles into structural aggregates.[16]

A promising option for tea-making involves the use of worm castings. Ann Lovejoy, a well-published garden expert living on Bainbridge Island, WA wrote that "worm compost differs from regular compost in being finer in texture, with a more complex nutrient base."[17] I have found vermi-compost to be a reliable and convenient ingredient for compost teas.

Compost teas can be applied as either a foliar spray or a soil drench depending on plant needs and environmental conditions.[18] In general, foliar sprays have gained recognition for their ability to help plants fend off insect damage.

According to the ATTRA web page, foliar feeding, can also be substantially more successful at transporting nutrients directly to a plant than soil applications, where the plant absorbs nutrients through the rhizosphere and also shares the nutrients with microorganisms in the soil. It would be interesting to see what effect foliar feeding versus soil drenches had a on plant nutrient response using teas made from on-farm inputs. Foliar feeding might be a more immediate way of treating the plant, whereas soil drenches help overall soil structure and longterm sustainable nutrient cycling, which are priorities in biodynamic fertility management.

Compost Tea Recipes

I have worked with three main recipes derived from various sources and personal concoctions: high bacteria compost tea, worm castings tea, and nettle tea:

~Compost-based teas are made using the Soil Soup Blender.™ Compost is filled into a mesh stocking and suspended in the 20-gallon container with the aerating machine running for two to three days. Supplemental ingredients can be added during the first twenty-four hours.

~High Bacteria Tea: Four pounds of compost, one ounce of molasses, twenty gallons of water, half gallon of nettle soup.[19]

~Worm Castings Tea: Two to four pounds of castings (remove worms while harvesting), one ounce of molasses, twenty gallons of water. Worm castings are recommended as the staple in the farm's tea making operation. They are easily available and a more reliable source for nutrients than general

compost sources. It would be interesting to try cow manure as well, although this type of high-nutrient tea might burn the plant.

~Nettle Tea: Two pounds of nettles (*urtica dioica*) soaked in five gallons of water. Nettle tea is an effective nutrient supplement and a deterrent of pests, especially on roses, currant bushes, and fruit trees.[20] After about a week of soaking, this mixture is strained and diluted with water (1:5 ratio) and either applied directly to the plants using the watering can or mixed in a compost-based tea. It will smell horrible, but it is effective as a mid-season nutrient supplement and a prophylactic spray against most fungal attacks. According to the journal *Biodynamics*, it is especially useful on cucumbers in order to help them maintain their deep green foliage.[21]

Application Procedure

Once the tea is ready, it can be transported in buckets and applied using a watering can. Application is a judgement call, but Elaine Ingham recommends one quarter gallon/three inches of plant for foliar sprays and five gallons/acre for soil drenches.[22] The watering can is actually more effective and efficient at applying the tea than the sprayer, although the sprayer has a finer nozzle which might be more appropriate in some foliar applications. It is important to look at the size of the crop, the overall moisture level of the soil, and the damage done (if this is a pest issue) when deciding how much to apply. When working with teas, always be aware of what is planted and growing in the garden in order to maximize nutrient inputs. Vegetable crops should get a dousing of tea at least every two weeks and it is

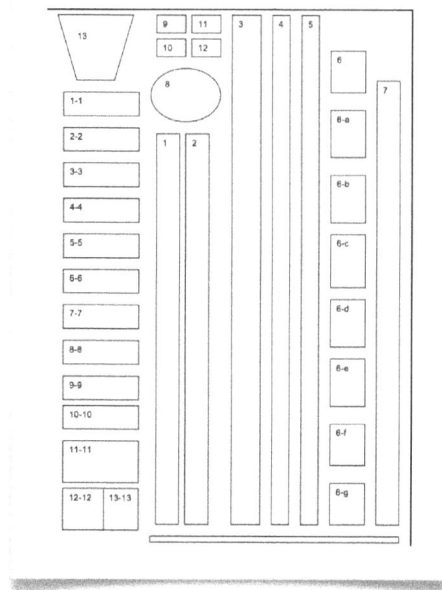

Garden Map

especially important to supplement plants during their development stages. In some cases, mature plants near harvest should be fertilized as well.

The Compost Tea Application Chart and Map

The Application Chart and Garden Map offer a general framework for the compost tea system. On a week-by-week basis, record where a tea is applied, use the map as a guide, and keep a record of batch ingredients in a journal or notebook. On the chart, specify which batch was used in the weekly slots. This system has proven helpful in maintaining the fertilization sequence throughout the summer (May through September). Also a good record of plant responses to certain tea types can be incorporated into journal entries and will serve to develop and refine this process.

Compost and Compost Tea Reports

Compost Monitoring Report: Our goal for the summer was to generate 10-15 tons of compost for the following season. This report will describe the activity of the first nine piles assembled over the summer. Six of these piles were made normally (Piles 1-6), while three were inoculated with Pfeiffer Biodynamic Compost Starter (Piles 7-9). Average

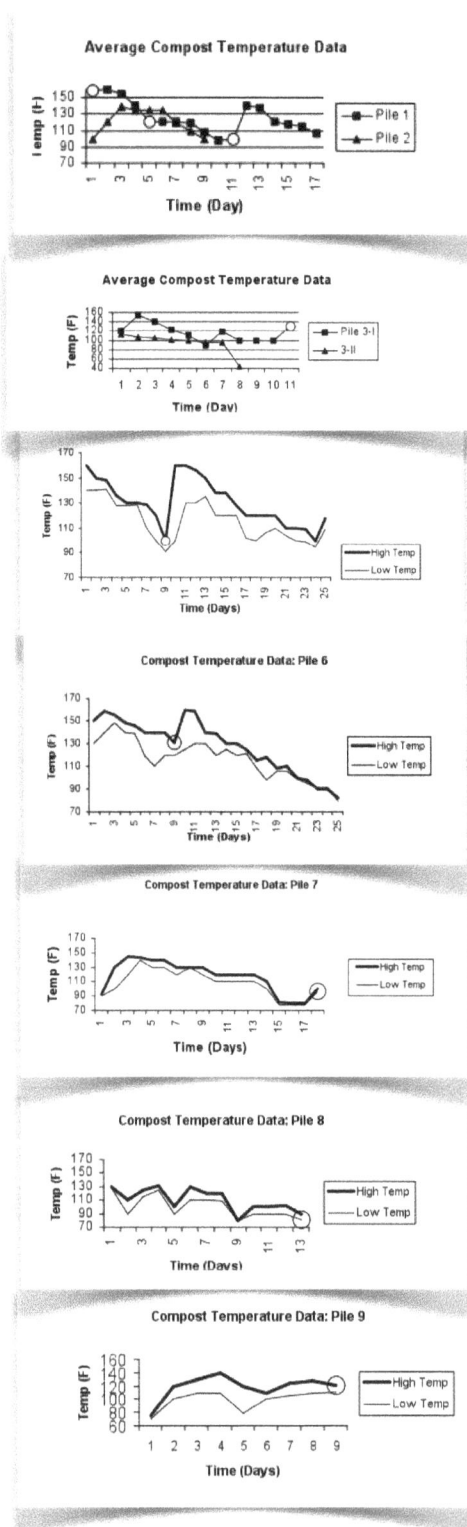

Average Compost Temperature Data

Average Compost Temperature Data

Compost Temperature Data: Pile 6

Compost Temperature Data: Pile 7

Compost Temperature Data: Pile 8

Compost Temperature Data: Pile 9

temperature data was collected for Piles 1-3 while data for Piles 4-9 shows the high/low temperature range. Initially, I tried to measure exactly the amount of ingredients going into the piles. However, with the desired carbon/nitrogen ratio of 30:1 in mind, experience shows that an estimated combination of green and brown materials in layers, plus ad hoc additions of blood meal, manure, grass clippings, or soil, as needed, will ensure a well-digested pile. Therefore I have not included the specific ingredient data in this report.

Pile 1 heated up very quickly and was turned within a twenty-four hour period (the circle indicates that the pile was turned). I added soil at this point to cool the pile down and it stayed at a good thermophilic temperature. At the second turning of this pile, I added a bucket of grass clippings. Pile 2 became relatively small (3'x4'x3'), so I combined it with Pile 1 after the temperature dropped to 90 degrees on Day 9. At this point the combined pile rose to 130 degrees in a forty-eight hour period.

Pile 3 was constructed on June 20 with both grass clippings and sawdust and the pile heated up quickly, but dropped in temperature rather rapidly after this point. Following this decline, I

Tea #	Date Made	Date Applied	Ingredients				Crops	Conclusions
			Cpt	H20	Ml	Nt		
1	5-18	5-16	4lbs	20g	n	n	Cabbage/Broccoli	No significant change
2	6-1	6-16	4lbs	5g	1oz	1qt	Carrots	10% germination
3	6-16	6-18	4lbs	5g	n	n	Lovage, kohlrabi, peas	Peas grew substantially
4	6-22	6-24	4lbs	5g	n	n	Sweet peas, artichokes	No significant change
5	6-26	6-28	n	5g	n	2lbs	melons	Dead
6 worm	7-12	7-10	4c	20g	n	n	cucumbers, kale, onion, basil, pea row, pole beans	good, green growth pole beans and cucumbers are thriving, basil slow growth
7	7-14	7-24	n	5g	n	2lbs	cucumber (1/3)	No difference. All look healthy
8 worm	7-24	7-26	8c (3 lbs)	20g	1oz	n	beans, beets, cucumbers, peas	beans/cucumbers-rapid, green foliage
9 worm	7-28	7-30	8c	20g	1oz	n	tomatoes (orch), kale, kohlrabi, basil	Slow growth basil, tomatoes fruiting, kale-flea beetle
10 worm	8/5	8/7	8c	20g	1oz	n	kale, artichokes	Flea beetle still persisting
11 worm	8/7	8/9	8c	20g	1oz	n	Herb garden	good re-growth on rosemary, lovage

Cpt-Compost Ml: Molasses Nt: Nettle

Compost Tea Chart

added even more green material because there was still a lot of un-decomposed sawdust in the pile. The pile did not seem to respond to this addition, so I rebuilt it and added a blend of blood meal, bonemeal, and greensand (N-P-K amendments, respectively) to help heat that pile. This change is indicated by 3-II on the graph. Pile 3 was then moved to the "curing" area. Both piles 1&2 (combined) and 3 were left to cure during the first two weeks in July. After this period it became clear to me that the piles needed some attention and neither were at the curing stage quite yet. What I learned from this was that when constructing a hot pile, there needs to be a good balance of carbonaceous and nitrogenous materials. Since I had been continuously adding material to both piles as I turned and monitored them, they were unfortunately still unfinished after a month's time. In addition, I had just been recording the average temperature for these piles.

For piles 4-9, I began recording temperature ranges, which gives a better picture of microbial activity throughout the whole pile. In some instances I noticed that some of the piles were only partially active. Good turning and watering practices can resolve this problem, but it also indicates that the layers were not evenly distributed. When piles 1, 2 and 3 were moved to the curing area, I added a dusting of blood meal (fifteen per cent nitrogen) to each pile. This supplemental nitrogen helped to break down semi-decomposed pieces of material. Piles 4 and 5 were made from a large amount of chicken bedding, kitchen scraps, and grass clippings. Both piles shrank substantially, so I combined them into one pile. Pile 4/5 heated up immediately, but its temperature dropped substantially in the first week. Since it was late in the summer, the pastures were drying up and leaving the "green" material relatively low in nitrogen. I sprinkled multiple handfuls of blood meal onto the piles in order to balance the carbon/nitrogen ratios, and enhance decomposition during curing.

Pile 6 consisted of straw and grass clippings from around the fences on the farm. It initially heated up to the thermophilic stage and then dropped slightly. On day 9 it was turned and heated up very well, and then declined steadily in temperature, as expected.

I built all three biodynamic piles (7-9) in similar fashion, using straw, garden waste, kitchen waste, cow manure, blood meal, and some grass clippings, and broke the sod underneath them in order to encourage interactions with the soil environment. All three piles rose in temperature slowly (see graphs 7-9), and never heated beyond 130 degrees. This gradual temperature rise was apparent in all of the biodynamic piles and did not seem a problem in terms of pile activity. In fact, it may be an indicator of microbial populations that are different from those in the regular piles. In general, biodynamic piles are more uniform in temperature than other piles, and they do not heat up as fast. It is not certain whether the Pfeiffer Compost starter is causing this difference among the piles, or whether the location is the cause. Being on a cement slab prevents aeration and contact with soil microbial life, but it does prevent nutrients from leaching into the ground.[23] It is interesting that both biodynamic piles have consistent temperatures from top to bottom, but do not heat up as fast as the piles in the compost area. Presumably there is different microbial life active in the BD piles and that would make a major difference:

"Research at Washington State University (WSU) by Lynn Carpenter-Boggs and John Reganold found that biodynamic compost preparations have a significant effect on compost and the composting process… Biodynamically treated composts had (lower) temperatures, matured faster, and had higher nitrates than control compost piles inoculated with field soil instead of the preparations, and it demonstrated that biodynamic preparations are not only effective, but effective in homeopathic quantities."[24]

Piles 7-9 were turned and combined and two more biodynamic piles were created, as well as a compost pile of slaughter offal. Their location did not offer a lot of shade and the piles became quite dry. A good turning with water and blood meal should do the trick, but future piles in this area might need a tarp cover and more frequent watering.

Compost Tea Monitoring Report

My initial project goal was to assess and compare the effect of compost tea applications on various organically grown crops in the Main Garden area (see chart above). However, this was problematic since I did not have access to technical equipment in order to determine the content of the teas, and it was difficult to find adequate room in the garden for control plots. Therefore my methodology was mostly qualitative, observational, and practical. I applied various recipes of compost tea to garden and orchard crops, and monitored any physical or productivity changes that arose after the tea was employed. This information was then incorporated into the development of a specific system for supplying nutrients to garden and orchard crops throughout the season.

Overall I did not see any dramatic changes in plant growth due to tea application. However, all of the plants that were given tea have grown deep, green foliage and produced a good harvest. The garden looked green. In two instances, smut on the onions and flea beetles on the kale, I applied tea to deter pathogens and pests, respectively. In both cases the tea did not seem to have a major effect, and manual controls proved more effective.

In general I found the worm castings tea, with a little molasses added, to be a reliable nutrient supplement. I usually brewed a batch for two days and applied it immediately. The cucumbers and beans have responded well to this type of tea, as well as the tomatoes in the orchard. I used the worm tea on the multiple rows of peas in the garden, but it was hard to tell if it affected them since they were near the end of their growing period.

The nettle tea seems to be an effective alternative, and I did not have any negative reactions to this type of tea. The melons died soon after I applied this tea, but I am pretty

sure that was a result of poor seed stock and
nursery maintenance. For compost tea, juvenile
nettles should be harvested from June to mid
and late July, before flowering.

Notes

———————————

[1] Kirschenmann, Fred 1997. "On Becoming Lovers of Soil," in: J. Madden (editor), *For All Generations: Making World Agriculture More Sustainable*. Glendale, WA.

[2] Reganold, John 1989. "Farming's Organic Future," *New Scientist* (June 10).

[3] Jackson, Wes 2002. "Natural Systems Agriculture: a Truly Radical Alternative," *Agriculture Ecosystems and Environment*, 88: 111-117.

[4] Høk, Johanna 2002. "Nutrient Recycling and Composting on S&S Homestead Farm," unpublished internship report; Huntington, Brian 2000. "Soil Health and Fertility," unpublished internship report.

[5] Høk, ibid.

[6] Coleman, Eliot 1995. *Four Season Harvest*. White River Junction, VT.

[7] Jenkins, Joseph 1999. *The Humanure Handbook*. Grove City, PA.

[8] Parnes, Robert 2013. "Soil Fertility: A Guide to Organic and Inorganic Soil Amendments." (https://soilandhealth.org/wp-content/uploads/01aglibrary/010189.fertle%20soil%20revise.pdf.) Retrieved June 12, 2021.

[9] See "Economic Valuation of S&S Homestead Farm." (https://docs.google.com/spreadsheets/d/1Na6dxEs-3AfhDFISvr7Jnoh5eRsGdoy5Qi5r0TfglC0/pubhtml.) Retrieved June 123, 2021.

[10] Jeavons, John 1995. *How to Grow More Vegetables…*Berkeley, CA.

[11] Edwards, S. 1998. "Vermicompost," *Biocycle*. July: 63-66. Updated by Adhikary, Sujit 2012. "Vermicompost, the story of organic gold: A review," *Agricultural Sciences* 03 (07): 905-917.

[12] Ingham, Elaine, et al. 2001. *Compost Tea Manual*, Corvallis, OR.

[13] Merrill, Richard and John McKeon 2002. "Compost Tea: A Brave New World," *Organic Farming and Research Foundation Information Bulletin* No. 9 (Winter).

[14] Reich, Lee 2016. "The Jury is Still Out on Compost Tea: Compost Tea is Currently Hot in the Gardening World, But Will it Also Move Beyond Fad Status?" *Fine Gardening* (Issue 107).

[15] Bess, Vicki 2000. "Understanding Compost Tea," *Biocycle* (October).

[16] Merril & McKeon, op. cit.

[17] Lovejoy, Ann 2001. "Compost Tea," *Seattle Post Intelligencer* (March 4).

[18] Ingham, op. cit, 4.

[19] Ibid.

[20] Thun, Maria 1999. *Gardening for Life: The Biodynamic Way*. Stroud, UK, 45-47.

[21] Grotzke, Heinz 1999. "Growing Cucumbers," *Biodynamics Journal*, no. 222: 13-17.

[22] Ingham, op. cit.

[23] Campbell, Stu 1998. *Let it Rot*. Pownal, VT, 93.

[24] Diver, Steve 1999. "Biodynamic Farming and Compost Preparation," *Attra (February)*, 5; see also: Carpenter-Boggs, Lynne 1997. *Effects of Biodynamic Preparations on Compost, Crop, and Soil Quality*. WSU, Crop and Soil Sciences. PhD Dissertation.

Nutrient Balancing on S&S Homestead Farm
Johanna Hök, 2002 & Henning Sehmsdorf, 2021[1]

The objective of this study was to develop a model nutrient budget for S&S Homestead Farm. Specifically, the project considered the cycling of NPK from soil, forage, hay, and manures excreted by livestock in pastures and feeding areas. Emphasis was put on the transformation of manure into compost, and how compost could be brought back to the fields. The model was to be used for assessing the sustainability of raising dairy and beef cattle, and sheep for meat without fertilizing with external inputs.

Context
Qualitative measures of S&S Homestead Farm practices reveal much about current soil fertility. On the basis of daily observations over four months, we concluded that the soil was healthy because there was a diversity of nutrient-rich forages in the pastures, and the animals were healthy. There were no apparent deficiencies of macro- or micronutrients in the soil. However, the animals — feeding on fresh grass in summer and hay in winter — could over time deprive the soil of such nutrients. Since about 5,000 pounds of meat left the farm annually, additional nutrients might have to be added to the soil to prevent deficiencies. The

study investigated how nutrients could be returned to fields and gardens by capitalizing on the special structural, biological, and chemical qualities of compost, and the fact that compost can be produced from materials available on the farm. On a biodynamic farm, one of the main goals is to create a self-sufficient organism independent of external inputs; therefore the availability of nutritional resources to make compost is of critical importance.

Because of time constraints, the study focused on the cycling of NPK, rather than on the whole range of macro- and micronutrients. Ninety-six per cent of total nutrients needed for growing plants are obtained through photosynthesis as plants use solar energy to harvest elements available in air and water. In organic practices, the remaining four percent can be secured from compost, bone meal, manure, wood ash, nitrogen from legumes and other green manures, and nutrients from certain herbs and weeds in garden beds and fields, and from household wastes.[2]

Until the time of the study, farm-produced compost had been used only in vegetable production. The previous year's compost was mixed directly into the soil in late summer when summer crops were removed. The remaining compost was bagged and used next season as potting soil for seedlings raised in the greenhouses, and as amendment when planting seedlings. Whenever there had not been enough compost to prepare winter beds, purchased organic fertilizers such as bone meal, blood meal, and green sand were used as soil amendments. Throughout the summer, organic fish emulsion was occasionally applied to boost seeds and newly planted seedlings. The underlying goal of the present study was to replace all purchased inputs by farm-produced composts.

Compost has many beneficial qualities lacking in other organic fertilizers. Compost has two main functions: First, it improves the structure of the soil, by making the soil easier to work, enhancing aeration and water retention, and resistant to erosion. Second, compost provides nutrients for plant growth, and its organic acids make nutrients in the soil available to plants. Fewer nutrients leach out in a soil with adequate organic matter,[4] and it hosts more microorganisms, because organic matter is the basis of the soil food web.[5]

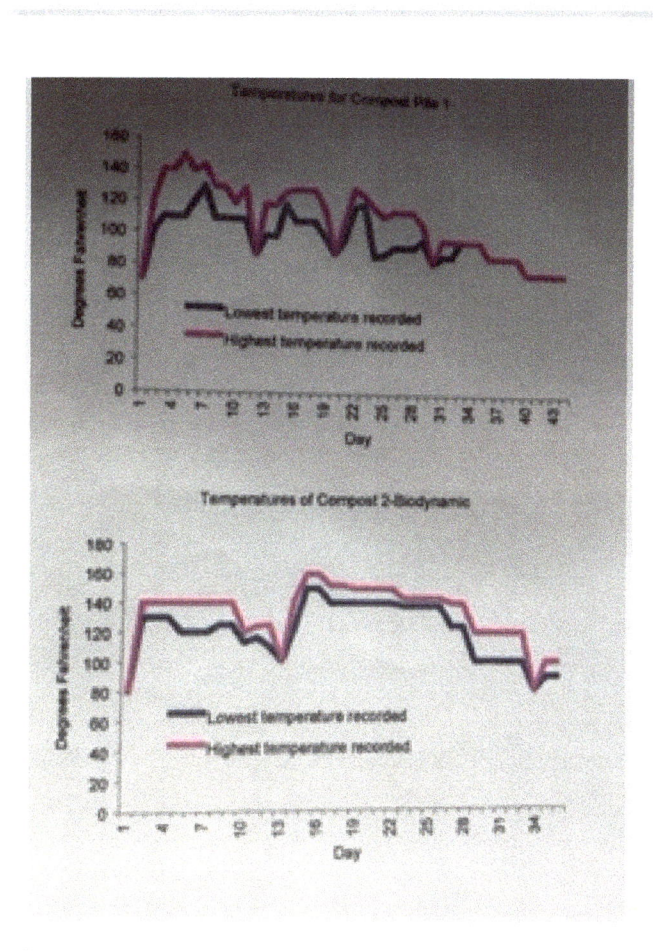

Temperature Cycles of Organic vs. Biodynamic Composts

On S&S Homestead Farm, we manage compost piles using organic and biodynamic practices, including the use of certain botanical ferments (biodynamic preparations). Composting is an aerobic process requiring oxygen.

No fertilizers of any kind have ever been applied to pastures or hay fields. However, the pastures get constant nourishment from the manures and urine of grazing animals. It has been estimated that grazing animals recycle ninety to ninety-five per cent of plant nutrients back to the pasture.[3] Since the hay fields had not been fertilized with manure, these fields would be the first areas to be considered for compost fertilization.

Microorganisms are responsible for the decomposition of organic material. As the microorganisms feed on oxygen and organic material, they exhale carbon dioxide and generate heat. Temperature is therefore a readily accessible indicator of the decomposition process. There are three main groups of microorganisms acting in a compost pile: bacteria, fungi, and actinomycetes. Different types of microorganisms act at different temperature intervals. Generally, the desired temperature interval for a compost pile is 120-160 degrees. Above 120 degrees pathogens are killed and below 160 degrees beneficial bacteria proliferate and do their work. In the cooler parts of the pile (below 100 degrees), macroorganisms such as earthworms, insects and nematodes are also active in the breaking down process.[6]

Method

An established model for general nutrient cycling in ecosystems was used to identify key areas for the recycling of NPK. The role of the sun was recognized in the nutrient uptake process directly dependent on light. Because of the limited scope of the project, and the overarching concern that the pastures and hayfields on the farm might be depleted of nutrients over time, the study focused on grazing animals and their excretions that can be composted. The model for the nutrient budget was divided into two sections. Section I describes composting inputs available on the farm, and the results of the composting experiments. Section II describes the nutrients leaving the farm in the animals and crops harvested, and discusses various alternatives for a balanced nutrient budget on S&S Homestead Farm, including certain farm-produced inputs not considered in the original compost study.

I. Inputs & Results

Specific microbial activity is essential to compost quality. The larger the number of beneficial microorganisms, the better the compost. Beneficial organisms flourish in relatively wet but oxygen-rich environments, while pathogenic organisms thrive under anaerobic conditions. Because it was a cheap and easy test to perform on site on a regular

basis, we used temperature as the key indicator of microbial activity.

Temperature is an indirect measurement of microbial action because such organisms radiate heat as they digest. During the "heating" (thermophilic) stage, the temperature was measured regularly in all piles until ambient temperatures were reached, usually after four to six weeks, when the piles reached the "curing" (mesophilic) stage, during which compost stabilizes, humus forms, and earthworms enter the pile from below.[7]

It was interesting to note that the biodynamically prepared compost piles consistently reached their peak temperature during the second heating cycle, i.e. after the pile had been turned once, rather than in the first heating cycle, as in the non-biodynamic piles, and temperature levels remained stable over time, indicating a slow and steady process of decomposition.

Turning the piles to aerate them or add water had a marked effect on the composting process, with a new heat cycle following after turning. Aeration means better aerobic conditions; however, with every turning, a certain amount of nitrogen is lost from the pile in the form of ammonium (NH_4). The stable temperatures in the biodynamic piles meant that less turning was required to stimulate decomposition. Building the piles with the tractor front loader created more heterogeneous and less well aerated piles than piles made by hand, but the work was accomplished faster.

Over the spring and summer, we assembled a total of eight experimental compost piles from available beef and dairy cattle, sheep, pig and chicken manure and litter, hay, green and woody plant materials, and kitchen wastes.

The intent was to calculate the NPK produced in these piles, made mostly by hand. Three of the piles were treated with the biodynamic preparations 502 (yarrow), 503 (German chamomile), 504 (nettle), 505 (oak bark), 506 (dandelion), and sprayed with preparation 507 (valerian).[8] The average finished volume of the piles was eighty-five cubic feet, and their average weight was estimated as 4,480 pounds, or two and a quarter tons. Using standard tabulations, we calculated the nitrogen content of the compost materials as follows: chicken litter 8%, pig litter 3.1%, sheep litter 2.7%, cow litter 2.5%, fresh grass clippings 2.4%, hay 2.1%, garden weeds, 1.5%, kitchen wastes 1.4%, woody debris (sawdust and plant stems) 0.09%.[9]

A well-prepared compost pile is estimated to produce fifteen to thirty pounds of N/ton, five to ten pounds of P/ton, and thirty pounds K/ton.[10] By that estimate, the almost eighteen tons of compost produced provided between 268-536 pounds of nitrogen, 89-179 pounds of phosphate, and 536 pounds of potash.

Even at the lower end of NPK production, these amounts would be sufficient to balance the nutrient withdrawals due to meat sales at current levels. The practical question would be how to spread that compost in the pastures and hayfields.

Leaves, Wood Chips, and Biochar

Over the years since completion of the original study, the farm has accessed other available nutrient sources, including deciduous leaves, wood chips and biochar. Besides beauty, shade, and food for honeybees and other pollinators, the approximately ninety maples, dogwood, flowering and fruiting apple, cherry, plum, pear, and peach trees on the farm provide more than twenty tons of composting materials in their leaves.

Agricultural research has shown that contributing that amount of leaf matter (dry weight) per acre would add on average 400 pounds of nitrogen, 40 pounds of phosphorus, and 150 pounds of potassium, as well as substantial amounts of calcium, magnesium, sulfur, boron, iron, manganese, chloride, sodium, copper, zinc, cobalt and nickel.[11]

Years of leaf application cause no decrease in the soil pH. Because the carbon content of leaves relative to nitrogen is about 50:1 (compared to 25:1 of compost), leaves worked directly into the soil cannot release their nutrients immediately for use by the plants, which is why we prefer to compost them for a year or two combined with hay, manure and other materials of high nitrogen content. Other proven methods of using leaves to increase soil fertility is to shred them and place them on top of crop beds to break down naturally over the winter and during the next growing season. Worms transfer this layer down into the soil.

A third method is to use leaves in worm bins constructed from hay bales placed on the ground and covered with a tarp. Over the course of a year, ground-dwelling worms convert the leaves into perfectly balanced plant food.[12]

Once we purchased a 1-Hp wood chipper, we discovered much better use for tree and brush trimmings than burning these materials in brush piles. The carbon to nitrogen ratio of wood is ten times greater than that of leaves (500:1), and therefore requires composting for optimal use. Returning seven and a half tons of composted wood chips to crop beds and fields provides an estimated forty-five pounds of nitrogen, ninety pounds of calcium, thirty pounds of potassium, and 7,500 pounds of carbon per acre.[13]

We make wood chips from clearing underbrush, felling diseased trees, or collecting blow-down from violent weather events common in the age of climate change. In the last couple of years, the annual production of wood chips amounted to an estimated fifteen tons, most of which ended up in compost piles and windrows.

In order to clear out forest underbrush as protection against fire, the San Juan Islands Conservation District mounted on-farm demonstrations of biochar production in small-scale, homemade kilns. Biochar is produced through pyrolysis or gasification by heating biomass under reduction of oxygen.[14] Biochar can be used as a soil amendment or to increase habitat for microorganisms.[15] A replicated study carried out on ten farms in San Juan County showed that bean fields treated with biochar increased total carbon by thirty-two to thirty-three percent, soil-available NH_4^+ by forty-five to fifty-four percent, soil-active organic N by forty-eight to hundred-ten percent, and active inorganic P by twenty-nine percent, while increases in phosphorus, iron, manganese, and zinc were shown in the nutrient density of the harvested dry beans.[16] On S&S Homestead Farm, we produce approximately four tons of biochar annually. We either apply the biochar directly to pastures and hayfields by spreading it with shovels from a truck bed, or we systematically incorporate it into our compost and, occasionally, work it into leaf and grass mulches applied to crop beds. We have not yet done soil tests to assess the results of these various treatments.

Animal-Based Inputs

Using standard tables and formulas,[17] we calculated the average amounts of NPK per pound of animal live weight excreted in manure solids and urine on pasture (grazing animals: cattle and sheep), or on bedding (confined animals: pigs and chickens), over the course of the year. In 2002, there were fourteen beef cattle (one bull, six heifers, six steers, one bull calf), one lactating dairy cow, twelve sheep, three pigs, and twenty-five chickens on S&S Homestead Farm, with an estimated total live weight of fifteen tons. The simple formula for computing the nutrient content in pounds for each animal kind was X (average amount of manure and urine annually per pound of live weight), multiplied by Y (weight of animals expressed in pounds, i.e. tons divided by 2,000), and multiplied by Z (average of N, P, or K per pound in solid or liquid manure according to animal species). The results are shown in table 1. To build the eighteen compost piles described above, we withdrew an estimated total of eighty pounds N, sixty-two pounds of P, and twenty-three pounds of K in chicken, pig, and cow manure and bedding. As noted above, the total of almost eighteen tons of compost produced provided between 268-536 pounds of N, 89-179 pounds of P, and 536 pounds of K.

	Nitrogen (lbs)	Phosphorus (lbs)	Potassium (lbs)
Cattle	1,076	433	630
Sheep	74	65	34
Pigs	16	22	6
Chickens	41	27	14
Total:	1,207	547	724

Table 1. Nutrient Value of Animal Manure and Urine

II. Nutrient Balances

As the farmers consume meat, dairy products, fruit, and vegetables, or sell them to the community, nitrogen and other nutrients leave the farm. The cattle and sheep graze, or are fed hay, on about sixteen acres, consuming about twenty tons of fresh forage in summer (fifty-five percent grasses, thirty-five percent legumes, and ten percent forbs),[18] and an equal quantity of hay in winter. Since the protein content in the hay is estimated to be about eight and a half percent, the amount of protein leaving the fields in forage and fodder per season is about 3,400 pounds. or 544 pounds of N.[19] However, the manure and urine excreted by the animals grazing in a balanced rotational system support the regrowth of the forages at such levels that nutrient balances remain intact.

Another question is what levels of nitrogen leave the farm in the carcasses of butchered animals? In 2002, the farm sold 4,657 pounds of beef and 720 pounds of lamb,[20] for an equivalent of 148 pounds of N, which is considerably less than the levels of N produced by the grazing animals themselves (a total of 1,207 pounds N, or 75 pounds N per acre). Soil tests taken in 2009 by Washington State U. Department of Crop & Soil Sciences found the nitrogen content of the pasture to have risen to 133 pounds/acre.

Nevertheless university farm consultants suggested adding another sixty pounds/acre of synthetic N in order to increase forage production to four tons/acre, a goal that would exceed farm requirements of two and a half tons/acre. The consultants' recommendation reflects the industrial notion that scaling up production is a good thing since it presumably leads to increased profits. This is, however, not always true. In our case, production increases would also necessitate expansion of hay storage and other infrastructure, bigger machinery and increased labor — all at substantial cost. Besides, a biodynamic farm is inherently committed to growing natural and social capital, rather than growing financial capital in the form of profit. Ecological health, resilience and long-term

economic stability of the farm are primary production goals. In this context, we were interested that the soil tests showed an increase of soil organic matter from previously three percent to nearly ten percent in the hay and forage fields, and twelve percent in the vegetable sites. The increase of soil organic matter shows that the soil food web is active and healthy.

Concluding Observations

Based on the results of the 2002 study, the farm implemented the following changes. Systematic composting of animal manures and bedding, plant and household wastes, leaves, wood chips, and biochar has become a regular production goal, netting farm-produced fertility valued at well over $6,000 annually.[21] However, while the original intent of the study was to develop a budget of compost as the primary source of returning fertility to hay fields, we came to the conclusion that the most cost-effective use of this farm-produced resource was to dedicate all of it to improving the soils in gardens, orchards, and crop fields, and to replenish the fertility of the hayfields through targeted grazing, rather than through mechanical compost applications requiring the purchase of a compost spreader and considerable labor. (The San Juan Islands Conservation District experimented with a shared-use compost spreader rotating between farmers in the county, but found the investments for purchase and maintenance of a quality spreader too high, and the logistics of moving the tool from farm to farm on neighboring islands insurmountable.)

Instead, we extended the practice of rotational summer grazing to winter feeding hay in the fields, by moving cattle and sheep through the hay fields on a monitored basis for systematic fertilization, while avoiding excessive pugging.[22] Typically, after cutting hay in June, the fields rest over the summer, during the typical drought, recently made more severe by climate change.

When the rains return in September, the mowed fields respond with about six inches of re-growth of grasses, legumes and forbs by the end of the month. Mowing is an essential part of pasture renewal. During the summer the grass typically stands at a height of two to four feet, which means that the cattle and sheep trample a fair portion of the forage. We have found that mowing the tall grass (as well as any thistles, burdock, rushes, or roses left behind by the animals), turns this biomass of plants and manure into excellent mulch that is rapidly metabolized into the soil, while removing the over-story that would otherwise shade the low-growing legumes and forbs.

By the end of October, the fields grazed by the cattle rotationally over the summer have regrown into a thick carpet of lush forage. This year, we are following the cattle (which have by now been moved into the hayfield) with two dozen sheep, who will fatten in the field in time for slaughter around Thanksgiving. They will further contribute to the fertilization of the field, while also breaking the parasite life cycles in cattle.[23]

The farm is meeting its long-term goal of becoming a self-sufficient organism.

Notes

[1] Based on an unpublished study, "Nutrient Recycling and Composting," by farm intern Johanna Hök, 2002; updated and amplified by Henning Sehmsdorf, 2021.

[2] Jeavons, John 2006. *How to Grow More Vegetables*…Berkeley, CA, 47-57.

[3] Russel, James, et al. 2018. *Pasture Management Guide for Livestock Producers*. Ames, Iowa. https://store.extension.iastate.edu/product/Pasture-Management-Guide-for-Livestock-Producers. Retrieved October 19, 2021.

[4] Jeavons, op. cit.

[5] Cogger, Craig 2015. "Soil Management for Small Farms," 7. https%3A%2F%2Fs3.wp.wsu.edu%2Fuploads%2Fsites%2F2079%2F2015%2F06%2FSoil-Management-for-Small-Farms-WSU.pdf&clen=286323&chunk=true. Retrieved October 21, 2021.

[6] See Parnes, Robert 1990. *Fertile Soil*, Davis, CA; Campbell, Stu 1998. *Let It Rot!*, Pownal, VT; Jenkins, Joseph 1999. *The Humanure Handbook*, Grove City, PA.

[7] Shaffer, Bob 2010. "Curing Compost: An Antidote for Thermal Processing." *Acres USA*, Vol. 40, No. 11. https%3A%2F%2Fwww.soilwealth.com.au%2FimagesDB%2Fnews%2FNov2010_ShafferCompost.pdf&clen=408973& chunk=true. Retrieved October 21, 2021.

[8] For further discussion of the biodynamic preparations and their intended effects on the soil, see Johanna Hök 2002. "Medicinal Plants on S&S Homestead Farm,"below. See also Goldstein, Walter, et al. 2019. "Biodynamic Preparations, Greater Root Growth and Health, Stress Resistance, and Soil Organic Matter Increases Are Linked," *Open Agriculture* 4: 187–202.

[9] See Jenkins, Joseph, op. cit, 56.

[10] Parnes, Robert, op. cit.

[11] Heckman, Joseph R. et al. 2004. "Plant Nutrients in Municipal Leaves," *Fact Sheet 824*, Rutgers U., New Brunswick, NJ. chrome-extension://efaidnbmnnnibpcajpcglclefindmkaj/viewer.html?pdfurl=https%3A%2F%2Fsustainable-farming.rutgers.edu. Retrieved October 25, 2021.

[12] Angst, Gerrit et al. 2019. "Earthworms Act as Biochemical Reactors to Convert Labile Plant Compounds Into Stabilized Soil Microbial Necromass," *Communications Biology* 2: (article) 441. https://www.nature.com/articles/s42003-019-0684-z. Retrieved October 25, 2021.

[13] N.a. 2019. "Nutrient Management for Recycled Orchards," *U. of California Agriculture and Resources*. https://orchard recycling.ucdavis.edu/nitrogen-management. Retrieved October 25, 2021.

[14] N.a. 2018 "Biochar Is a Valuable Soil Amendment," *International Biochar Initiative*. https://biochar-international.org/biochar/. Retrieved October 25, 2021.

[15] Noraini, Jaafar, 2014. "Biochar as a soil amendment and habitat for microorganisms," U. of Western Australia. https://research-repository.uwa.edu.au/en/publications/biochar-as-a-soil-amendment-and-habitat-for-microorganisms. Retrieved October 25, 2021.

[16] Gao, Si 2016. "Locally produced wood biochar increases nutrient retention in agricultural soils of the San Juan Islands, WA," *Research Works Archive*. http://hdl.handle.net/1773/36699. Retrieved October 25, 2021.

[17] Parnes, op. cit, 133-135.

[18] For details of the forage mix, see Reeve, J.R., et al 2011. "Sustainable Agriculture: A Case Study Of a Small Lopez Island Farm," below.

[19] See Merrill, A.L. & Watt, B.K. 1973. "Energy Value of Foods: Basis and Derivation." *Agriculture Handbook* No. 74, ARS, USDA, Washington DC.

[20] For protein levels in beef and lamb, see Ashbrook, Frank G. 1973. *Butchering, Processing and Preservation of Meat: A Manual for the Home and Farm*, New York.

[21] S&S Homestead Farm Production 2016. https://docs.google.com/spreadsheets/d/1Na6dxEs-3AfhDFISvr7Jnoh5eRsGdoy5Qi5r0TfglC0/pubhtml. Retrieved October 29, 2021.

[22] For details of rotational winter feeding, see Henning Sehmsdorf 2021. "Ecological Livestock Raising," below.

[23] Retrieved October 25, 2021.

Seeds of Life

Tasha Wilson, 2005[1]

"All that is born is born of *anna* (food). Whatever exists on Earth is born of *anna*, lives on *anna*, and in the end merges into *anna*. *Anna* indeed is the first born amongst all beings."[2]

Food is the basis on which all life depends. We live in a time where food is processed, pasteurized and genetically modified. Rarely do we know who grew our food and where, what chemicals were added, or even whom we support when purchasing our foods. Today, the global food system is monopolized by some ten corporations.[3] This includes the seed industry, an industry people tend not to know very much about, but is crucial to the future of world food security and farmers everywhere.

Today most of the world's population is concentrated in non-industrialized countries, and seventy per cent of that population earns a living by food production (versus two percent of population in industrialized countries). For centuries, Third World farmers were successful in cultivating crops, and through natural selection helped develop species that are hardy and diverse. For example, Indian farmers developed 200,000 different varieties of rice

through selective breeding. New Guinea developed more than 5,000 varieties of sweet potatoes. This was made possible by farmers' freely sharing seeds with each other, another tradition that goes back centuries. A farmer who wanted some of his neighbor's seed would trade for it by offering an equal amount of his own seed in return.

Not only was seed shared, but farmers could plan what they were going to plant the next season by monitoring how seeds did in their neighbors' fields, learning from others' mistakes and successes. This was an effective and sustainable method.

Unfortunately, the way the seed industry is set up today, farmers can no longer continue this time-honored method of seed propagation. In 1987 the Dag Hammarskjold Foundation organized a meeting on bio-technology called the "Laws of Life." The meeting addressed the emerging issues of genetic engineering and patenting. The meeting made clear that giant chemical companies (such as Monsanto) were "repositioning" themselves as "life science" companies. Their goal was to control the agricultural industry through patents, genetic engineering, and mergers. Today just ten corporations control thirty-two percent of the entire commercial seed industry valued at an estimated $23 billion. These same corporations control one hundred percent of the market for genetically engineered, or transgenic seeds, as well as the global agrochemical and pesticide market. The resulting monopolization of the industry led to establishing the World Trade Organization (WTO) and international treaties like the Trade Related Intellectual Property Rights Agreement. This agreement, through the WTO, makes it possible for a corporation to patent the genetics of a seed and monopolize it by claiming it as private property.

Critics have decried this legalized practice as a form of biopiracy.[4] Corporations claim they have "discovered" seeds and genetic materials actually used by indigenous people for centuries. The result is corporate monopoly over the seed and criminalization of seed-saving and -sharing. These laws maximize profits for big corporations, with no regard for small-scale farmers. Local markets for seed and food are deliberately destroyed and replaced by monopolistic global markets.

Among the 250,000-300,000 plant species of today, a minimum of 10,000-50,000 are edible by humans. Around 7,000 of these are farmed for food. Out of these 7,000, just thirty species provide ninety percent of the world's caloric consumption and only four species, rice, maize, wheat, and soy, provide the bulk of calories and proteins consumed by the world's population through global trade. When you take into account that all these crops are grown from seed and that ten corporations control twenty-five percent of all commercial seeds and one hundred percent of all genetically modified seeds, you realize why the monopolization of the seed industry is such a big problem.

Traditionally, local growers cultivated diverse crops by conserving plant breeds and seeds through natural selection, and ensuring their continued use. But as global markets began to replace local markets, monocultures began to replace diversity. For example, China originally cultivated about 10,000 different varieties of wheat; by the 1970s that number had fallen to 1,000. The U.S. at one time raised over 7,000 varieties of apples; now more than 6,000 of those have disappeared.

Industrial agriculture promotes monocultures in order to centralize control over the production and distribution of food.

It is easier to have patents for thirty varieties of corn if there are only seventy varieties available in the market. Vandana Shiva refers to this ultimate control as "food totalitarianism." Aside from the loss of diversity, another problem is the hybridization of the seeds replacing local varieties. In order to secure patents, corporations claim seeds as "their inventions." To do this, they breed "new" hybrid seeds that are actually crosses of original breeds, so they can then claim any seed with similar genetics as stemming from their "invention."

Unfortunately, hybrid seeds are vulnerable to pests and require pesticides that the original seed did not, because natural selection had bred resistance. But the fact that farmers are forced to buy pesticides in order to save their crop is a huge gain for the seed corporations, who are also the purveyors of agrochemical products. Because of the costs of pesticides never needed before, more and more farmers have to buy on credit. In many cases crops fail as a direct result of pests or large-scale seed failure. Farmers throughout the world who plunge into debt, commit suicide. In India, many have killed themselves by drinking the pesticides.

Vandana Shiva references a study comparing traditional polycultures with industrial monocultures, showing that polycultural systems can produce one hundred units of food per every five units of input, whereas industrial monocultural systems require three hundred units of input for every hundred units of food.

Corporations like Monsanto maximize profits by undermining traditional agricultural practices. In 1998, the USDA, in collaboration with Pine Land Company, announced "Terminator Technology." This patented invention permitted the creation and sale of sterile seeds. Making the seeds sterile meant selectively programming the plant's DNA to kill its own embryos. It was designed to force farmers to buy new seeds from corporations each year.

Monsanto bought the patent on the terms of five per cent of sales going directly into the pockets of the USDA. Monsanto planned to commercialize the technology on the open market. At least seventy-eight countries applied to be owner/licensees of patented seeds.[5]

Fortunately, qualified officials who studied the new technology announced publicly that there was a serious possibility Terminator seeds might spread their genetics to surrounding food crops. As a result, the Consultative Group on International Agriculture Research, the world's most important agriculture research system, stated that they would not use "Term Tech" in its breeding work. Yet corporations strongly argued that the technology was necessary for them to recoup their investments. Finally, in response to international outrage, Monsanto stated in October 1999 that it would abandon their plan to commercialize "Terminator Technology."

Not all government agencies are turning a deaf ear to the injustices being done to traditional growers and their communities. One particular

project through the WSU Agriculture Department, in collaboration with USDA and USAID (United States Agency for International Development), has sites in both the U.S. and in Malawi.[6] The project is focused on dry beans.

Beans are a great crop to raise because they are self-pollinating. This means that the harvested seeds can be saved for planting the following year. This is especially significant for small-scale growers because large-scale seed companies will often not fill orders for them, since bigger orders are more cost effective.

Dr. Carol Miles of WSU heads up this project, both in Malawi and in the U.S. In the summer of 2004, I was given the chance to participate in one of the trials in Washington.

The goal of the project was to test breeding lines, rather than varieties of beans. The difference is that varieties are what is sold to the consumers, whereas breeding lines (also referred to as parent lines) are what these varieties stem from. In the bean business everyone generally uses the same parent lines, but they want different qualities carried through to different varieties, depending upon where the varieties are being grown. For example, a variety of beans grown in Malawi, whose parentage is Pinto, would have different characteristics than beans of the same parentage grown in Washington, because of differences in elevation, heat, and precipitation. These characteristics would be tested in Dr. Miles' bean trials.

The parent lines we were testing were all main-market classes. Main-market classes are what you find in the grocery store: navy beans, pinto beans, black beans, and Great Northern beans, for example. Dr. Miles was able to get these beans from a USDA breeder who works with mainstream agriculture. Even though the size of the order for the trials was smaller than an average order for an industrial farm, the breeder provided the beans, because the results would benefit him. Before breeding lines can be released, they have to be tested.

Another objective of the same breeder was the cultivation of a breeding line superior to the ones already available. The trials were sited on five small farms in Washington, including Vancouver, Moses Lake, and Lopez Island. At each farm the same breeding lines were planted, a total of eight different breeds, with four standard market varieties.

A standard market variety is a bean that has already been approved, released into the market, and sells successfully. The purpose of the main-market class is to provide a comparison for other breeding lines. You may wonder why small farmers would want to participate in the project that may help their competition, large scale farms. But the project does benefit small-scale farmers as well, because they participate in growing the beans and can observe the different stages of development, and draw their own conclusions. After the harvest, they keep a portion of the beans for their own consumption, or for planting the next year.

Commercial agriculture works on a large scale producing crops all across the country. In order to compete with other large scale corporations, they cannot look for perfection in their crops. Instead they have to produce a crop that is versatile and will do well in many places. By contrast, small-scale farms can focus on perfecting a crop and are able to compete in the market by offering the same products, but of higher quality. The bean trials help small farmers perfect a breed.

The project is repeated in Malawi, where it faces certain challenges. Because the culture is deeply rooted in tradition, farmers are resistant to outside influence and the idea of new crops being introduced is problematic. Malawis are a communal people. The project takes place only on six farms, but all villagers come and go, inspecting and critiquing the project. This is done not only out of curiosity, but also from tradition. The practice of observing your neighbor's fields and exchanging news with him is very old.

Through this tradition the farmers learn about different crops and use this information when deciding what to plant for the next year.[7] The project uses a breeder in Malawi who has beans saved from previous harvests, which are planted together with the breeding lines from the U. S., and the two are compared. As in the U. S. trials, the farmers are allowed to keep a portion of the harvest to eat or plant. Both the U. S. and the Malawi trials are beneficial to local farmers and the agriculture industry.[8]

In the summer of 2004, I was given the opportunity to participate in the Lopez Island dry beans trial to take place on Henning Sehmsdorf and Elizabeth Simpson's farm, S&S Homestead. This was the second such trial done at the farm. The previous summer, Dr. Sehmsdorf and Dr. Miles had run a dry-beans trial, growing heirloom beans. This year Dr. Sehmsdorf agreed to participate in the breeding line research, which would begin in May of 2004.

At the time, Dr. Sehmsdorf was teaching an Agricultural Science Class and offered the project to his students. I had been working at the farm every Saturday, in addition to the class, and was excited to get involved with the bean project. I met with Drs. Miles and Sehmsdorf to lay out the schedule for the project and learn what my role would be. Dr. Miles had decided that we would plant on May 18, so that the beans would be ready for harvest the first or second week of September. The Agricultural Science Class would help in the planting. My job would begin a couple of weeks later. I would come three days a week, and stay up to three hours, recording observations and maintaining the beans. The information I collected were the date of fifty percent emergence, the date of first flowering, the date of fifty percent flowering, plant height at fifty percent flowering, and the general condition of the plant stand. I was also responsible for weeding between the rows of beans, and keeping a journal of the project.

On May 18, the Agriculture Science Class met at the farm to begin planting the dry beans provided by Dr. Miles. The beans, main-market classes and standard-market variety, were to be planted four different times each in ten-foot plots. This required five and a half rows, each row measuring hundred and fifty feet. Before planting, we had to make furrows in the ground with pick-axes and clear out any rocks that would hinder the beans' growth.

We planted approximately hundred and sixty beans for each ten-foot plot. Each individual plot was labeled with a number, rather than the breed name. This is the standard method, because referring to a bean by number rather than name, prevents bias in the data collector.

After the beans were planted, I met with Drs. Sehmsdorf and Miles to outline my schedule. I would need tri-weekly visits to the farm in about three weeks. The first data that I would need to collect would be fifty percent emergence. This would occur when about sixty sprouts were showing in a plot. The timing would be different for each plot and not necessarily the same even for beans of the same breed. Three weeks after May 18th, I began to visit my beans. It was very exciting because there had been so much change. The miracle of growing amazed me.

Looking at the sprouts, I was filled with wonder. To think that these tiny, fragile things would grow into sturdy plants capable of providing essential food was amazing. To know

that I would have some part in that, made me proud and determined.

The date of fifty per cent emergence for the first plot passed and was closely followed by others. I settled into a steady rhythm of weeding, tending, and recording. Last summer, Dr. Sehmsdorf had decided to build a six-foot fence around the perimeter of the field. Unfortunately at this point the fence was not finished and we were forced to resort to other measures of precaution. We spread the farm dog's discarded hair throughout the patches as a way to scare the deer away and then added plastic netting over the fragile seedlings. Soon though the fence was done and we were able to remove the netting.

Another obstacle was the dry, hot weather we were experiencing. Fortunately when we planted the beans, Dr. Sehmsdorf had us lay drip hoses in the bottom of the furrows, so that we would have an efficient way to irrigate the plants if there was a lack of rain. S&S Homestead Farm supplies water to its land through a sustainable system. The water source is a pond dug when they first started the farm. The pond is filled with rain water distributed over the farm using a solar-powered micro-irrigation system.[9]

As the summer went by, I recorded more and more data: date of first flower, date of fifty percent flowers, plant height at fifty percent flowering in centimeters, and then finally the plant stand. Plant stand is a measurement of average quantity, in which I measured two and a half feet in from both ends of a plot and then counted the number of plants in the five feet of middle section left.

I developed a technique for the weekly weeding: I would begin with a pick axe walking up between the rows and hacking at the roots of the weeds that had sprung up. I then collected the weeds in a wheelbarrow which I emptied on the compost pile. Then I weeded the actual rows, a much more precise task. I used a trowel to get to the roots of the tougher plants, making sure to avoid the roots of the beans. Most of the weeds were mustard, which the bugs appeared to love. The bugs never went after the beans, perhaps because of their developed taste for the mustard. There were always lots of ladybugs and no trouble with aphids, which I am sure was due to the ladybugs.

Things went very smoothly, and about the middle of July the plants began to bear fruit. The pods were long and beautiful. Some were flecked with purple and startling shades of green. I could not resist the temptation to try them and I was not disappointed. They were sweet and crunchy and when I bit into them, intense and fresh flavor burst in my mouth. It made me wonder how anyone could ever eat anything not just off the vine. As September approached and the summer came to an end, I found myself with not very much to do.

The time had come for the beans to dry out so that when we harvested them, we could collect the seeds to be analyzed and then replanted. All was going well, but then about a week before the scheduled harvest, a buck deer got into the field and had his way with our precious plants. Dr. Simpson walked down early one morning and caught him in the act. We were shocked; as far as any of us knew deer were not supposed to be able to jump six-foot fences. Dr. Sehmsdorf could testify that in his forty years of farming, no deer had ever got into his orchard enclosed by a six-foot fence. We were crestfallen. Fortunately there were enough plants left for Dr. Miles to collect and analyze and some beans left for the farm to keep. The day of the harvest dawned bright and sunny and as we all assembled to pull up the plants, I had mixed feelings. I was glad that the project was over and grateful for what I had learned, yet also sad because it was the end of

something I had gained tremendously from. We worked fast and within an hour had harvested the middle five feet of each plot that Dr. Miles told us to gather from.

Some of the plots had been so badly damaged by the deer that there was simply nothing left to collect. We left two and a half feet at both ends of each plot to be gathered later by our Agriculture Science Class for the farm. Then Dr. Miles left to catch the ferry, taking the beans with her to her lab in Vancouver. There many tests would be done to analyze the success and faults of the breeding lines. These conclusions would be shared with the breeder to determine what breeds to use as parent lines. My part in the project was complete.

The one thing that all people in the world have in common is food. Food is our life source. Food sustains us, it is what we cannot do without. In our world today, it is easy to forget that the food you eat did not just appear in your grocery by magic. It began with a seed. Something so small that if you dropped it, you would probably never see it again. Yet from that tiny organism comes our source of life.

We cannot forget the farmer who tends that seed and is responsible for what it becomes. Yet in much of modern agriculture, the seed is compromised, abused without our consent and without our knowing. If you take the time to examine the growth process, you cannot help but marvel. It is truly a miracle. We must not let ourselves forget that, or the results will be disastrous. Fortunately today there is a counter balance to the industrial food system. Projects like Dr. Carol Miles' bean trials are paving the way for the survival of local markets and small-scale farmers.

Recently India's sacred tree, the Neem Tree, was patented by an American company, Thermo Trilogy. The Neem's seeds have a fungicidal quality that indigenous people have

known about for centuries. Yet Thermo Trilogy
was able to claim that they had discovered this
trait, and so it became illegal for people to use
the seeds without paying fees to Thermo
Trilogy. Last week the patent was revoked, for
the first time in history, as a means of
protecting traditional knowledge and
practices.[10] Actions are being taken, there are
people who care, the future is not as bleak as it
may appear. We must not forget, it is the
farmers that grow our *anna*, it is the seed that
anna comes from, and that we are all connected
through *anna*.

Notes

[1] Senior thesis submitted in partial fulfillment of graduation requirement at Lopez Island Public School, Lopez Island, WA.

[2] Ancient Indian Upanishad, quoted in Shiva, Vandana 2000. *Stolen Harvest: The Hijacking of the Global Food Supply,* Cambridge, MA, 5.

[3] See Simpson, Elizabeth 2007. "Pastoral versus Industrial Food Production," below.

[4] See Vandana Shiva, op. cit.

[5] See http://www.monsanto.com/monsanto/layout/about_us/default.asp. Retrieved 7/15/2021.

[6] Miles, Carol A. et al. 2005. "Dry Bean Variety Trial Comparison WSU Vancouver REU & Moses Lake". *Sustainable Seed Systems*. WSU. http://SystainableSeedSystems.wsu.edu/. Miles, Carol A. 2003 "Growing the Dry Bean Market". WSU and Bean/Cowpea CRSP Work to Expand Viability of Beans in Africa and at Home. WSU. http://aenews.wsu.edu/Aug03AENews/Aug03AENews.htm#.

[7] Carol Miles, personal communication, March 16, 2005.

[8] Http://SustainableSeedSystems.wsu.edu/.

[9] See Sehmsdorf, Henning 2004 (2021). "Solar Micro-Irrigation and Energy Production,"below.

[10] Hellerer, Ulrike & K.S. Jarayaman 2000. "Greens persuade Europe to revoke patent on Neem tree…," *Nature*, vol 45: 266-267. See also Organic Consumer Organization 2005. "Biopirates Lose Patent on Seeds of India's Sacred Tree, the Neem," *Organic Bytes*, vol. 52: 1. www.organicconsumers.org/patent/neemtree030905.cfm. Retrieved July 14, 2021.

This Isn't a Show Garden!

Lisa Murgatroyd, 2009[1]

"I arise in the morning torn between a desire to improve (or save) the world and a desire to enjoy (or savor) the world. This makes it hard to plan the day" (E.B.White).[2]

In my year at Lopez Island Elementary School, as the first garden-to-cafeteria program designer and educator, the best decision I made came the night before the first day of school: to always choose what would encourage personal investment by staff and students. This is the story of where that decision, and the hard work of everyone involved, led us. That decision became a thousand other decisions that year. I have learned that my investment of time, emotion, or action is the path to meaningful ownership. When you feel ownership in something, whether it is a love relationship, a fight for civil rights, or dedication to a daily practice, that something becomes inseparable from your being.

I was certain that the new garden-to-cafeteria program would secure its future only when the whole elementary school felt it was theirs. We needed to reach a place in the school culture where the work of the program was housed on the fingertips and lips of every student and staff member, instead of on my makeshift desk, smooshed between the copy room and the teachers' lounge. From the beginning I had to believe that everyone could be inspired in some way. How? Where to start? How to gain access to the teachers' and students' time and make it time well spent? What would success mean to us? These were questions we worked through all year.

On the first day of school, I was a complete rookie. I didn't know what needed to be done. I didn't know the school or the community. I had never been a teacher or created a program, but here I was. It was baptism by fire. To my dismay, in endearing jest, the principal told me later that I had the look of "a deer in the headlights" the first few weeks of school.

I walked into his office the first day and he told me two important things: "Find the seams in the culture of this school and see how your program can fit into them. You have to do that to make this work." And, "We don't have a lot of time to work with here. We have to teach a lot of other things, too, like the standards." I took this to heart in figuring my approach. I knew to expect this in a public school: pressure to meet the standards, but not enough time or resources to do it. Teachers would be overworked and class time would be full before the garden-to-cafeteria program asked for anything. If the program was to survive, it had to become what was taught in school, not just something extra that took time away from everything else. The garden, farm, working with food in class, had to become a lens through which the standards in every subject could be taught.

To own the project, the teachers had to have a hand in designing it. So I went about assessing their needs, and the more effort I made to involve them, the more willing they were to make time for me. Including everyone in making decisions was messy at first: it's difficult to meet the desires of so many different individuals. But this approach paid off big time in the end. After preliminary meetings with each teacher and then collaborative meetings with the whole group, we created a structure for the program that worked.

Our final plan, after several weeks of trial and rehashing sessions, defined my role: ~ I would act as a resource to teachers, providing ideas and support for them to use the garden, food, and natural world as a lens to teach their subject standards; ~ I would create and facilitate classes using the garden or food to develop specific themes they were teaching (i.e. learning fractions or the history of Lewis and Clark); ~ I would help the kitchen staff incorporate student-harvested garden produce into their meals; ~ I would co-ordinate weekly harvesters, prefigure harvest amounts, and find volunteer vegetable washers; ~ I would provide weekly L.I.F.E. (Lopez Island Farm Education) lessons, incorporating the standards in interdisciplinary ways, while teaching students growing techniques and how to enjoy fresh, healthful food.

We scheduled care of the school garden together. The six beautiful, kid-friendly raised beds would be cared for by all the classes on a rotating basis, an exercise in cooperation and shared responsibility. Each class would harvest or water for one week. We completed planting and other garden work during my weekly classes with students, and on other days, when teachers felt compelled to get students outside and working.

Every chance I got, I encouraged students' sense of ownership in the garden: "You guys, this is your garden to create and love. We can either make it beautiful or let it lie. In the end it is up to us to figure out how to do it together." We involved the whole elementary school in determining the crops we would grow and where we would put them, preparing the beds, growing the seedlings, planting, and care. When it rained or was too cold, we worked in the classroom learning about food or cooking. Everything was an experiment, an adventure in teaching, learning, and growing food.

We expanded the garden so that each homeroom had its own garden bed. I acted as support for teachers on their projects, encouraging their investment and creative impetus. Even the teachers who had never grown a plant in their lives took to their class gardens with vivacity and boldness. They created salad gardens, flower gardens, and even a French-fry and catsup garden with a carpet of cover crop growing beneath.

After some experience with teaching their students in the garden, the teachers told me that they noticed that the students who had the hardest time focusing in the classroom excelled in garden lessons. This was exciting, though not surprising. University of Illinois studies show that the test scores of children who participate in outdoor learning, actively doing instead of listening to a lecture or reading a textbook, rise in every discipline, a whopping twenty-seven percent in the sciences.

With new opportunity, there was also

confusion. We had notable adventures in harvesting, for example. In one miscommunication, immature garlic was pulled up instead of chives being cut. Whole sections of chard and kale plants were pulled up instead of individual leaves being harvested, and cabbage plants cut instead of lettuce! This needed to happen. It was an educational garden, and served its function beautifully. I loved witnessing the process of personal discovery that kids and teachers experienced. They were reconnecting themselves to food.

Day after day, a new moment in the garden or classroom would speak to someone, and they would become alive with wonder. We found that with some creativity, we could fit gardening and food into any lesson. In math we learned ratios and fractions by making potting soil, about data tables and how to make line graphs by measuring the growth rate of garden plants over time, to make bar graphs by counting plant families, and about pie charts and percentages by experimenting with the composition of our garden soil.

In social studies we examined how farming is done all over the world and learned about what foods grow in different climates. We learned the geography of countries while mapping the migration of apples, cheese, and other foods across the world. To learn about colonial times in America, we made foods from ingredients that were important in Wampanoag and Pilgrim agriculture. We put on skits depicting the history of different important seed crops.

Using apples as a model for the Earth's spherical shape, we figured exactly how much of the apple skin, representing the crust of the Earth, could actually support human life through the growing of food. The kids were amazed at the tiny ration of apple skin, smaller than a fingernail, on which we could actually grow food! We enriched the language arts curriculum by integrating into it visits to S&S

Homestead Farm, a diversified sustainable farm and champion of local food education.

Students learned how food is grown there, visited the animals, milked the cow, and explored the interconnections of life on the farm. Back at school, they wrote stories and poems about their experiences. My favorite was a story about the chicken that wanted to go to kindergarten. During the students' visit, the chicken had actually hopped on the school bus and was waiting for them on a seat when they returned. They created poems that read like a beat poet night in San Francisco: "Milking the cow, squeezing a hot dog," "Chickens squawking like a woman screaming, their eggs, white circle smooth."

For a science class learning about the solar system, we explored the farm on a "tasting adventure route," making different stops to explore the role of the sun in growing food. Back at school, students tied their experiences together in a song they wrote about interconnections on the farm that are powered by the sun. Several teachers told me that science was their most challenging subject to teach in the classroom. However, from introducing the world of microorganisms through compost, to the composition of air through photosynthesis, to investigating seasonal rhythms through garden observation journaling, the possibilities for teaching the sciences in the garden were endless.

We didn't let these opportunities get away. After all, the ideas about the natural world we were studying were functioning all around us in the garden. It felt magical. Throughout the year, we found that the kids loved participating in anything to do with food. If they had grown it, they would eat it, guaranteed. At 9 am, children would ask for thirds on turnips when we cooked them together to complement the children's story, "The Enormous Turnip."

In wellness class, when learning about vitamins, the kids would ask the teacher, "Can I have another vitamin-A snack please?" Every day in the garden students would eat kale and Brussels sprouts right off the plants. It wasn't long before the school chef and I teamed up. If we could prepare a healthy food item in one of my classes that she wanted to introduce to the menu, chances were that the students would take it from the lunch line because they now knew that food.

When the chef had tried to serve hummus the year before, there was a lot of "weird" and "yuck" heard, and she was left with a bowl full of hummus at the end of lunch. She tried again this year, and, after we made hummus in class with parsley from the garden, they ate it right up. Now it is a standard item on the menu. We realized that we had struck a gold mine in child psychology for getting kids to eat well. I used the same approach with the kitchen staff, offering myself as a resource to meet their needs. From the beginning the staff welcomed this challenge, their fire growing as the year progressed. They were empowered in their unique role as agents of change in students' lives.

Lopez Island School has a Title I designation. This means that over fifty per cent of the students qualify for free breakfast and lunch at school, the bulk of their daily nutrition. Cafeteria staff attended workshops entitled Rethinking School Lunch. Through culinary classes they taught students to make and enjoy delicious, healthy foods. By the end of the school year, the staff had also researched how and where they could purchase local and organic foods. Due to their work and dedication, with the help of grant monies, the cafeteria is going organic, and the students love the fresh, flavorful food.

Since the kitchen started the year on a budget that would support only "heat and serve,"

many would have said it couldn't be done. But they did it. By spring, the "garden as lens" idea was everywhere in the elementary curriculum, and soon the teachers were taking it on themselves. They taught nutrition by cooking recipes to complement a book the class was reading, used the garden to create math puzzles, and illustrated erosion by growing plants in different densities.

The wellness class learned about nutrients through a wall-sized fruit and vegetable rainbow they had made. Art classes learned drawing by sketching their growing plants in natural fiber diaries they had created. When the music teacher jumped on the bandwagon, it felt like we had reached a tipping point. They were making it their own.

In mid spring, when the garden climbed out of its winter shell, the whole elementary school serenaded its return in a beautiful concert based on gardening and growing. They sang, "Inch by inch, row by row, gonna make this garden grow, gonna mulch it deep and low, till the rain comes tumbling down." There were flower costumes, butterflies, and scarecrows. Then came the encore, "The Weeds Came Back" to the tune of "The Cat Came Back." A gang of wild third grade boys dressed as weeds danced across the stage. The grand finale was a perfectly timed percussion piece in the garden, using every kind of garden tool. Students danced around the garden to the beat holding lunch trays and harvesting into them. The show ended with the students balancing the trays on their heads to indicate "a balanced meal." The crowd of parents and community members went wild; it was glorious. The Spirit was catching everyone.

The librarian created a year-round display of garden books for kids and another of teaching resources. The superintendent began making presentations about our garden-to-cafeteria program at school conferences. The principal attended a nation-wide Sustainability and Education retreat to focus his own work on the program. A Farm-to-School Institute was held, attended by educators from around the country. The school board considered funding part of the program's needs for the following year, a big step on a shoestring budget. Even the school secretary was giving me updates about the veggies she had refused all her life but was eating at school now. Children came home from school or from visits to the S&S Homestead Farm eager to eat vegetables and to plan and start gardens at home.

Every time we cooked in class at school, we sent the students home with the recipe, so that students could apply what they learned to their lives outside school. Parents told me they were making the recipes with their kids and began to give me suggestions about new ones we could consider for future food lessons. The principal who had called me a "deer in the headlights" was now saying that he felt the culture of the school was changing because of the work we were doing. By the end of the year, he was a dear friend to me. Teachers said they could feel the school changing too, and they were excited by the project, refreshed.

One parent told me that the L.I.F.E. program was the reason their kindergartener attended the school. Truth be told, the Lopez School garden did not look like a show garden that year. The plant-spacing imperfections, late plantings, uneven watering, densely planted seeds, and over-ripened fruits were not the marks of neglect, however, but the marks of learning children. All over the quirky garden was the evidence of motor skills being stretched to new limits, children taking responsibility for nurturing their own plants, and interesting features that showcased their creative design and thoughtful planning, such as pioneering experiments in milk jug greenhouses and tomato trellising. The students who worked in the dirt all year, the kids who

helped with extra garden watering during recess, the teachers who stayed up late developing ideas and preparing to integrate food and garden into their classes, the kitchen staff who discovered new sources and preparations of food, the principal and superintendent who extended their hours to attend conferences and meetings -- this garden was now theirs!

Notes

[1] Photo credit: *Life Garden Program*. https://www.lopezislandschool.org/cms/one.aspx?pageId=500997. Retrieved July 15, 2021.

[2] Quoted in Shenker, Israel 1969. "E. B. White: Notes and Comment by Author," *New York Times* (July 11).

THE STAFF OF LIFE?
The Culture of Gluten Intolerance As Seen Through the Eyes of a Homestead Baker

Henning Sehmsdorf, 2014[1]

When I was a child in Germany, my grandmother baked all our bread. It took four days—refreshing the sourdough starter on the first day, mixing the sponge on the second, kneading the dough on the third, shaping the loaves on the fourth, and letting them rise four or more hours before baking the bread in a wood-fired oven. One morning, when I came into the kitchen and saw the glistening loaves lying on the counter, my grandmother told me that the cat had had her litter in the warm, rising dough during the night. But Grandmother had scraped off the mess and baked the loaves anyway. "Bread is the staff of life," she said. "It is sacred!"

We still bake all the bread we eat in our household, mostly long-fermented sourdough rye made from grain raised on our farm, and we supply loaves to our whole-diet CSA which, besides bread, includes vegetables, fruit, meat, cheese, and eggs. Remarkably, we have found that people suffering from gluten intolerance can safely eat our bread without gastric discomfort, presumably because the gluten protein, which gives the bread its shape and elastic structure, is broken down into harmless peptides easily digested in the human gut.

These two vignettes, one describing an unforgettable childhood experience, the other a current food practice in our home, illustrate the making and eating of bread as a cultural practice of deep significance for bodily and Spiritual health.

Historically, the consumption of bread has served as a marker of community identity. Some 3,000 years ago, in Homer's *Odyssey*, the hero in exploring the shore of the land of the Cyclops, anxiously asked his fellow seafarers, "Are they bread eaters?"[2] In other words, are they human? And a millennium later, Jesus taught his followers to pray to God to "give us our daily bread," and to celebrate the presence of the divine in the world through eating bread in Holy Communion.

With this as background, it strikes me as a sign of a profound cultural shift to read the following on the Rodale website recently: "It's not every day that we send out such a special announcement! We're thrilled to bring you this opportunity to pre-order *Wheat Belly Total Health*. With *Wheat Belly*,[3] Dr. William Davis gave millions of people the ability to take back control of their health. He enabled them to reverse years of chronic health problems by removing grains from their daily diet."

The cultural implications of the Rodale announcement are staggering. For one, bread is no longer seen as sacred and therefore becomes exchangeable for any nutritionally equivalent food stuff. For another, the recommended response to gluten intolerance is total avoidance of any grain, instead of inquiry into the causes of why so many people are no longer able to tolerate a food that has been central to the human diet and culture since civilization began. What Rodale should be asking is: what changes are needed in the way we produce bread for this food to regain its historic role as the sacred "staff of life?"

Here are some of the relevant facts: roughly one third of people with European ancestry carry predisposing genes for gluten intolerance. However, in traditional food cultures very few people ever developed any symptoms of the disorder. In other words, genetic predisposition is not sufficient to cause the problem. And yet,

in the U.S., gluten intolerance and celiac disease have quadrupled since WWII. At least one per cent of the population are now affected, with similar numbers recorded in Europe. Why?

Here is the second relevant fact: in traditional food cultures grain was prepared for human consumption by long fermentation to enhance digestibility and deepen flavor. Rudolf Steiner would refer to the cultural knowledge embedded in fermentation as "intuitive science." In the contemporary industrial food system, by contrast, intuitive ways of knowing have been replaced by reductive analysis and technologies. In regard to bread production, this shift has led to profound changes not only in how bread is regarded by producers and consumers, but also how grain is grown, and how it is processed into a loaf of sliced bread.

The history of wheat (*triticum aestivum*) as a domesticated food can be traced back about 8,000 years. The tradition of fermenting wheat to create leavened bread arose in Egypt around 1700 B.C. About the same time Slavic, Celtic, and Germanic peoples developed rye (*secale cereale*) as their principal grain, commensurate with the cooler climate and shorter growing season of their region. Unlike their wild ancestors, and some domesticated grains such as einkorn, emmer, and kamut, both wheat, and to a lesser extent rye, carry a small chain of peptides as part of the gluten protein that can potentially cause digestive intolerance and disease, which traditional cultures learned to circumvent by fermentation.

Thus wheat and rye became the "staff of life" in many cultures in the Mideast, Europe, and in America. Bread became the most important food staple, and therefore sacred. It held this preeminent place until the rise of industrialized food systems in the nineteenth and twentieth centuries. As early as the mid-nineteenth century, bakers in the U.S. and England began applying industrial methods to speed up bread production, which soon stripped this food of its sacred significance. In 1910, seventy per cent of all bread was still baked at home in America, but by 1921, when "Wonder Bread" was introduced,[4] home production had declined to less than thirty per cent. By the 1950s the "continuing mixing method" was developed. By this method, still in use today, a slurry of ingredients travels on a conveyor belt to the oven and to the finished loaf in as little as three hours. This process all but eliminates the time-consuming and labor-intensive fermentation process requiring extended periods of rest, during which the bacteria in the yeast become active, and the gluten is broken down into harmless peptides. To stimulate the necessary rising action of the bread, chemicals distributed in a soy filler are added along the way: ascorbic acid, hydrochloride, and sodium meta bisulfate to soften and strengthen the gluten; ammonium chloride and phosphate to feed the yeasts; amylase to break down starches into sugars; and protease to improve the extensibility of the dough.

As the methods of producing bread changed, so did the methods of producing grain. Grains, and especially wheat, were hybridized to increase net output, as well as raise gluten content to ensure adequate rising action during reduced fermentation. The grains were systematically treated with chemical fertilizers, pesticides, and herbicides, and they were genetically modified to control the explosion of weeds and plant disease under monocultural, industrial-scale production systems.

However, a significant cultural counter shift occurred in the early 1970s, when "back to earth" idealists cried out for healthier, safer, and more natural foods. For their bread, growing numbers of consumers demanded something better than the mass-produced, chemical-laced, and nutritionally deficient product sold in supermarkets. Many small-scale bakeries sprang up to meet this emerging market, albeit by making imitation sourdough breads while continuing to use chemical additives in ready-made mixes. Nevertheless, as the Bread Conference demonstrates, there is growing awareness that bread is more than mere food, that it is sacred, and that bodily and spiritual health are related.

Webster defines "sacred" as something that is dedicated, set apart in a place of honor, holy by association with the divine, and therefore entitled to reverence and respect, not to be profaned.[5] When the sacred is surrounded by ritualized traditions, it becomes religion. The place of bread in Holy Communion speaks to its sacred nature.

Michael Pollan elaborates this connection in his bestseller *Cooked: A Natural History of Transformation,* where he describes the story of Western civilization as pretty much the story of bread. He sees bread as the alchemical transformation of the fourth of the classical elements, which is air:

"Symbolically, air is not nothing. Air elevates our food, in every sense, raises it from the earthbound subsistence of gruel to something so fundamentally transformed as to hint at human and even divine transcendence. Air lifts food up out of the mud and so lifts us, dignifying both the food and its eaters. Surely it is no accident that Christ turned to bread to demonstrate his divinity; bread is partially inspired already, an everyday proof of the possibility of transcendence."[6]

The imperative to heal the earth, physically and Spiritually, is the foundation of biodynamics, of sacred agriculture. We are reminded of St. Paul's injunction to the Romans that, to do the work of the Spirit, we must "not conform to the world."[7]

As with all imperatives, meaningful implementation begins at home. It begins with caring for the soil on our farms and in our gardens. When it comes to bread, it begins with growing our own grain, or sourcing it from a responsible neighbor and baking our own long-fermented sourdough bread, or finding a baker who does.

Speaking with Paul, we must learn to say no to a commercial culture that puts profit before the health of the consumer and of the world around us. It is our own implicit collusion with that culture that causes the suffering from gluten intolerance and celiac disease. We must reject the technological fixes offered by commercial culture: the gluten-free foods, the grain substitutes, and the pharmaceutical remedies.

Biodynamic beekeeper Günther Hauk says it well in *Towards Saving the Honeybee* when he calls for a "tremendous shift in paradigms," in order to "restore our ailing bees to a level of vitality in which they can, once again, begin to flourish and thrive." What ails the bees is, at its base, the same thing that ails sufferers of gluten intolerance. Both suffer from pervasive cultural failure. As Hauck says, what is needed in both instances, is a fundamental "shift in attitude from a functional, profit-oriented, mechanical approach to a Spiritual, organic one, on the basis of heartfelt reverence."[8] Only when bread once again becomes the sacred staff of life, will the suffering end.

Notes

[1] Presentation given at the Bread Conference held at WSU Mount Vernon Research Center, September 5-6, 2014.

[2] *Odyssey*, book IX. Compare 'Lotus Eaters help define Meaning of Humanity" (http://hawkodyssey2011.blogspot.com/2011/05/lotus-eaters-help-define-meaning-of.html#. Retrieved July 9, 2021).

[3] Davis,William 2013. *Wheat Belly 30-Minute (or Less!) Cookbook: 200 Quick and Simple Recipes*. Kutztown, PA.

[4] "Wonder Bread's 100th Anniversary" (https://www.wonderbread.com/anniversary. Retrieved July 9, 2021.

[5] 1956. *Webster's New Collegiate Dictionary*. Springfield, MA, 744.

[6] Pollan, Michael 2013. *Cooked: A Natural History of Transformation*. New York, NY, 250.

[7] Romans 12:2: "And be not conformed to this world: but be ye transformed by the renewing of your Spirit, that ye may prove what is that good, and acceptable, and perfect, will of God."

[8] Hauck, Günther 2002. *Toward Saving the Honeybee*. San Francisco, CA, 60.

Part IV. Growing Animals

Ecological Livestock Raising
Henning Sehmsdorf, 2021[1]

Imagine this scene: A woman, my wife Elizabeth, and a cow walking quietly toward a barn. The grass is green and lush, and the early morning sun rising above the trees to the East bathes the scene in a luminous aura of light. The cow's name is Loveday. She is one of our two Jerseys, and she is on her way to be milked. It's around 6:30 in the morning, in May.

The scene is the very image of ecological livestock raising. Basically, what we do on our farm is harvest the sunlight.[2] The energy of the sun makes the grass grow, which feeds Loveday, who gives us about four gallons of milk every day, which we drink or make into butter and cheese. Milk from a grass-fed cow is much sweeter than milk from a grain-fed animal, because it contains high levels of natural sugars. Loveday's milk also feeds the chickens, turkeys and pigs. The sun rays shining on the skin of the cow are converted into vitamin D. We consume the milk, butter, and cheese raw, which means that all these products are alive with enzymes which are the catalysts without which we cannot metabolize our food.[3] We also do not homogenize the milk, because homogenization interferes with the absorption of the milk calcium into the bloodstream.[4] The only feed any of our two dozen beef and dairy cows get is green forage in summer, and hay in winter. Cattle, like sheep, are ruminants, which means that they can produce all the proteins and other nutrients they need from grass through the process of gradual mastication, also known as chewing the cud. They are fed no grain, and so they also don't produce the deadly E-coli strain 0157:H7; nor do they require any other supplements, medications, antibiotics, or wormers.[5]

Livestock Raised on S&S Homestead Farm
The livestock we raise include beef and dairy cattle, sheep, pigs, chickens, turkeys, bees, cat and dog — a fair sampling of once wild animals domesticated by human beings over tens of thousands of years to create a relationship that altered both the animals and the farmer:

"The domestication of animals went to the bottom of a primeval relationship, altering it at the roots. Man as a farmer sees himself and his flocks and herds with different eyes from those with which the hunter watches his quarry in the forest…The ploughing oxen team and the herded sheep are the creation of the farmer in a sense that the wild goats and cattle which the hunter shoots and traps are not."[6]

Beef Cattle

Cattle were domesticated by 4000 B.C. in the Fertile Crescent of Mesopotamia and spread from there into Europe and eventually to America.[7] Their wild progenitors — known in German as *Aurochs* (primordial oxen) — continued living in the untamed forests of Europe until the seventeenth century, when the last recorded live specimen died in 1627, in Poland. In 2018, I came across a life-size monument of that last wild *Aurochs* in a forest in Eastern Germany, and a splendid beast it was, standing at least six feet at the shoulder, and its horns the same span from tip to tip. From skeletal remains, its live weight has been estimated around 2,500-3,000 pounds, about twice the weight of most modern cattle. No doubt, in domesticating the primordial ox, man had seen fit to reduce the animal to manageable size.

Laurie Carlson suggests that the domestication was mutual. The culture of pastoralists and sedentary farmers evolved as hunters stopped pursuing their prey in the forest. Instead they cleared trees to grow animal fodder, including wild oats and rye, in enclosed pastures and meadows.[8] Both species benefited from the symbiotic relationship. Humans provided the cattle with feed and protection from predators. The cattle provided wealth, meat, milk, skins for clothing, bones for toolmaking and art, dung for fertility, fuel, and building materials. The relationship also created a balance of trust and mutual dependence. Today's farm animals are in effect the creation of the farmers and would not exist without their central place in society's food system. Without their manures, farms would have to rely entirely on synthetic inputs — with all the negative ecological and health effects of such artificial fertilizer.

On S&S Homestead Farm, for a quarter century we have bred beef cattle by crossing English Angus with Simmental, an Alpine breed. The Simmental are particularly efficient in converting grass cellulose to high-quality protein, without the need for supplementation with grain or other feed concentrates. The only mineral we supply is selenium, which is deficient in the soils of Western Washington. If not supplied in the feed, the lack of selenium causes white muscle disease (essentially the animal dies from heart failure).

The Simmental are large-framed animals that reach their phenotypical maturity at about three years of age. We slaughter the beef when they are thirty to thirty-six months old, usually at the end of the Spring flush, after the animals have had a chance to finish on lush grass and are in prime condition. The bulls often achieve a dressed weight up to 2,000 pounds.

We do not segregate the bull from the beef herd, which means that we do not control when calves are born. However, we find that the calves usually are birthed in time for weaning to occur naturally when the grass starts growing in March. Because the bull is always with the herd, he is mellow, does not challenge our fences, and fertility is invariably one hundred per cent, making artificial insemination unnecessary.

Dairy Cattle

In 2010, my wife and I decided it was time for the farm to add dairy production. We bought a bred Jersey cow from a registered breeder on a neighboring island, for several reasons. Jerseys are relatively small, and they produce milk with high fat content. They are also known for their gentleness. Equally important was that most Jerseys produce A2 proteins which affect the body differently from A1 proteins found in conventional milk sold in the U.S.[9] When A1 protein is digested in the small intestine, it produces a peptide called beta-casomorphin-7 (BCM-7). The intestines absorb BCM-7 and pass it into the blood stream. BCM-7 has been linked to stomach discomfort and lactose intolerance. By contrast, the structure of A2 protein is more comparable to human breast milk, and to milk from goats, sheep, and buffalo.[10] To make sure that the Jersey cow we acquired produced only A2 protein, we had her DNA tested by using a hair sample.

Dairying is as old as the domestication of bovines, perhaps even older. In Switzerland, archeologists discovered utensils used for straining milk and rennet dating from 6000 B.C.,[11] but they were probably used for milk from sheep or goats rather than bovines.

In most cultures the world over, the lactating cow has been revered as divine. In ancient Egypt, *Hathor*, the cosmic cow, was worshipped as the Great Mother. Her body was the heavens and the Milky Way spewed from her udder.[12] In Nordic cosmic history, *Aupumla* ("rich in milk") nursed the sleeping giant from whose body the world was shaped. In licking the primordial ice, she brought forth the first male gods.[13]

For us, our Jersey cow, whom we named Loveday, became the queen of the farm for fifteen years. We initially cross-bred her with our beef bull to produce meat that was superior in taste and fat marbling, but eventually raised a Jersey bull calf to breed our own dairy line. Unlike their sisters, Jersey bulls become aggressive by the time they are about two years old, even though we leave them with the herd. Last year, for example, when Ares, the current dairy bull, started to lift me up by my belt to get me out of the way, I knew it was time to put him to the knife. We replaced him with a purchased bull calf fastened on our lactating milk cow, who raised him together with her own heifer calf. Both calves are now of breeding age and will hopefully continue our dairy herd of maximally two lactating cows and their offspring, plus a bull.

Sheep

The domestication of sheep antedates that of cattle by several thousand years, and next to dogs are among the first animals to be tamed by humans. In ancient Mesopotamia, wild mouflon *(ovis gmelini)*, shaggy-haired animals with large curving horns in the male, were primarily raised for meat, milk, and skins. Woolly sheep began to be developed around 6000 B.C., and imported to Africa and Europe via trading. The hair fiber of primitive sheep was harvested by plucking rather than shearing, a trait that survives in modern hair sheep, such as the Katahdin, which are bred strictly for meat production. Other breeds, such as the East Frisian developed in the island chain off the North-German coast, were cultivated mostly for milk. Today, however, the most popular breeds are dual purpose, producing both superior meat and a sheared wool fleece, while the milk is left for raising the lambs.

Normally, our annual sheep flock consists of half a dozen ewes, a ram, and one to two lambs per ewe, born in early spring. As with the cattle, we do not segregate the ram from the flock for most of the year. But we do separate the lead ram and the male lambs from the females after summer solstice until breeding time in November. Unlike cows who (like humans) ovulate every month, sheep are solar animals, meaning that the ewes start ovulating when the sun stands in the zenith at mid-summer.

Unlike most conventional sheep producers, we do not want to lamb until the Spring grass is lush and plentiful, which in the San Juan Islands occurs in late March or April. Lambing in Spring rather than in the middle of Winter means that it can safely take place outside rather than requiring the protection of a barn. The lambs do not need heat lamps, or other supplementary gear and veterinary inputs to survive. We make sure lambing occurs close to the house, where we can keep an eye on them. This helps to protect them from predation by crows, ravens and eagles, and we can lend a hand in case of a breech birth.

When we started building our sheep flock in 1994, we began with purebred Suffolk, which provide high quality and large meat carcasses. But we found that the overbred ewes were poor mothers. A percentage of the lambs were born with double eyelids, requiring surgical correction to prevent blindness (by cutting a slit in the skin below the lower eyelid to pull it back from the eye as the skin healed). So we cross-bred the Suffolk ewes first with Romney, then with Polypay (a mix of five breeds) and finally with Churro. This produced lambs that are vigorous, with no birthing problems, free of genetic defects, and resistant to internal parasites.

The churro (whose name means sheep in Spanish) were developed by the Dené (Navajo) from stock brought to the New World by the Conquistadors in the sixteenth century. Churro are hardy animals, very fertile, and sport multicolored coats. However, they are also relatively small and flighty, and not nearly as trusting as European-descended sheep, and we found them more difficult to manage. For these reasons, we continued our breeding program with French Cheviots, Ramboulliet, and other breeds, with the overall goal of hybrid vigor and health. Although we raise sheep mostly for meat, we also love the beautiful skins. Unfortunately, there is only a limited market for sheep skins, and tanning costs are high. The mixed-wool fleeces of our sheep are not desirable for most spinners; however, we have often harvested the wool for teaching purposes. Apprentices and public school students have learned to shear, rove, card, spin and knit farm-produced wool into socks, mittens, and other textile and fiber art products.[14]

Pigs

Recent studies of mitochondrial DNA and nuclear genes from wild and domestic pigs have shown that the domestic pig (*sus domesticus*)

originates from the Eurasian wild boar (*sus scrofa*).[15] Separate wild subspecies in Europe and Asia existed as early as 500,000 years ago, and were domesticated independently about 9,000 years ago. Wild boars continue to thrive outside of human habitation, and some farmers deliberately interbreed the two species for hybrid vigor. My German brother-in-law, for example, regularly hunts wild boar, and when he comes across a litter of wild piglets, he takes them home to his farm for just this purpose.

We usually raise no more than three to four pigs per year, for the simple reason that on our limited acreage, we cannot grow enough grain to produce more pork. This is regrettable because pork raised on a diet of barley, raw milk and whey, vegetables and fruit tastes entirely different from commercial pork grown on feed concentrates. The meat is sweet and succulent, and absolutely delicious. A ham smoked in our wood-fired oven, for instance, does not require any glazing for baking, and you wouldn't insult the delicate flavor with mustard or any other condiment. Home-smoking the bacon also means that you can avoid the use of nitrates that turn into harmful nitrosamines in the body. Because of the limited number we can raise, we do not breed the pigs on the farm. Instead, we buy them as piglets when weaned at about six weeks of age.

Over the years we have raised American Landrace pigs known for their high carcass weight in the ham and loin, Hampshire and Duroc crosses of British origin, prized for their high muscle content and lean meat, and Mangalitsa hogs, an Old World breed indigenous to Hungary. Unlike most pigs bred in the last fifty years for lean meat, Mangalitsa provide plentiful fat and marbling, and thus more flavor. For the last few years, we have settled on an American heritage breed called Guinea hogs. These hogs were originally imported from West Africa and the Canary Islands in conjunction with the slave trade, as documented in 1804 by Thomas Jefferson, himself a slave owner. Guinea hogs for a long time enjoyed great popularity among homesteaders for their abundant and firm fat, tender meat and fine hams, as well as for their gentle nature. We use their rendered lard in pie crusts and bread dough, as a flavorful alternative to olive oil for frying, and to butter spread on our sourdough rye bread.

Chickens

Chickens (*gallus gallus domesticus*) were domesticated around 2000 B.C. from the red jungle fowl (*gallus gallus*) native to Southeast Asia and China. Like most other birds, including some feathered dinosaurs, chickens are a species of Theropoda, a large and diverse group of animals with hollow bones and three-toed limbs. Today scientists generally hold that birds did not evolve from dinosaurs. In fact they are the direct descendants of the few dinosaurs that survived the Cretaceous–Paleogene extinction event, when an asteroid perhaps ten miles wide smashed into the Gulf of Mexico causing most large animals to disappear from the face of the earth. A group of smallish therapods on the southern continents survived the disaster — the birds. One bird family that evolved in Asia was the red jungle fowl, and chickens are the domesticated subspecies of that bird.[16]

For many centuries chickens were bred for cockfights and ceremonial purposes. Between 400-200 B.C., the Hellenistic world discovered chicken as a food, and since then chicken has become the most widely consumed meat in the world.[17] Today, there are many chicken breeds, some grown for show, some for fighting, but mainly for eggs and meat.

Over the years, we have tried breeding our own chickens, with limited success. For the sake of economic efficiency, most kinds of chicken had

the brooding instinct bred out of them, privileging egg and meat production over time spent on hatching eggs. This also means that farmers are dependent on professional hatcheries using artificial brooding machines for replacement birds. The one breed that still broods eggs are the Bantams, but even they are not completely reliable.

Similarly, genetic alterations of broiler chickens make it nearly impossible for small-scale farmers to raise meat birds naturally. Cornish Cross, for example, have been bred to achieve carcass weights of ten pounds in as little as two months,[18] so that they can be slaughtered while the meat is still tender. But due to the accelerated growth rate, the chickens' immune systems are so weakened that growers routinely put antibiotics in their feed, which end up being ingested by the consumer.[19] Therefore over the past years, we have opted for slow-growing dual purpose chickens such as Orpingtons, Speckled Sussex, Rhode Island Reds, and Bielefelds. We have bought day-old chicks from commercial hatcheries (and watched many of them die in transit through the Post Office). Most recently we have successfully brooded out fertilized Bielefeld eggs in a commercial incubator. We have accepted that these chickens do not provide good broiler carcasses, but are good layers and make excellent broth birds.

Turkeys

The American Livestock Conservancy states that "all domesticated turkeys descend from wild turkeys indigenous to North and South America." The turkeys are said to descend from velociraptors that "were about the size of a domesticated turkey, being only about three feet tall and six feet long, with most of the length coming from the tail and weighing in at around twenty to thirty pounds fully grown."[20]

Turkeys (*meleagris gallopavo*) were first domesticated 2,000 years ago by Mayans in

Central America and by Aztecs in Mexico. In 1519, the Conquistadors took them to Spain, from where they reached England by 1524. "On a continent where fine dining still included eating storks, herons, and bustards, the meaty, succulent turkey was a sensation."[21] Several European varieties were developed over the next century. When Pilgrim settlers brought turkeys back to New England in 1620, they were surprised to discover wild turkeys. These cross-bred with the European strain to produce the Heritage Turkeys, as they are now known.

Commercial breeders in Washington State and Oregon, between 1920s-1950s selected for turkeys that showed rapid weight gain and broader breasts, with massive muscle on both sides of the keel bone. The breeding practice, however, produced animals too large to mate without human intervention. The bulkier muscle keeps males from mounting females effectively. Given the lower fertility rates, artificial insemination is now required to produce viable eggs.

On S&S Homestead Farm, we raise Heritage Turkeys, which mate naturally with expected fertility rates of seventy to eighty per cent. The turkeys have a productive lifespan of five to seven years for hens, and three to five years for toms. They can live outside, making indoor housing unnecessary.

Requiring twenty-six to twenty-eight weeks to reach market weight, Heritage Turkeys develop a strong skeleton and healthy organs before building muscle at growth rates comparable to commercial varieties before the 1920s. Our biggest challenge is to keep eagles and raccoons out of the turkey yard, especially when the hens are brooding and vulnerable to predation. This is the only time we keep them in an enclosed space, where for twenty-eight days the hens sit quietly staring into space, as they wait for their poults to hatch.

Bees

Honeybees (*apis mellifera*) are essential to the health of the ecosystem, both on the farm and in the larger natural environment. They play a decisive role in flower, vegetable, fruit and nut reproduction.[22] While the more than 4,000 non-honey bee species in North America, as well as bumblebees, wasps, beetles, butterflies and other insects, also contribute to pollination, honeybees pollinate an estimated 170,000 plant species. 40,000 of these would face extinction without honeybees.

There are eight species of honeybees, six of European origin and two of Asian. Honeybees were originally imported from Europe to America in the 17th century. In a single year, one honeybee colony can gather about forty pounds of pollen and 265 pounds of nectar, processed into honey by the bees. Bees' wax is used for making candles, lip balm, soaps, and cosmetics. Pollen has culinary and medicinal uses, providing high levels of amino acids, triglycerides, phospholipids, vitamins, macro- and micronutrients, and flavonoids.[23]

Honeybees evolved over millions of years. The earliest recorded bee was found in Myanmar, encased in a piece of amber dated to 100 million years ago. Domestication of bees no doubt began with humans harvesting honey from wild hives, in probable competition with wolves, bears, skunks, bee-eating insects, birds, and spiders.[24] Depictions of humans collecting honey from wild bees date to 10,000 years ago. Beekeeping in pottery vessels began about 9,000 years ago in North Africa. 4,500 years ago, honey was stored in jars found in the tombs of Egyptian pharaohs.

Beginning in the 18th century, German beekeepers built movable comb hives (skeps) to allow honey harvest without destroying the colony.[25] Today, most bees are kept in rectangular boxes equipped with pre-fabricated, plastic combs. Langstroth hives are preferred by industrial beekeepers who move the boxes from field to field along established seasonal routes. Biodynamic beekeepers, however, prefer pear-shaped Top Bar Hives, straw skeps, or hives from hollowed-out logs, to "allow bees to develop in accordance with their true nature."[26] Organic shapes made from non-synthetic materials mimic natural environments. Biodynamic beekeepers encourage bees to construct perennial colony nests by building their combs from wax they extrude from their own bodies.

On S&S Homestead Farm, we keep bees for ecological rather than for commercial ends. Lopez Island has a short nectar-flow season which limits feed sources for bees. Therefore, we do not harvest any of the honey stores the bees need to survive in the non-growing season. During the winter, we supplement bee feed with a blend of sugar, chamomile tea, thyme and salt to mimic the organic substances bees naturally receive from nectar. We observe the bees through a window in the side of the Top Bar Hive, but do not open the hive.

We also do not intervene with the queen's mating naturally by flying to the seasonal drone congregation in the sky, where she breeds with up to twenty-five feral males. This natural practice guarantees genetic diversity in the hive, in contrast to commercial "re-queening" with an artificially inseminated queen produced in a laboratory for genetic control.[27]

By not clipping one wing of the queen bee (as is done in commercial settings), we encourage spring swarming of the colony. The mother queen, after producing a new queen, vacates the hive with half of the worker bees, but leaves behind all the accumulated stores of nectar, honey, and pollen in support of the new generation led by her daughter.

Because of our hands-off approach, we do not find it necessary to treat the bees with oxalic acid or any other organic or synthetic biocide to control Varroa or other phoretic mites. Instead, we build place-specific immunities by providing a clean and naturally supportive environment for the bees. So far, this approach has worked, and our bees survive year after year. Our approach to bee health parallels our practices on the farm as a whole. The health of the soil, the plants and animals, as well as our personal health, has essentially been supported for over half a century by the balanced, natural environment of the farm, without substantive allopathic or pharmaceutical interventions.

A century ago, Rudolf Steiner warned that honeybees would vanish from the face of the earth, unless agriculture shifted its focus from industrial bee management for profit, with potentially disastrous consequences for natural and human ecology.[28] The dreaded phenomenon of bee colony collapse in our time confirms the accuracy of Steiner's prediction. Steiner forecast ecological consequences of the bees' disappearance beyond the loss of commercial honey or pollination services. Of greater significance is the loss of life support provided by the bees in their symbiotic exchange with flowering plants. Steiner described how in siphoning nectar from the flower, the bees replace the nectar with formic acid, which is the same substance found in ant and wasp venom. Bee venom is a silica acid containing enzymes, amino and other acids, a total of sixty-three components necessary not only for flowers, but for all of organic life. Without a constant influx of bee venom, the flowering plants would eventually go extinct.[29]

The biodynamic perspective on beekeeping emphasizes their "eusocial (from Greek *eu*=good + social) character, meaning that bees exhibit an advanced level of social organization, in which a single female or caste produces the offspring, and non-reproductive individuals cooperate in caring for the young and the hive as a whole.[30] The eusocial behavior of bees expresses what Goethe would surely describe as an archetype (*Urphänomen*) of selfless love operating in the natural world. In her widely acknowledged book on beekeeping,[31] Jacqueline Freeman depicts bees as teachers of wisdom needed for human society to evolve toward the same level of mutual responsibility. As farmers who see the bees' enterprise as a sacred task, we agree.

Dog & Cat

Dogs (*canis familiaris*) are domesticated descendants of now extinct grey wolves.[32] Dogs were the first wild animals to be tamed by hunters, even before the advent of agriculture about 14,000 years ago. The most recent common ancestors of dogs and wolves date to about 30,000 B.C. In that long span of domestication, dogs have evolved into many subspecies and breeds. On the farm, we prefer dogs of strong herding and guarding instincts, notably German shepherds and Border Collies, and we consider them farm animals rather than pets, although they are strongly attached to us, and we to them. They live outside the family home, which benefits their health and orients them to their guard function. The wolf-like German shepherd was developed as recently as 1899 for herding sheep on the village commons. Strength, endurance, intelligence and fearless loyalty are outstanding characteristics of the German shepherd. We remember Ursa taking on a rambunctious bull to protect my wife, or flying into an eagle clutching a chicken.

The Border Collie, which originated from the Anglo-Scottish border region is another highly intelligent, energetic, and tireless dog, also bred for sheep control, but with a different focus. While the German Shepherd was bred to accompany large herds over extensive community pastures, the Border Collie served to move smaller contingents of sheep in and

out of enclosures as instructed by the farmer. We are fond of Molly, our Border Collie mix. However, she is afraid of sheep (or any large animal) and is uncharacteristically dumb. She makes up for insufficient intelligence with endless sweetness and a pronounced funny streak.

We note that the Latin word *cattus,* Old English *catt, catte,* Dutch *kat* and German *Katze,* French *chat* all relate etymologically to *chattel,* meaning controlled assets, including slaves, bondsmen, money, land, or income. This is interesting because a typical cat is not an animal anyone can fully control, let alone enslave. Cats (*felis silvestris lybica*) own themselves, and they allow humans to feed and love them. Wild cats apparently domesticated themselves during the late Neolithic age (around 1900 B.C.) by settling near agricultural communities in the Fertile Crescent, where grain harvests and stores attracted mice and rats. Unlike dogs, cats by and large remained genetically unchanged by domestication, so that the DNA of domestic and wild cats remains the same even today.[33]

On the farm, we usually maintain one cat who, like the dog, lives outside the house. We don't breed this cat; he or she usually comes to us from some neighbor blessed with too many kittens of uncertain parentage. The cat hunts rodents in exchange for a little feed and lots of affection. For the cat's protection, we are careful not to use poisons in controlling vermin like rats, mice, or predators such as raccoon and mink. Almost every day, the cat will demonstrate its hunting skill by bringing us the remains of its prey. Sky, our current cat, is a neutered tom tabby. He is very social, a good mouser, and a talker who likes to follow us on walks around the farm, and even farther afield.

Methods

The commitment to ecological livestock production means to preserve the animals' ability to grow in "the image of nature." It amounts to harvesting the sun through the animals and through the feed we grow for them. To accomplish this, we make breeding choices supported by our methods of raising the animals. We feed green forage and hay to the ruminating animals; grain, potatoes, vegetables, fruit, organ meats and other offal to the pigs, chickens, and turkeys. The bees provision themselves from flowering plants. The cat gets a large part of his food from hunting, and the dog eats the leftovers from our table. Feed crops are supported by the soil fertility provided by the animals and cover crops, and by biodynamic sprays funneling cosmic energies to the farm organism. Fundamentally, the biodynamic farm operates as an integrated, self-organizing, self-supplying, and self-correcting system powered by the sun, but closed to market-based inputs as much as possible.

Rotational Grazing and Haying

Rotational grazing is a system of sectioning portions of green forage in a pasture during the growing season. In winter, cattle are fed cured hay, but the principle of controlling animal movement by a rotating fencing system is the same, and the underlying goal is the same. Whether feeding forage or hay, the farmer uses a movable electric fence to control animal movement, stocking rate, and density.

From late March through the end of November, we move two dozen cattle across ten acres of pasture by dividing the field into eight strips. The strips are separated by movable tape fences electrified to a voltage between five and ten Kv pulsing through the wire every twenty seconds. If an animal touches the wire, the impulse travels through its body to the ground and back to the charger. The animal recoils from the shock and the circuit is broken until another animal touches the wire. Only then is electricity actually consumed and leaves the system. The system is

inexpensive to run, only a couple of dollars in energy costs per month for a fifty-acre farm. The herd is introduced to a strip when the grass is about six inches high, and moved to the next strip when the forage has been grazed down to an average of two inches.

Because of the stocking rate and density (twenty-four animals per one and one quarter acre), the herd tends to stay in place instead of moving around and wasting forage by trampling. Confined to a limited area, the animals consume everything. Instead of preferentially selecting sweeter legumes, they also feed on tougher grasses such as tall fescue, which play an important role as roughage in the ruminant diet. If the herd had access to the entire ten acres at the same time, the animals would return to the more delectable forage species and graze them down repeatedly. The legumes could not replenish nutrient stores through photosynthesis and instead would draw on reserves in the roots until these are exhausted, and the plants die.

This is why it is easier to overgraze a larger pasture than a smaller one allowed to rest between grazing intervals. We move our herd across the ten acres eight times in the course of the summer. Each strip is grazed for about a week — more or less, depending on weather and growth rate — and then rests for eight weeks before the herd returns to it. During the time the herd spends in each strip, most of the forage is consumed, and what is left behind is a dense layer of manure. By the time the herd comes back, the grass has regrown to about six inches, and night crawling worms (*lumbricus terrestris*) and beetles have metabolized the cow pies into the soil, leaving no habitat for parasites to invade the digestive system of the grazing animals. During the winter months, we use the same system, but in reverse. Animals are fed hay by rotating them through a non-growing pasture or empty crop field. They leave behind well fertilized sod as they excrete while feeding beneath the tape fence in front of them. Because we move the temporary fence in ten-foot increments every few days, the entire field is covered with nutrients to support next year's pasture or a succession crop planted in the field.

We feed the hay daily in fifty-pound square bales small and light enough that my wife and I can easily lift them without a machine. There are economic, nutritional, and ecological reasons why we prefer daily feedings of small bales to feeding large round bales once a week or so.

The production and distribution of 1,200 pound round bales requires large and expensive equipment, including a fifty-horse power tractor, at least. Our tractor produces only eighteen HP at the PTO, which is just enough to operate the old sickle bar I restored. I rake the freshly cut hay into windrows with an equally ancient hay rake, and bale it into square bales to produce an average of twenty tons of hay annually. We make sure that cutting, raking, baling and stacking the hay in the barn loft takes no more than two-three dry days in late June or early July, thereby preserving the full nutrient content of the cured hay. At the current cost of about $300 per ton for good hay out of the field, home production saves a great deal of money, making small-scale beef and dairy production financially viable.

The nutritional reason for feeding daily becomes clear when you break open the bale in the field and inhale the sweet smell of fresh hay not leached by rain or sun, and therefore nutritionally dense and palatable. Laying down the hay on the opposite side of the electric fence means that the cattle waste little by trampling. We estimate that rotational summer grazing, which extends the season by at least six weeks per year, and daily winter feedings of hay, save perhaps up to fifty percent of hay, in comparison to more mechanized methods

intended to save labor by maximizing machine use.

Moving the cattle frequently in winter prevents the typical soil pugging that occurs when heavy animals are fed from a stationary round feeder supplied once a week or even less often.

Grain and Feed Supplements

Beef cattle can usually maintain themselves even on dry grass during protracted summer droughts, which are becoming more common on Lopez Island due to climate change. We extend milk production by supplementing the dairy cattle on late summer pasture with wheelbarrows of whole corn plants, comfrey, sugar beets, kale, cabbage, chard, grape, ivy, and bean vines, and with succulent weeds such as lambs quarters, clover, bindweed, burdock, and milk thistles. Hay feeding does not begin until November.

On the two-acre crop field where we often winter the cattle, we plant barley in spring. One of our interns calculated the total amount of nitrogen deposited by the cows at 128 pounds per acre. We need half that to grow the grain, with enough left over to support the forages recolonizing the field from the seed left in the soil from the previous year. We repeat the cycle in winter.

The barley, usually about two tons, is harvested by a neighbor, who owns a small combine. He charges about as much as the grain is worth on the commodity market. Nevertheless, the process is worthwhile, both economically and ecologically. The barley provides organic feed for chickens and pigs at the price of non-organic grain. Because the grain is grown on the farm, we can be sure of its quality, and we avoid importing weeds and other invasive plants with grain purchased from another site. The field yields not only grain but also straw. At the current price for square bales sold at the local feed store, the straw is actually worth more than the grain, and is a valuable farm resource. We use it for animal bedding, as well as for mulch in vegetable and fruit production. By returning unused nutrients from the winter sacrifice area in the form of animal feed and straw, we avoid leaching excess nitrogen into the groundwater we drink or give to the animals.

Nutrient Cycling[34]

Cows excrete over ninety per cent of the nutrients they take in from grass and hay and convert into proteins and vitamins. Their manure is full of enzymes and bacteria. Collected and piled in a heap, the excrements turn into a potent source of soil fertility. If our milk cow calves in winter, we keep her and the calf in a stall in the milk barn and daily collect their manure in a compost pile under the barn roof. The pile heats up to between 120-160 degrees, and pathogens in the manure are killed, while beneficial bacteria and fungi flourish. The metabolic activity of the microorganisms fixes the nitrogen, phosphorus and potassium (NPK) in the fecal matter, so that they cannot leach out in water or volatilize into the air.

By the middle of March, we move the manure pile to open ground. We cover it with straw to keep the summer sun from drying it out. As the internal heating stops, the pile cools, and earthworms move in. We usually leave the pile alone for several months. Eventually the entire pile is digested by worms, leaving behind their castings. The worms mineralize the organic matter in the manure, producing a high quality, balanced fertilizer. By the time the compost is put on vegetable beds, it is crumbly and moist, but no longer wet.

Worm castings are also found in the open field, wherever animals have been grazing. The night crawlers (*lumbricus terrestris*), as their name implies, come out at night to feed on animal droppings. We calculate that between the cows and sheep, the ruminant animals on the farm produce about 90,000 pounds of dry matter (DM) in the form of manure per year. Only a small portion of that is captured in the barn; the rest stays behind on the ground in the pastures. It is metabolized into the soil first by macro-organisms such as earthworms and beetles, then by microscopic soil organisms. There are as many one billion microorganisms in each teaspoon of a healthy soil well supplied with organic matter and free of toxic fertilizers, pesticides, or herbicides.

Some bacteria and fungi are decomposers that break down the organic residue pulled into the soil by the worms. Other bacteria and fungi immobilize nitrogen (N_2) by storing it in their bodies. Protozoa and nematodes feed on these, releasing plant-available nitrogen (NH_4+). Others live in myccorhizal association with roots and deliver nutrients (such as potassium) and water to the plant. Other bacteria, fungi, nematodes and micro-arthropods are parasitic and pathogenic, which means, for example, that they feed on roots and other plant parts and potentially cause disease. But to the degree that soil organisms are fundamentally in balance, the forces of growth and decay, life and death support each other in the food web below our feet.

Another method to capture nutrients in ruminant manures is through cover cropping. After feeding the animals rotationally in a crop field, the field is planted in rye, vetch, clover, field peas, or fava beans. Timing is critical. If the seed goes into the ground three weeks before the last frost date, there is enough growth to direct-seed barley into the cover crop before the end of May. Following the barley harvest in summer, the cycle is repeated, either by planting a fall cover crop, or by feeding the animals rotationally in the field during the winter.

A third method of nutrient cycling involves biodynamic preparations.[35] Eight herb and plant substances (chamomile, nettle, yarrow, dandelion, valerian, and oak bark, plus silica crystals) are fermented over the winter buried in various animal sheaths (cow intestines, mesentery, stag bladder), and in clay pots.[36] At planting time, the fermented preparations are potentized in water and sprinkled on the soil, or they are sprayed on grain or hay fields throughout the year. They are also inserted into compost piles.[37] These fermented substances function in soil and compost like homeopathic remedies do in the human body. Applied in minute doses, they regulate metabolic processes to enhance energy and nutrient exchanges, and generally strengthen the micro-organic life in the soil and the health of the plants. For best results, the biodynamic preparations should be produced on the farm where they are to be used. They cycle the life energies of the farm organism. But they can also be purchased commercially. Since 1985, the Josephine Porter Institute has provided the preparations for farmers or gardeners who do not raise ruminant animals.[38]

Water Cycling

Early on, we dug a retention pond to capture winter run-off water. The pond is about 150 feet across, fifteen feet deep, and holds about 750,000 gallons of water. Most of the water comes from the roofs of two barns. During the rainy season (November through April), the water runs down gutters into cattle tanks equipped with overflow pipes funneling the water into a cistern. The full cistern overflows into open swales to carry the water by gravity into the pond about a thousand feet away. During the drought season (July through September), the water is pumped back uphill to the cistern, where a shallow well pump pressurizes the water (forty psi) for use in irrigation. Photovoltaic panels provide the energy to pump the water back uphill into the cistern, and a small computer regulates the flow of the water to the orchard, and to a two-acre vegetable field.

The solar-driven catchment system supplies most of the non-potable water needed for the farm operation. Our hand-dug wells from which we draw water for drinking, cooking, bathing, and laundry, rely on aquifer reserves of unknown quantities. A benefit of using pond water for irrigating the vegetable and fruit crops is that it is soft, unlike hard ground water. Rainwater is more easily transported in the plant capillaries which supports plant productivity and health. The animals, on the other hand, benefit from the added calcium and other minerals, and therefore are given groundwater to drink.[39]

The quantities of water from the pond and wells are not sufficient to irrigate the pastures, hay or crop fields. We water the vegetable beds, but must rely on rainfall to support forage, hay and grain production. On Lopez Island no more than fifteen inches per year fall during periods of heavy rain in winter, alternating with long drought periods in summer. When it rains, it is imperative to hold as much of the water as possible in the soil, instead of allowing it to run off and into Puget Sound.

The key to water retention in the shallow glacial soils of the island is the humus they contain. Humus is the organic matter in soil derived from microbial decomposition of plant and animal substances. It consists of about sixty per cent carbon, six per cent per cent nitrogen, and smaller amounts of sulfur and phosphorus. When we first started farming in 1970, the levels of soil organic matter on the whole farm were less than three percent. Fifty years later, after animal droppings had worked into the soil over many seasons, the humus content has grown to an average of twelve percent, and even higher in the intensely managed vegetable beds.[40] This organic matter acts like a sponge, holding rain water in place during the rainy session and releasing it gradually during drought. As a result, there is little of the massive ponding observable during the winter in many fields on the near mainland, wherever soil humus as been depleted due to mechanical and chemical farming practices.

On our farm we see do not see the depletion of soil fertility common on grasslands on Lopez Island, wherever animals were eliminated when farming collapsed in the County during the 1950s. Local farmers could not compete with industrialized agriculture in Eastern Washington made possible by the construction between 1933-1942 of the Grand Coulee Dam on the Columbia River, built to produce hydroelectric power and provide irrigation water. In the absence of animals on hay fields and pastures, humus levels rapidly declined and with them the capacity to retain and cycle rainwater. In ecologically managed agricultural enterprises, the presence of ruminant and other farm animals is an irreducible requirement for effective nutrient and water cycling.

Ecological vs. Industrial Livestock Raising[41]

In concluding this essay, I turn to the comparison of two farms described in Michael Pollan's recent best-seller, *The Omnivore's Dilemma*.[42] These two farms are comparable in size — about 500 acres — and in the livestock they raise, but differ in production methods and ecological impacts. The Naylor Farm in Iowa produces cattle, pigs and chickens on corn and beans, while Polyface Farm in Virginia raises the same livestock, but does so mostly on grass.

The Naylor Farm annually imports calves, weaner pigs, and chicks which they raise in CAFOs (Confined Animal Feeding Operations) on feed grown mono-culturally (meaning that a single species, such as beans or corn, is cultivated in each field). Polyface Farm breeds perennial livestock and feeds them poly-culturally on diversified, perennial stands of grasses, forbs, legumes and wildflowers.

The Naylor Farm methods are mechanical and rely on heavy machinery running on fossil energy. Polyface Farm methods are biological and are fueled mostly by solar energy harvested through forages and the animals.

The Naylor Farm relies on global markets to supply synthetic fuels and fertilizers, and to distribute its products. Polyface Farm purchases chicken feed from a neighbor nearby, and markets its products locally.

The Naylor Farm represents the industrial prototype of farm production as a manufacturing process. The farm functions as a machine converting purchased inputs of seeds, animals, fossil fuels and petrochemicals into outputs of food and fiber. Inputs are brought from faraway factories and refineries, and outputs are distributed worldwide through the global trade system.

The industrial farm has an economic orientation, reducing product unit cost through mechanization, labor minimization, specialization, and disproportionate development of enterprises. Self-sufficiency is not a typical objective of an industrial farm: feeds, energy and fertilizers are imported, and farm production is dictated by market demands.

Inadvertently, the industrial farm also produces waste streams conventionally disregarded when calculating the bottom line. Manure, nitrogen and pesticide run-off pollute groundwater, streams and oceans. Heat and exhaust from heavy machinery dissipate into the atmosphere in the form of lost energy and greenhouse gases. Indirectly, industrial farming causes secondary chemical pollution, for instance, from mercury, a by-product of polyvinyl chloride (PVC). There is growing evidence that highly mechanized systems of food production permanently degrade soil, water and genetic resources to such a degree that, as non-renewable resources become limited in the near future, the prospect of attaining even the modest yields of pre-industrial agriculture may be dim.

By contrast, Polyface Farm represents the prototype of ecological farms based on the concept of food and fiber production as a biological and social process.

Such a farm is self-organizing, self-correcting, self-sufficient and self-capitalizing. Polyface Farm depends mostly on solar energy harvested through diversified, perennial forages to feed livestock born, bred and slaughtered on the farm. Its products are distributed to nearby markets. Because production relies on biological rather than chemical or mechanical processes, thermal energy losses and carbon dioxide pollution are minimized, and manures and animal offals are recycled as local fertility.

On ecological farms, economic stability is achieved by efficient labor input, balanced combination of enterprises, diversification, and highest possible self-sufficiency in regard to fertility, feed, and energy.

It is important to note that organic rather than chemical production does not necessarily make a farm ecological rather than industrial. As defined by the USDA Organic Standards, organic production simply means that inputs are non-synthetic, and free of genetically modified organisms and radiation.
However, to the degree that food and fiber production is organized as a mechanical manufacturing process, an organically certified farm would still have to be considered industrial, and its impacts on the physical and social environment would be more or less the same.

Personally, I have doubts that large-scale farms, whether organic or conventional, can ever be fully ecological. Given declining oil supplies and global climate change, it is increasingly important to consider the ecological opportunities afforded by small-scale, diversified, integrated and self-sufficient food production. On Lopez Island, the price of diesel now tops $5/gallon. If only for economic reasons, it is necessary to make livestock raising as little dependent on fossil fuels as possible. At the Eco-Farm conference in California, one of the keynote speakers made the point emphatically that if America wants to feed itself thirty years hence, we will need to add forty million small-scale farmers to grow food and fiber on integrated and ecologically responsible farms, and supply the food needs of their own, local communities.

Notes

---

[1] Based on a workshop on animal husbandry held on S&S Homestead Farm, 2008.

[2] For a quantitative analysis of the solar energy embodied in the production of S&S Homestead Farm, see Haden, Andrew 2002, "Emergy Analysis of Food Production on S&S Homestead Farm," below.

[3] See Howell, Edward 1995. *Enzyme Nutrition*. New York, NY.

[4] See Schmid, Ron 2003. *The Untold Story of Milk*. Brandywine, MD.

[5] See Robinson, Jo 2000. *Why Grassfed Is Best!: The Surprising Benefits of Grass-fed Meats, Eggs, and Dairy Products*. Vashon Island, WA. See also Barnett, Tanya 2003. "Gratia Plena," below, and Sehmsdorf, Henning 2014. "Blood into Milk: Small-Scale Biodynamic Dairy" below.

[6] Sandars, Nancy K. 1985. *Prehistoric Art in Europe*. New York, NY, 171.

[7] Carlson, Laurie W. 2001. *Cattle: An Informal Social History*. Chicago, Il, 19.

[8] Carlson, op.cit., 21.

[9] A2 protein is mainly found in breeds that originated in the Channel Islands and southern France. These include Guernsey, Jersey, Charolais, and Limousin cows. See Arnarson, Atli 2019. "A1 vs. A2 Milk — Does It Matter?" https://www.healthline.com/nutrition/a1-vs-a2-milk#definition. Retrieved September 7, 2021.

[10] Metropulos, Megan. 2017. "The Benefits and Risks of A2 Milk." https://www.medicalnewstoday.com/articles/318577. Retrieved September 5, 2021.

[11] Carlson, op.cit, 246.

[12] Ibid, 47.

[13] Sehmsdorf, Henning 2020. *Myth & Tradition in Norwegian Literature & Folklife*. Lopez Island, WA, 28.

[14] See Perlman, Deanna 2014 "Wool Processing at S&S Homestead: Grow, create, teach, learn." https://sshomestead.org/research2/. Retrieved September 12, 2021.

[15] Giuffra, E. & J. M. H. Kijas, V. Amarger, Ö. Carlborg, J.-T. Jeon and L. Andersson 2000. "The Origin of the Domestic Pig: Independent Domestication and Subsequent Introgression," *Genetics* 154 (4): 1785-1791.

[16] Pennisi, Elizabeth 2018. "Quaillike creatures were the only birds to survive the dinosaur-killing asteroid impact". *Science*. doi:10.1126/science.aau2802. Retrieved September 13, 2021.

[17] Charles, Dan 2015. "The Ancient City Where People Decided To Eat Chickens." https://www.npr.org/sections/thesalt 2015/07/20/424707879/the-ancient-city-where-people-decided-to-eat-chickens. Retrieved September 13, 2021.

[18] See "Typical Broiler Body Weights & Feed Regiments." https://www.pinterest.com/pin/484277766154093950/. Retrieved September 14, 2021.

[19] On the deleterious health effects both for farm animals and end-consumer from non-ecological commercial breeding programs and feeding regiments, see Elizabeth Simpson & Henning Sehmsdorf 2021. *Eating Locally and Seasonally: A Community Food Book for Lopez Island (and All Those Who Want to Eat Well)*, Lopez Island, 10-12.

[20] Reese, Frank & M. Bender, S. Beyer, J. May, D. Williamson, S. Pope 2021. "Definition of a Heritage Turkey." https://livestockconservancy.org/heritage-turkey-definition/. Retrieved September 14, 2021.

[21] Ibid.

[22] "The Benefit of Bees," *The New Agriculturist*. http://www.new-ag.info/00-5/focuson/focuson8.html. Retrieved September 15, 2021. See also National Research Council of the National Academy of Science 2007. *Status of Pollinators in North America*, Washington, D.C.

[23] Johnson, John 2021. "Bee Pollen: What to Know," *Medical News Today*. https://www.medicalnewstoday.com/articles/bee-pollen. Retrieved September 15, 2021.

[24] Isom, Cathy 2019. "Predators that Prey on Bees and Your Harvest." https://agnetwest.com/predators-prey-bees-harvest/. Retrieved September 15, 2021.

[25] See Tautz, Jürgen 2007. *Phänomen Honigbiene*. Heidelberg, Germany. (English translation: Sandmann, David C. 2008. *The Buzz About Bees: Biology of a Superorganism*, Berlin, Germany).

[26] LaPado-Breglia, Christine G.K. 2011. "Smitten with Bees," *Chico News & Review Archives*. https://www.newsreview.com/chico/content/smitten-with-bees/2978078/. Retrieved September 15, 2021. See also Wilhelmi, Christy 2013. "Biodynamic Beekeeping with Michael Thiele," *Gardenerd. https://gardenerd.com/blog/biodynamic-beekeeping-with-michael-thiele/*. Retrieved September 15, 2021.

[27] See for example: Burns, David & Sheri 2021. "Requeening a Beehive." https://www.honeybeesonline.com/requeening-a-bee-hive/. Retrieved September 18, 2021.

[28] See Sehmsdorf, Henning 2015. "Bee Colony Collapse and Biodynamic Strategies to Restore Bee Health," below..

[29] Steiner, Rudolf 1923-4. *Das Wesen der Bienen* (On the Nature of Bees), Dornach, Switzerland, 123ff.

[30] See Batra, Suzanne W. T. 1968. "Behavior of Some Social and Solitary Halictine Bees Within Their Nests: A Comparative Study," *Journal of the Kansas Entomological Society*, 41 (1): 120–133. See also Crespi, Bernard & Douglas Yanega 1995. "The Definition of Eusociality," *Behavioral Ecology*, 6: 109–115.

[31] Freeman, Jacqueline 2014. *The Song of Increase: Returning to Our Sacred Partnership with Honeybees*. Battleground, WA.

[32] See Bergström, Anders et al. 2020. "Origins and genetic legacy of prehistoric dogs", *Science*, 370: 557–564.

[33] Smith, Casey 2017. "Cats Domesticated Themselves, Ancient DNA Shows," *National Geographic: https://www.nationalgeographic.com/science/article/domesticated-cats-dna-genetics-pets-science*. Retrieved September 19, 2021.

[34] For on-farm intern research on nutrient cycling and composting, see Hök, Johanna 2002. "Nutrient Cycling and Composting on S&S Homestead Farm," and Gigot, Jessica 2003. "Compost Management," below.

[35] See https://www.zinnikerfarm.com/biodynamicpreps. Retrieved September 22, 2021. See also Sehmsdorf, Henning 2021. "Biodynamics: A Personal History of an Applied Idea," below.

[36] For the making and application of BD preparations, see Wistinghausen, Christian von, Wolfgang Scheibe, Eckhard von Wistinghausen, & Uli J.König 2000. *The Biodynamic Spray & Compost Preparations Production Methods, Booklet 1;* and *The Biodynamic Spray and Compost Preparations: Directions for Use, Booklet 2*. East Troy, WI.

[37] See Sehmsdorf, Henning 2008. "On-Farm Research: Biodynamic Forage Production," below. Also see Reeve, Jennifer R., Lynne Carpenter-Boggs, & Henning Sehmsdorf 2011. "Sustainable Agriculture: A Case Study of a Small Lopez Island Farm," *Agricultural Systems* 104(7): 572-579.

[38] https://jpibiodynamics.org/. Retrieved September 22, 2021.

[39] See Sehmsdorf, Henning 2004. "Solar-Powered Micro-Irrigation & Energy Production," below.

[40] See Sehmsdorf, Henning 2021. "Farming for Health: The Economics of Stewardship," below (p. 3).

[41] For further perspectives on an industrially-based food systems, see Simpson, Elizabeth 2007. "Pastoral versus Industrial Food Production," below

[42] Pollan, Michael 2006. *The Omnivore's Dilemma: A Natural History of Four Meals*. New York, NY.

Gratia Plena
Tanya Marcovna Barnett, 2003[1]

For years now, friends have told me about the simple beauty of the chapel of St. Ignatius at Seattle University. After a recent meeting on SU's campus, I decided it was high time to visit this house of worship. Alone inside, I slowly explored the chapel's nooks and sacred art, until I found myself drawn toward an impressive sculpture. The sculpture — carved from a rough, single block of marble — appeared to have a stream of milk cascading down its face. Its sculptor must have found a milk-colored ribbon within the marble, and smoothed and polished it so that it appeared wonderfully liquid and lactescent. The "milk" flowed from a bowl that the sculptor had carved out of the top of the marble block then covered with gold leaf. This tilted bowl seemed almost like a halo hovering above the sculpture. I sat for quite some time, drinking in the soothing sense that the sculpture evoked. And then, "she" began to emerge. Vaguely at first, then with increasing vividness, the outline of a life-sized, female human figure took shape within the milky ribbon. I grew increasingly entranced by the ghost-like figure. Whispering voices from the back of the chapel broke my trance, and I overheard bits of a conversation between recently arrived visitors: "A single piece of marble … do you see her silhouette? …it's called Gratia Plena … that means 'full of

grace'." I remembered "Hail Mary, full of grace," from the prayer based on Elizabeth's joyful greeting of the pregnant mother of Jesus. The woman before me seemed full of the milk of loving kindness; brimming with the Sustainer's life-giving grace poured out for all creation.

For me, this sculpture also seemed to celebrate the first tastes of God's nourishment as newborns, as mammals. When we drink our mothers' milk, every cell of our bodies is flooded with its life-giving mystery: sun, water, plants, and animals transformed in one body to become food for another. As infants we are "filled with good things" (Luke 1:53) necessary for growth. We too are filled with grace.

Over the two weeks since delighting in Gratia Plena, "she" had largely been out of my thoughts, but she reentered my mind when in mid-November I visited S&S Homestead Farm on Lopez Island, just north of the Strait of Juan de Fuca. As with the chapel, for years I had heard glowing descriptions of Henning Sehmsdorf and Elizabeth Simpson's homestead but had never taken the time to visit. However, I found it hard to resist Henning's recent invitation to come stay on the farm for several days and "eat food that will make you never want to return to Seattle."

When I reached the farm, Henning, Elizabeth, and two German shepherds greeted me. After dropping off my gear in their straw bale guesthouse, we met up again in their dimly lit milking barn. I sat silently beside the couple as they hand-milked a Jersey cow named Loveday — four hands working in harmony with her udder. I listened to the rhythmic sound of streams of milk rushing into a stainless steel bucket, one stream after another. This twice-a-day, hour-long ritual determined the pace at which everything else on the farm could move. This ritual also helped to determine the contents of each day's meal for Elizabeth, Henning, and eighteen other families on the island who relied upon Loveday. I found myself entranced by Loveday's seemingly endless flow of milk. At this point, my thoughts flashed back to Gratia Plena: milk flowing, God's love and sustenance made manifest through an earthly being, the hungry being filled. Elizabeth gently broke my trance as she handed me a wine glass filled with Loveday's warm, fragrantly pungent milk. As with my reflections around Gratia Plena, I felt awe in this earthy mystery that flooded my own cells, in this conversion of life through Loveday's body to become life for my own.

Over the course of my days on their farm, Henning and Elizabeth spoke of the "members" of their farm — chickens, pigs, sheep, cattle, dogs, a cat, the land itself, and each other — as "co-workers." They explained that every member contributed to making the farm work. Rather than being expendable farm "assets" and "resources," each animal received a name and the level of respect due to them as co-workers.

Likewise, Elizabeth and Henning did not treat Loveday as a resource, chattel, or a machine. Rather, because of Loveday's great responsibilities — supplying more than fifty people with milk every week; generating substantial financial income for the farm; replenishing the ground with her manure; and providing constancy and much joy — it seemed that she enjoyed a proportionally high level of respect. She grazed freely on acres of pesticide-free pastures, which were "rotated" to allow the land time to recycle her manure. Raised chemical-free as a calf — by Dominican Sisters on a neighboring island — her body continued to remain free of synthetic hormones, antibiotics, and other medicines. She ate organic hay, greens and vegetables from the garden, and barley, all grown on the farm. She also received grooming and lots of affection every day.

All of this attention, Henning commented, helped her to respond to her human co-workers as she responded to her own calf. As a result, she willingly "let down her milk" to them. She was neither a machine, nor a coddled pet. Instead, she was the heart of the working organism that was the farm. She was full of the tangible Spirit of God, and this Spirit flowed out of her body to be food for many.

With what I know about the dairy industry in our country,[2] I realized how rare it is for a cow to be treated as a co-worker rather than a machine — or, more accurately, a dispensable cog in a larger industrial machine. This industrial dairy machine works to produce the greatest quantities of milk with the smallest input costs (feed, land, buildings, etc.) For decades, farm policies in our country have supported and subsidized this mechanistic model, which requires placing large numbers of cows on as little land as possible. Over eighty-five percent of cows in our country now live in Concentrated Animal Feeding Operations (CAFOs). But such a mechanistic system is actually "too" productive: milk quantities far exceed consumer need. Our current oversupply of milk floods the national economy and threatens to supplant the remaining small-scale, sustainable milk production here and around the world. The problem of overproduction is so great that, for example, in 1985 our government paid over fourteen thousand dairy farmers to kill their cows and get out of the dairy business. These cows were "dispensable cogs" and their milk a waste product.

CAFO cows don't live like Loveday. She produces a quantity of milk that's natural to her body, about twelve pounds a day. After receiving stress- and disease-producing (e.g., mastitis, uterine disorders, enlarged internal organs) injections of recombinant Bovine Growth Hormone (rBGH), Loveday's CAFO sisters can produce up to a painful forty-nine pounds of milk. Loveday grazes freely and eats food that her body was created to consume; CAFO cows typically never taste fresh grass — they're confined to small areas and fed a highly concentrated diet that is foreign and stressful to their bodies. Loveday's manure is a blessing to the land; the enormous quantities of CAFO cow manure are almost always an air and water-pollutant, and environmental curse. Loveday can expect to produce her rich milk for at least twelve years; her CAFO counterparts usually "wear out," cease milk production, and are slaughtered within two years.

During my stay at S&S Homestead, Gratia Plena came to mind frequently as I reflected on these contrasts. Elizabeth and Henning's honoring of the land and animals as co-workers felt reminiscent of the way the Holy Spirit worked with and through Mary. Although Mary considered herself the "lowliest" of servants, she was the Spirit's honored co-worker; and through her, all generations could taste God's loving kindness. Indeed, Mary was a beloved partner in an act that could radically feed the world's deepest hungers.

It feels blasphemous to even consider the idea that the Holy Spirit might have treated Mary as a resource or machine. For me, Gratia Plena and Loveday both provide hopeful antidotes to regarding each other, animals, or any member of creation as mere resources. Both invite a rediscovery of the sources of God's daily grace. Both evoke the honor of all members of creation to become co-workers with Spirit. It's a choice between cheapened "gifts" and the fullness of grace. I hunger for the latter.

Notes

[1] Originally published in *Earth Ministry Newsletter*, 2003.

[2] Based on *The Humane Farming Association*, www.hfa.org.

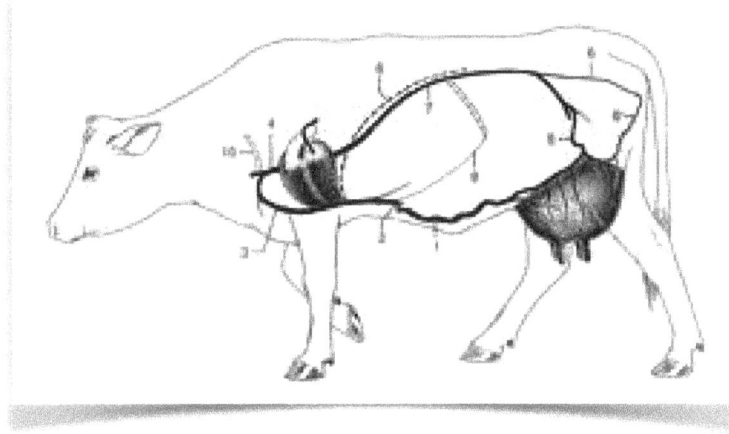

Blood Into Milk: Small-Scale Biodynamic Dairying

Henning Sehmsdorf, 2020[1]

At S&S Homestead Farm, we milk two Jersey cows, and our dairy is state-licensed to produce milk and make cheese. Why do we milk cows, and why do we bother to be licensed? The answer, in short, is that the cow is the ruminant animal par excellence. From her milk we produce the most healthful foods imaginable. However, the distribution of that food in our community is strictly regulated by the state. Without a license we can eat our own cheese, but we cannot sell it to our neighbors.

Ruminant Par Excellence

All domesticated ruminants, such as sheep, goats, and cows, turn the cellulose in grass into high quality protein, but cows do so to an eminent degree. Unlike the more flighty and nervous sheep and goats, cows are inherently inward-turned and meditative, as if their whole purpose in life was to chew their cud. Cows turn most of their energies on the process of digestion that produces the foods and other inputs essential to the biodynamic farm economy.

Monogastric farm animals, such as horses, pigs, poultry (and humans, for that matter), digest their food in a single stomach using enzymes and gastric juices secreted by the salivary glands, pancreas, and liver. Ruminants digest feed through fermentation by symbiotic bacteria in multiple stomachs, and through enzymes produced in the pancreas, abomasum, and intestines.

The rumen of a fully grown cow represents about eighty per cent of her digestive system, and most of the time contains about forty gallons of chime, that is, grass that has been suffused by digestive acids. It travels into the rumen by the esophagus and after multiple rounds of chewing, leaves it by way of the reticulum, which is the second stomach. The reticulum contains multiple fingers that "retain" any foreign object such as a piece of plastic string or a twig that might have been ingested by the cow as she plucks forage from the ground. After passing through the reticulum, the chime enters the omasum, the third stomach, in which feed particles are further reduced in size to enter into the abomasum, the fourth stomach, for final fermentation through enzymatic digestion. The omasum plays a major role in the absorption of water, electrolytes, volatile fatty acids, and minerals. In the abomasum, sometimes referred to as the "true stomach," protein from the feed and rumen microbes is passed into the bloodstream. What remains leaves the animal through the intestines in the form of manure.[2]

In contrast to the digestive system of the mature cow, in a newborn calf the rumen occupies only about thirty per cent of its digestive system. The abomasum occupies seventy per cent, with the reticulum and omasum as yet in rudimentary stages. The reason for this is that the newborn lives exclusively on milk. Eighty per cent of the proteins in cow's milk consist of casein, which coagulates into cheese by interaction with the enzymes chymosin, pepsin, and lipase in the lining of the abomasum. Over the next three months, as the calf starts nibbling grass in imitation of its mother, the rumen, reticulum and omasum gradually expand, and the abomasum shrinks to about seven per cent of the total digestive system.

Discovery of the enzymatic function of rennet in the calf abomasum gave rise to cheese production, as farmers learned to harvest the rennet by hydrating dried stomach pieces in whey or salt water. Today, much of the rennet used in making cheese commercially is derived from plant sources, notably thistle, artichokes, nettles, and other plants. However, it is claimed that animal rennet is better for long-aged cheeses, because plant-derived rennet may leave a bitter taste.[3]

Blood Into Milk

After the ruminant system of the cow has extracted the nutrients from the feed, these are absorbed into the bloodstream and travel to the heart, which pumps the blood back to the udder. Milk is secreted from blood in alveolar glands synthesizing leucocyte cells, amino acids, enzymes, vitamins, calcium, sugars and other minerals carried in the bloodstream. Each 500 ml of blood flowing through the udder produces one ml of milk.

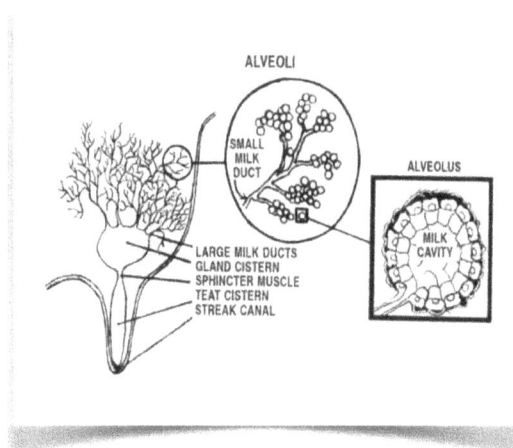

Image: H.D. Tyler, "Alveoli Duct System," 2022.

Scientists are satisfied with that description which, however, does not do justice to the miracle of transubstantiating blood into milk any more than the scientific definition of photosynthesis does justice to the miraculous process by which plants use sunlight, water, and carbon dioxide to create oxygen and energy in the form of glucose.

The udder has four chambers to which the four teats attach. Each chamber houses a gland cistern, where the milk collects. During nursing or milking, a sphincter muscle involuntarily opens or closes the canal of each teat in response to what the cow feels. She does not control the stream of the milk consciously, but if she feels the calf's presence, the sphincter muscle relaxes, and she "lets down her milk." Or, if she transfers her affection to the farmer because of the way she is handled, the sphincter muscle opens spontaneously. The only way the farmer can override this response is with a milking machine that forces the fluid from the udder mechanically.

Above each of the four milk cisterns in the udder, there are the alveoli which attach to the cistern like branches of a tree. Each branch supports milk cavities that are like fruit on the tree, and each cavity is filled with cells

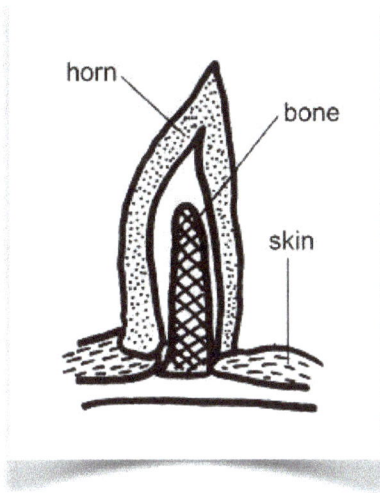

Image by Ruth Lawson. Otago Polytechnic.

surrounded by blood. This is where the transformation of blood into milk takes place.

The farmer who milks his cow by hand can feel the thick vein that runs from the heart of the cow to her mammary gland, which yields the precious liquid to the rhythmic motion of the calf's nursing or of the milker's hand.

The Cow Horn as Digestive Organ

Cows are also recipients par excellence of the cosmic energies from the sun that are brought to earth through photosynthesis. The forages on which the cow lives are the main vehicle by which these energies of the sun are captured, but not the only one.

Her skin, for example, turns solar energy into the essential vitamin D that makes milk such a vital food stuff. Beyond that, the cow's horns also funnel the energy of the sun into the breath of the cow, carrying cosmic energy into the animal through oxygen intake.[4] The cow digests the energy of the sun through her breathing!

The horn of the cow consists of an outer sheath of keratin — a protein substance akin to hair, nails or hoofs — surrounding a honeycombed, hollow bone core connected to the sinuses. Solar and subtle energies received by the sheath acting as a capacitor (like a mica stone)[5] are transmitted in the breath to the cow's bloodstream — energies flow in. By contrast, other ruminants — sheep, goats, deer — have solid bone horns that are not connected to the animal's sinuses and therefore cannot transmit outside energies to the breath. De-horned cows are deprived of the capacity to receive cosmic energies through the horns, which is, of course, why biodynamic farmers do not mutilate their cows by cutting off the horns, and why they consider the cow the essential biodynamic animal because of her enhanced ability to ingest the energies of the sun by means of her horns.

In this context, it is important to remember that breath is the most important nutrient supporting all organic life. Deprive yourself of oxygen for more than two minutes and your brain cells begin to die. After oxygen, water is the second most important nutrient, and most mammals can live without water for several days, and they can live without solid food for several weeks. All nutrients come to us from

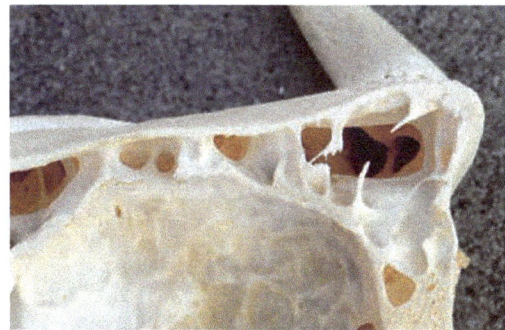

Photo by Kagfreiland. Demeter USA, 2016.

the cosmos in the form of light, air and water, all three transformations of the same solar energy ingested by the cow through her horns.

Dairy Products

Besides fresh milk, and its derivatives, butter and cream, the dairy products of S&S Homestead Farm include long-fermented hard cheeses, such as Cheddar, Fontina, and

Parmesan; soft cheeses, such as Quark (a German soured milk cheese), Mozzarella, Camembert, Queso Fresco (a Mexican cheese made from milk acidified with whey, lemon juice, or vinegar), Feta, Shangleesh (a Lebanese yogurt-based herbal cheese cured in olive oil), Paneer (a traditional soft cheese from India), fresh herbal cheeses, and Yogurt.

Making cheese essentially involves heating milk to different temperature levels (depending on the kind of cheese to be made); adding various bacterial cultures or starters; adding rennet to precipitate the separation of curds and whey; cooking, draining, pressing and aging the curds in conformity with traditional or new recipes.

Pasteurization & Fermentation

Raw dairy products are alive with enzymes. Enzymes are complex proteins catalyzing biochemical processes in the body. The relative absence of enzymes in many modern foods forces the pancreas to compensate by producing the missing enzymes, which in turn can lead to pancreatic cancer. Pasteurization systematically kills enzymes; in fact, the absence of enzymes is the test by which the effectiveness of pasteurization is assessed.

	Protein	Fat	Vit B2	VitB12	Folic Acid	VitA	Chol
milk	3.3g	3.5g	0.18g	0.41μ	6.4μ	31μ	11mg
yogurt	3.3g	3.5	0.18g	0.88μ	13μ	31μ	11mg
quark	13.5g	0.3g	0.30g	0.40	16μ	2μ	1mg
whey	88g	1g	0.15g	0.20	1μ	3μ	1mg

Nutritional Values of Dairy Products

Pasteurization also destroys fifty per cent of vitamin C, and eighty per cent of water soluble vitamins. Without vitamin D in pasteurized milk, the body cannot absorb milk calcium. However, synthetic vitamins (D2 and D3) are difficult to absorb and have been linked to heart disease.[6]

Milk pasteurization is achieved by heating the milk to 145 degrees for thirty minutes or, alternatively, to 162 degrees for fifteen seconds. For maximum health, we aspire to make our dairy products from raw milk, without pasteurization or homogenization. Homogenization is a process by which commercial milk producers for cosmetic reasons assure that fat particles do not rise to the top as in natural milk. However, this process interferes with the capacity of the fat globules to bind with essential nutrients such as vitamins A and D, and therefore we avoid homogenization. Dairy products we make from raw and un-homogenized milk include butter (room temperature), sour cream (heated to 72 degrees), cottage cheese (heated to 80 degrees), quark (heated to 85 degrees). In keeping with traditional recipes, most of our hard cheeses stay below 145 degrees. Fontina, for example, requires the milk to be heated to 102 degrees, and cheddar to 90 degrees.

On the other hand, yogurt and cheeses derived from yogurt, such as shangleesh, paneer, or queso fresco require heating the milk above pasteurization levels.[7] Yogurt is an ancient food appearing as early as during the Neolithic period, around 5,000–10,000 years ago, probably as a result of milk naturally souring in warm temperatures. Henning's mother, for example, routinely made yogurt (which she called "thick milk") by setting out raw milk in the sun in shallow bowls until it thickened naturally.

Today, commercial yogurt is mostly made by pasteurizing milk and then fermenting it with historical probiotic cultures originating in places like Greece (l. *acidophilus, s. thermophilus*), or Bulgaria (l. *bulgaricus*), or recently isolated from the intestinal tract of a healthy human (*L. rhamnosus*, 1983).[8] We use commercial yogurt as a starter culture for our own yogurt, which we then perpetuate by inoculating our pasteurized milk with a spoonful of the homemade product.

Fermentation in dairy production restores the biotic life eliminated by pasteurization and, in raw milk cheeses, achieves a healthful bacterial balance to eliminate pathogenic contamination. Fermentation also enhances nutrition, flavor, and digestibility. If it tastes good, it is also good for you. Our taste buds and our noses tell us that it is a wholesome food.

For a healthful diet, it is useful to compare the protein, fat, vitamin, folic acid, and cholesterol levels of various dairy products. For example, whey, a by-product of cheesemaking we use in cooking and baking, but also feed to pigs and chickens, exceeds the protein levels of milk and yogurt by a factor of almost twenty-seven, and that of quark by a factor of six and a half. On the other hand, the fat content of quark is one tenth of that of milk or yogurt, while the fat content of whey is one third of the two aforementioned products. The vitamin levels of the products mentioned here vary widely, with the vitamin A level of raw milk exceeding that of whey ten-fold, and that of quark fifteen-fold. And if you are conscious of cholesterol levels in your food, you might want to be aware that quark and whey contain one tenth of that substance than either raw milk or yogurt, while the cholesterol level in butter exceeds that of milk and yogurt twenty-five fold, and whey two-hundred fifty fold.[9]

Milk Parlor

Why A Certified Micro-Dairy, and How?

Obviously, S&S Homestead would not require a state license to supply itself with the high-quality foods and feeds produced by our micro-dairy. One reason for upgrading our operation to the level required for certification were the marketing opportunities not available without a license, such as selling cheese through local food stores. But as a farm dedicated to teaching agricultural sustainability, we also felt that certification would provide us with infrastructure to teach small-scale dairying to interns, apprentices, school classes and the general public at state-certified standards, and provide vocational training. In other words, the second reason for upgrading our dairy enterprise to state standards was to reassure consumers made wary of raw foods by ingrained (albeit uncritical) cultural habit, and to show how to produce such foods safely.

Given that we already had a fully equipped processing kitchen in our main barn, upgrading to a licensed facility permitting us to produce and process milk into long-fermented cheeses, was neither difficult, nor expensive.

We did not apply for a license to sell the raw milk, which would have required another level of infrastructure. Until the Washington Legislature made unlicensed cow share programs illegal,[10] we had supplied up to twenty families every week with quantities of milk, ranging from a quart to two gallons each. We provided glass canning jars to our customers, which they were supposed to return the following week in sterile condition.

Unfortunately, we found that almost no customer was able to meet that standard, and the jars came back looking and smelling like cheese, thereby potentially exposing consumers to dangerous pathogens. The only legal solution would have been to acquire an autoclave at a price of $15,000-20,000, an outlay not warranted by a small-scale dairy. Instead we opted for a license permitting the production of raw milk to be processed into hard cheese. Under the terms of the license, we age the cheese for at least six weeks to achieve bacterial balance through controlled fermentation, thereby eliminating the possibility of pathogenic contamination.

In making the changes needed to qualify for certification, we worked closely with a food safety inspector from the Washington Department of Agriculture, who gave us good advice on how to meet legal requirements adapted to our local conditions. We developed a Hazard Analysis Critical Control Points (HACCP) plan that detailed the steps from milking (in the milk parlor), to decanting and cooling the milk (in the milk room), to manufacturing the cheese (in the cheese room).

In the milk parlor (in the barn), we exchanged the old milk stanchions made from raw lumber with stanchions constructed from paintable and washable wood surfaces. We also installed a half-wall in the hay loft above to keep hay from falling into the parlor. But we convinced the inspector that instead of encasing the entire milk parlor in a closed room lighted and vented with electrical fixtures, the open barn structure was preferable because of the natural air flow and daylight already in place. In the milk room, we installed two sinks, a small one for the milkers to wash their hands when they entered from the milk parlor, and a larger one in which to decant the milk from the buckets to half gallon jars, which are placed into a small freezer converted with an external thermostat to keep the chill temperatures just above freezing (34

degrees).

In the opening to the adjacent cheese room, we installed a swing door so workers could move through without contamination between the rooms, mounted a second hand-washing sink just inside the door, replaced the five-gallon water heater with a unit ten times that size, installed a fan system to remove humidity from the room, painted the cement floor with a washable paint, placed the freezers and refrigerator on trolleys so that they could be moved for cleaning, and installed a small wine cooler as a cave allowing for temperatures around fifty degrees for aging cheeses. We also poured a cement slab in front of the outside door and capped the door with a small roof so that people could leave their shoes outside and enter the cheese room without bringing in rain or mud. We received our license in 2014 and have been inspected annually since then.

Dairy Production & Economics

Since we made all of the improvements ourselves, the total cost of earning certification amounted to $6,700, which we capitalized over fifteen years as an infrastructure expense, with the farm acting as lender to the dairy enterprise.

During its first year, the licensed dairy produced nearly $15,000 in fresh milk, butter, yogurt, soft and hard cheeses, and whey, with production rising and waning in rhythm with the seasonal grass growth. The values of various products were estimated by prices for organic dairy products in the local market.

In building our dairy herd, we did not have a good year in 2014. Abby, whom we purchased in 2008 as a two-year old, had produced a calf as she did every year, and she provided most of the milk that season. Circe, born to Abby two years earlier, and pregnant with her first calf, delivered by breech birth. We could save the cow, but not the calf, and because of injuries suffered during the difficult birth, Circe could not breed back. We had to slaughter her, which returned at least some value to the enterprise, but her loss cut milk production for the year in half. As if the loss of the second milk cow had not been enough, our Jersey bull, Freyr, got into a fight with the beef bull, broke his back, and died. Because of the unexpected suddenness of this accident, we were not prepared to butcher him, which dealt another blow to the economics of the dairy.

single cow, the small-scale dairy proved its fiscal viability, and earned an important place in the overall economy of the farm.

The labor cost involved in producing and processing the milk was calculated at $6,875, which is accounted for on both the income and expense side of the ledger. Since the farmers do not pay themselves a salary or take profit from the farm — but are compensated for their work in food, housing, capital gains, and other tangible and intangible benefits offered by life on the farm —— their effort is represented as part of the non-cash value streams flowing through the enterprise.[11]

Income	Cash	Non-Cash	Expenses	Cash	Non-Cash
Fresh milk		$11,736	Cap15yrs		-$180
V/a prod	$550	$2,592	Interest		-$54
Slaughter	$2,100		Infra15yrs		-400
Calf		$600	Interest		-$120
Compost		$1,420	Permit	-$50	
Labor		$6,875	Vet	-$40	
			Grain	-$350	
			Forage		-$1,920
			Hay		-$1,920
			Labor		-$6,875
			Overhead	-$400	
			Slaughter	-$450	
			Bull Loss		-$2,500
Totals	$2,650	$23,223		-$1,290	-$13,969
Net gain	$10,614				

Dairy Income and Expenses 2014

Nonetheless, the overall economic returns of the dairy were quite positive for the year. The value of the raw milk, which the farm household (including interns and apprentices) either consumed fresh or processed into various value-added products, yielded nearly $15,000. The income from the slaughter of one cow, the value of the newborn calf, and of the compost-based fertilizer raised the total income to more than $25,000 for the year. On the expense side of the ledger, capitalization (of the cow) and infrastructure costs amortized over fifteen years, amounted to around $750 for the year. Variable out-of-pocket expenses (permits, veterinary cost, grain, the value of green forages and hay), estimated overhead expenses (wear-and-tear, utilities, insurance, taxes), slaughter and loss of the bull, amounted toapproximately $15,000. This left a net gain of around $10,000 for the year. Besides the unsurpassed quality of the foodstuffs from a

The economic and health benefits of maintaining a micro-dairy are palpable and significant. As important is the place of the cow herself in the overall life of the farm. Milking, and the care of the cow and her calf, begin and end our day. Our awareness of the complex and subtle processes in her body remind us daily of similar transformations in the soil, plants and animals within the microcosm of the farm organism, and deepen our perception of the place of the farm in the whole of the cosmos. The lactating cow represents the essence of the biodynamic farm.

Notes

[1] Based on a paper presented at the Tilth Conference in 2015.

[2] For a general description of the digestive process in the milk cow, see van Loon, Dirk 1976. *The Family Cow.* Montgomery, IL.

[3] See https://backyardgoats.iamcountryside.com/home-dairy/vegetable-rennet-vs-animal-rennet-in-cheese/. Retrieved September 28, 2021.

[4] See Steiner, Rudolf 1993. *Spiritual Foundations for the Renewal of Agriculture.* Junction City, OR., 71. See also König, Karl 1982. *Earth & Man.* N. p., 273-278; Poppelbaum, Hermann 1992. "Horns & Antlers," *Star & Furrow* 78: 11-17; Ritter, Georg 1971. "Horns & Antlers," *Star & Furrow* 47: 26-29; Schad, Wolfgang 1977. *Man & Mammals: Toward a Biology of Form.* Garden City, NY, chapters. 6-7; Lovel, Hugh 2000. *A Biodynamic Farm: For Growing Wholesome Food.* Austin, TX, 46.

[5] For the use of mica in capacitors, see Cole, Jason 2001. "Micas," U. of Waterloo Earth Sciences Museum. https://uwaterloo.ca/earth-sciences-museum/resources/detailed-rocks-and-minerals-articles/micas. Retrieved September 30, 2021.

[6] See Fallon, Sally 2001. *Nourishing Traditions: The Cookbook that Challenges Politically Correct Nutrition.* Washington, D.C., passim. See also Howell, Edward *1946. Food Enzymes: Health and Longevity.* 3rd ed. 1985. Twin Lakes, WI.

[7] For recipes of cheeses and other dairy products made on S&S Homestead Farm, see Simpson, Elizabeth & Henning Sehmsdorf 2021. *Eating Locally & Seasonally: A Community Food Book for Lopez Island (and All Those Who Want to Eat Well).* Lopez Island, WA, 113-122.

[8] https://en.wikipedia.org/wiki/Lacticaseibacillus_rhamnosus. Retrieved October 1, 2021.

[9] The nutritional information incorporated in the table above was derived from Elmadfa, Ibrahim, Waltraut Aign & Erich Muskat 2002/2003. Die Grosse GU Nährwert-Kalorien-Tabelle (The Comprehensive GU Nutrition & Calorie Tables), Munich, Germany. Quoted in Kühne, Petra 2003. "Milk as Part of Our Food," Biodynamics 245: 35-37. The nutrient amounts are stated in micrograms (µg), milligrams (mg), or grams (g), per 100 g of product.

[10] Washington State allows the sale of raw milk by licensed sellers, and the licensed sellers must meet stringent legal standards set forth at RCW § 15.36 et seq. and by the Washington State Department of Agriculture. See https://www.foodsafetynews.com/2009/11/skirting-the-law-with-cow-share-agreements/. Retrieved September 29, 2021.

[11] For a fuller discussion of "livelihood" and budget formation, see Sehmsdorf, Henning 2005. "Self-Sufficiency of a Small Family Farm;" and Sehmsdorf, Henning 2021. "Farming for Health: The Economics of Stewardship," below.

Bee Colony Collapse and Biodynamic Strategies to Restore Bee Health[1]

Henning Sehmsdorf, 2015

As reported by The New York Times[2] and Reuters,[3] according to the annual survey published by Bee Informed Partnership, a consortium of research laboratories and universities, in May 2015, approximately 5-6,000 beekeepers managing about 400,000 colonies in the U.S. reported losing forty-two per cent of their bees during the previous twelve-month period. This is an increase of almost twenty-five per cent above the losses reported in the two years before, and the second highest since surveys began in 2010. Most ominous was the fact that honeybee deaths during last summer were even greater than during the winter, for the first time ever. Poor nutrition, the Varroa mite and pesticides, among them especially neo-nicotinoids, were listed as possible causes for the disappearance of the bees.

Cynically, Bayer Crop Science, the biggest manufacturer of neo-nicotinoids, hailed the results of the survey because winter bee losses seemed to have stabilized at relatively lower rates, even though any losses higher than eighteen per cent are generally considered economically unsustainable for commercial

pollination services, estimated to be worth ten to fifteen billion dollars per year.

What is Bee Colony Collapse (BCC)? Bees can die from a variety of causes: disease, natural disasters (sudden weather changes, storms, fires), lack of food, human interference (exterminators, destruction of natural habitat, toxins). However, BCC is a radical and new phenomenon first called by that name in 2006 when it was discovered in North America and in Europe that whole colonies were simply disappearing, leaving behind intact hives, combs filled with brood, and plentiful food stores.

The main reason governments, industry and scientists are currently paying attention to BCC is because of the economic importance of honeybees as pollinators. According to the Agriculture and Consumer Protection Department of the Food and Agriculture Organization of the United Nations, the value of global crops relying on honeybee pollination was estimated to be close to $200 billion in 2005. Shortages of bees in the U.S. have increased the cost to farmers renting them for pollination services by up to twenty per cent.

Many causes are blamed for BCC: pesticides, primarily neo-nicotinoids; infection with mites (Varroa and Acarapis) and other pathogens; genetic changes; loss of immune sufficiency; loss of habitat due to mono-cropping and competing uses (urbanization and commercial development); malnutrition; and changes in beekeeping practices especially in the U.S. (mobile beekeeping). Most assessments, however, ignore the quintessential role of bees in maintaining the ecological health of the entire planet.

A notable exception to this narrow perspective can be found in the work of certain evolutionary biologists such as Jürgen Tautz[4] and his colleagues in Germany. Another springs from the development of biodynamic bee sanctuaries in Europe and North America. Ninety years ago Rudolf Steiner warned that the honeybees would disappear from the face of the earth with disastrous consequences for natural and human ecology, unless agriculture shifted its focus from industrial bee management and production for profit.[5] The dire facts surrounding BCC make Steiner's prediction and call for action ever more urgent today.

Jürgen Tautz characterized honeybees as one of most successful species in the history of evolution. Honeybees co-evolved with flowering plants dependent on them for their own survival. For ca. four and a half billion years life developed on earth according to fixed rules of reproduction. Three and a half billion years ago single-cell life forms emerged capable of extracting matter and energy from the environment for reproduction and genetic variability. Six hundred million years ago multi-cell life formed building on the principles of specialization and cooperation. Flowering plants have existed for hundred thirty million

years, and thirty million years ago bees evolved to achieve the superorganism guaranteeing the immortality of its genetic substance through specialization within the colony.

In 1793, German botanist Conrad Sprengel discovered the role of honeybees in flower reproduction. Two generations later, Darwin not only confirmed Sprengel's insight experimentally — by covering flowers with nets to exclude insects, with predictable results — but showed that flowering plants had evolved to compete for pollination by offering different levels of pollen and nectar quality and temperatures. Plants adapted to remove sensitive sexual organs to the interior of the efflorescence as protection against wind, weather and predation by some classes of pollinators (such as the rose beetle who simply eat the sexual organs). Flowers developed optical and aromatic qualities to increase their attractiveness to honeybees in particular.

Thus, while honeybees are not the only pollinators — butterflies, flies, beetles, solitary bees, wasps, bumble bees also contribute to pollination — none perform this task as effectively and with the same level of mutual benefit for the plants and pollinators, as do honeybees.

The eight species of honeybees (*apis mellifera* = honey bearing bee), of which six are of European origin and two Asian — in contrast to no fewer than 4,000 different species of non-honey bees in North America alone — pollinate eighty-five per cent of all flowering plants, ninety per cent in the case of fruit trees. Altogether, *apis mellifera* pollinate 170,000 plant species, and of these 40,000 indispensably so. In other words, the plants would be doomed to extinction without honeybees. This statistic makes clear the profound role and significance of honeybees to plant ecology world wide. A single bee colony can visit one million flowers every day over a land area of four hundred square kilometers, with each individual bee visiting 3,000 flowers per day, at distances of ten kilometers in straight flight.

In the early 19th century, German beekeeper Johannes Mehring came to the insight that any given bee colony constitutes a single being (*Einwesen* or *Bien*). Worker bees function as the metabolic organs, the queen as the female and the drones as the male reproductive organs. The comb in which the bees raise their brood and store their food, functions as the bone-like cell structure comparable to the spine and skeletons of vertebrates. Later in the century, American entomologist William Morten Wheeler, on the basis of his work with ants, coined the term "superorganism" to describe the self-organization of all state-building hymenoptera (winged insects).

Jürgen Tautz points out that bee colonies share certain characteristics with mammals, which makes them evolutionarily superior to most vertebrates. Mammals have low reproduction rates — so do bees (one new colony on average per year); mammals produce mother's milk in specialized glands — bees feed their brood with substances produced in glands in their mandibles; mammals gestate offspring in a physical uterus — bees gestate their young in a social uterus (comb); mammals produce a body temperature of 35 C — so do bees; large-brained mammals excel among vertebrates in cognitive capacity — honeybees exceed most vertebrates in their capacity to learn and communicate with each other in the superorganism.

How is the bee superorganism actually organized? How do the thousands of individual bees in the hive know what to do? How are decisions made and communicated? What motivates worker bees to forgo reproduction, leaving the function and privilege of genetic transfer to the queen? As a Darwinian biologist, Tautz emphasizes that while bees act "as if" a central intelligence informed their behavior, the driving force behind their actions is instinct working through the genes. Decision making processes are decentralized and emerge phenomenologically as bees interact through smell, touch, and vibrations communicated through the cell walls of the comb.

Rudolf Steiner, the founder of Biodynamics, took a different view. What Mehring called the "Bien," Steiner referred to as "Spiritual Bee," by which he meant an overarching conscious intelligence informing all of nature. Different animal species are informed by "group souls" in a hierarchy of Spirit that has its pinnacle in the Creative Logos, as described in Genesis and the Gospel of John.

Similarly, ancient Greek philosopher Aristotle argued that "the whole (which) is greater than the sum of its parts"[6] derives from the formative forces shaping the elements of earth, fire, air and water into natural phenomena. Every natural thing embodies an "indwelling purpose" (*telos*), which it expresses in the course of its life-cycle.

Thus the bee superorganism on the basis of information flows makes decisions beyond what individual bees are able to decide on their own. Environmental conditions in the colony shape the characteristics of individual bees and direct their actions and behaviors. For instance, the pheromones constantly emitted by the queen bee retard the reproductive capacity of the worker bees. However, when the queen dies or is lost in mating flight, the absence of her pheromone stimulates some worker bees to lay unfertilized eggs. These eggs pupate into males that carry the genes of the colony beyond the hive to guarantee the survival (or immortality) of the larger bee community.

On average, *apis mellifera* live in colonies numbering 20,000 in winter, 50,000 in summer. Most honeybees are now domesticated, residing in wooden caves provided by their human keepers. While the smallest of domesticated animals, the honey bee because of its pollination capacity, is of the greatest economic and ecological significance to the farm. During the summer, bees collect nectar to make honey, pollen to produce a protein-rich food, and propolis from buds, fruits, leaves and flowers for use as sealants in hive construction and for medication. The bees transport the nectar in a special honey belly at back of their trunk, the pollen on their hind legs, and the propolis in a pouch on the bee mandibles.

The bees also build sexagonal cells from wax sweated from six glands located on their abdomens. In the cells the bees store nectar they transform into honey by fanning, ferment pollen by mixing it with enzymes, and they use the combs as brood chamber, and as communication device.

All worker bees are sterile females, while the male drones function exclusively to fertilize the single nubile female — not the queen in their own colony, but one encountered during feral mating with queens from other hives. Every colony has only one queen, recognizable by her long hind body. The queen lays up to 200,000 eggs per summer. The eggs metamorphose into larvae that pupate in their cells. Until they are fully grown, they are fed royal jelly (a secretion produced in the workers' hypopharanx), while the queen is fed royal jelly for as long as she lives.

Fertilized eggs produce females, unfertilized eggs produce males. The worker bees live a total of four to six weeks, cycling through sequential tasks: cleaners, builders, brood nurses, guardians and, finally, collectors. In summer the bees produce a few young queens by constructing larger, round queen cells, rather than the hexagonal cells which produce worker bees and drones. Bee colonies multiply by swarming. The old queen leaves the hive with a large portion of the colony. Bees survive the winter as a complete colony, contracting to a tight cluster and warming themselves through wing vibrations, with the required energy supplied from honey stores. Bees sting when the colony is threatened, but not when approached gently. If the beekeeper is around the bees all the time, and they recognize his voice and scent, they do not behave aggressively.

The communication system of honeybees is a marvel of special interest. Rudolf Steiner held that bees communicated in the dark hive by taste and smell rather than by sight. Jacqueline Freeman, a biodynamic farmer who maintains a bee sanctuary in southern Washington, describes bee communication with each other and the outside world — including the beekeeper — as a vibrational hum like an inverted Om. The bees don't actually hear the hum — they do not have ears — but feel it in their bodies. According to Freeman, the hum of the bees is communicated to human observers as language: they teach the wisdom needed by humanity to reach the next level of evolution.[7]

Jürgen Tautz employed time lapse photography and microchips to measure and document how bees experience the world through sight and smell, and communicate with each other. The knowledge bees require to survive in competition with other organisms is based on genetic predisposition, experiential learning, and communication. How do bees learn? Colors, for example, do not exist outside of the perceptual world of living beings. Colors are created in the nervous system in response to electromagnetic waves such as light, depending on sensory capacities, and on the significance of color for the survival and reproduction of the organism. Bees instinctively prefer blue and yellow, but learn to distinguish between color levels. The two compound eyes of bees each feature 6,000 lenses producing as many simultaneous images, while the human eye features a single lens producing a single image. Bees see ultraviolet light, i.e. short waves, while humans see long wave light. Bees therefore see black where humans see red. In the course of evolution, flowering plants have developed reflective four-fold patterns on their coronas, which mirror ultraviolet light visible to bees but not to humans, which means that bees can see optical patterns invisible to the human eye. In speedy flight (thirty km/hour), bees do not see color at all, that is, their color sense is shut off until they reach their target, at which point it is turned back on. Bees see in slow motion, while for humans rapid movements appear blurred. For bees, moving objects remain distinct, which means that they can see swaying flowers, drones following the queen, and the like, clearly.

Likewise bees possess a sense of smell a thousand-fold more powerful than the human, made possible by olfactory organs seated in individual cells located on the underside of their feelers (as are sensilla for touch, humidity and temperature). Thus flower aromas attract bees over long distances, and bees use air currents streaming toward them to identify targets. Unlike butterflies, flies, and other pollinators, honeybees stay with the flower they began their daily work with. They quickly learn the patterned combination of color and aroma to identify a particular flower. A single exposure to an aroma suffices for a bee to identify a target; color and patterns require three to five exposures.

Experiments demonstrate that the learning capacity of bees matches that of the lower vertebrates, including mammals. They are able to distinguishing abstract conceptual pairings such as right-left, symmetric-asymmetric, equal-unequal, more-fewer. Bees are able to abstract rules and apply them to new situations. Tautz was able to teach his bees signs, thereby

enabling them to negotiate unfamiliar labyrinths equipped with such signs. Bees are able to identify different localities and time schedules with specific decisions and activities, such as planning a work schedule to harvest different quantities of nectar at different times and sites, depending on the productive capacity of the flower target.

Are these responses not signs of "bee intelligence?" By its very nature, the bee super-organism is tied to a stable locality, and most bees stay at home much of their lives. But to harvest materials and energy flows, bees must venture into a hostile world to find flowers at great distances, and find their way back to the colony. For their orientation, bees make use of geographic aids using sights and smells from trees and bushes as landmarks. In practice flights they map the hive surroundings, using celestial aids such as the sun and polarization patterns of bundled sunlight on overcast days. Assessing the position of the sun and polarization patterns includes awareness of compass changes arising from the earth's diurnal rotation.

The bees' sense of time also allows them to gauge the time of day when flowers are open and producing nectar, and to schedule their visits accordingly. If a food source is exhausted, it is struck from the bee's memory. On the other hand, if bad weather prevents harvesting a known food source, bees are able to remember the location for up to a week. If a bee finds a plentiful food source, she returns up to ten times to memorize the quickest and straightest route. Having determined the route, she returns to the hive and informs her co-workers through her dance.

Bee master Karl von Frisch (1886-1982) observed that if a food source is within fifty to seventy meters from the hive, the bee performs a round dance. If farther away, the bee performs a waggle dance, throwing her lower abdomen from side to side at a rate of fifteen times per second, alternating from side to side. Then the bee runs in an arc back to the starting point of the waggle dance and repeats. Next time she runs in the opposite direction, back to the starting point, and repeats again, the whole dance cycle being carried out in an area of no more that two to four centimeters on a selected place in the comb.

Time lapse video recordings have shown that the circular movement is actually an illusion created by the rapid body oscillations produced by the five wing muscles pulsing, while the wings are uncoupled. The frequency of the pulses correspond to 230-237 wing beats per second, moving the body forward as a leg is lifted to gain better footage. As the leg is lifted, the rim and wall of the cell respond with a felt oscillation.

Groups of worker bees surrounding the dancer imitate the dance exactly. The dancer indicates the direction of the food source by positioning herself at an angle, where one vector points to the position of the food source, the other vector to the position of the sun indicated in the dark hive by the deviation of the sun's position relative to gravity (the cells of the combs being precisely perpendicular). When the informed worker bee emerges from the hive, she factors in the position of the sun to establish the direction of the food source. A newly recruited worker may need thirty times as much time to find a signaled food source than the experienced worker bee does. Therefore, mixed groups of up to ten bees fly together, the experienced bee circling the target with swooping flight maneuvers and emitting an aromatic substance from a gland located at the tip of her abdomen.

According to Tautz's research, in the world of honeybees, reproduction occurs in two ways: sexual mating and swarming. In most animal species, sexual reproduction couples individual males and females to produce offspring, which in turn couple for reproduction. In the case of bees, however, only the queen couples. She mates with up to twenty drones from among the thousands encountered on her maiden flight to a gathering place (*lumen*) several hundred feet in the air. In the course of the summer, the queen produces two to three daughters, one of which replaces her in the old nest when she swarms, taking part of the colony with her. Queens reproduce for several years, the drones only once, which means that numerically speaking offspring production is extremely low among bees. In most other species, the numerical relationship between male and female is the opposite: few males can fertilize many females. Where the reproductive rate is low, genetic survival requires protective care of the young from birth to sexual maturity (as in the case with humans), provided by the mass of sterile female worker bees.

Reproduction by swarming occurs when the old queen leaves the hive with about seventy percent of the worker bees as soon as the young queen is ready to reproduce. Her bridal gift includes thirty per cent of the worker bees (the youngest and oldest), plus ready-made combs filled with honey and pollen. In moderate climates like ours, swarming normally occurs between April and September, leaving behind enough brood to replace the departed workers.

Preparation for swarming is indicated by the construction of queen cells at the lower edge of the combs, or in the middle of the brood nest, colonized with eggs. When the first queen larva is developed sufficiently to pupate, the cell is lidded, and the old queen leaves the hive a few days before the new queen emerges. Accompanying worker bees fill their crops with honey to last for ten days, during which time the swarm must have found a new abode. Just before swarming, the bees emit high frequency vibrations and scratch and bite the queen to make her move. A bee fall emerges from the hive and collects near the hive in a tight cluster, sending out scouts to look for a new home. If the remaining population is not strong enough for further division, the workers destroy the remaining queen cells and repeat the cycle later.

The remaining daughter superorganism develops its own genetic signature by the new queen's mating with multiple drones on her maiden flight. Even a colony that occupies the same nest over time, alters its genetic makeup with each new generation. The superorganism is the same and yet different. The primary swarm around the old queen retains its genetic identity until the old queen is replaced. For the queen bee the cycle from embryonic development to reproductive mating is one month. However, normally a whole year passes before a new queen is produced, extending the life cycle of the queen bee to nearly twelve months.

The queen bee continually lays eggs which develop into sterile females, unless the worker bees start building queen cells and feed the eggs laid there with royal jelly, thereby manipulating the generational succession through their action. By dividing into daughter colonies, the superorganism effects a different and simplified life cycle rhythm, bypassing the four life cycle phases of the individual organism from egg to embryo, pupa and adult.

Genetically, sexual reproduction is the indispensable requirement for evolution. The honeybees avoid death of the superorganism through reproduction by colony division, while also retaining genetic evolution through sexual reproduction by the mortal queen.

The immortality of the superorganism is made possible by the continual replacement of its members, every four weeks to twelve months of its worker bees, every three to five years of its queen. Drones live two to four weeks. At a population of 50,000 and a daily death rate of 500 (one percent), the entire colony is replaced within four months, except for the queen. A new queen, however, changes the genetic make-up of the colony entirely; this happens either when a new queen is raised to enable colony division by swarming, or when an emergency queen is raised in case of need.

Honeybees leave the autonomous world of the hive to collect matter and energy for their survival and to support annual colony increase. The sun provides energy to plants to produce organic substances harvested by bees. Flowering plants and bees support each other in the most important task (*telos*) of all living organisms, which is reproduction. Through pollination the bees accomplish the sexual reproduction of plants. The "fruits" of bee colonies are the production of branch colonies. In this sense the sexual organs of the super-organism (queen and drones) are the "seeds" of the bees.

We have noted the honeybees' productivity as pollinators, but what about bee productivity in regard to nectar, pollen, propolis and wax? Tautz estimates that only fifteen percent of worker bees can carry nectar and pollen simultaneously. The rest are specialists, carrying one or the other, or performing other tasks in the hive. Five to twenty per cent of out-fliers are strictly scouts informing the colony of new food sources or a new hive cavity. A strong colony produces an average of three hundred kilograms of honey in a summer, eighty-five per cent of which is consumed by the bees to energize temperature regulation in the hive, with fifteen per cent left for winter storage or harvest by beekeepers. On average, a bee returns fifty times as much energy to the nest as she uses per flight, about fifty kilograms per bee lifetime. One milligram of honey is calculated to provide twelve joules of metabolic energy.

Temperature regulation in the hive is indeed one of the most critical tasks carried out by designated heater bees who insert their upper torso into open cells and, with their wings uncoupled, rev up their wing muscles to create heat transmitted to surrounding brood cells.

Amazingly, the social roles of the honeybees are modified by temperature levels in the brood stage. Bees raised in cooler temperatures tend to be employed inside the hive, those raised in maximum temperatures of thirty-six degrees C tend to focus on outside work, nectar collection, scouting, communication (dance), and display higher levels of learning capacity. Summer bees live an average of four weeks and tend to be raised at higher temperatures. Winter bees can live up to twelve months and can become active in the second season as collector bees; they are raised at lower temperatures. Temperature regulation in the hive also includes cooling, as needed. Selected worker bees can be seen standing at the hive entrance fanning air into the opening. Inside, other workers fan to transport the fresh air to the center of the hive.

What about wax production? The honeycomb is an integral organ channeling matter, energy and information throughout the super-organism, effecting the homeostasis of the colony. Worker bees spend ninety per cent of their lives carrying out various tasks on the comb. Twelve-eighteen day old worker bees produce most of the wax from eight glands on their bellies, kneading the wax with their mandibles while adding glandular secretions. When building a comb in a new location, a colony builds an average of 100,000 cells and invests seven and a half kilograms of honey in metabolic energy.

The comb starts with wax randomly placed on the roof of the cave and developed into a hexagon of geometrically perfect cell walls exactly 0.07 mm thick, angled at exactly 120 degrees, and sloped slightly toward the cell floor. The distance between neighboring combs is eight to ten millimeters, allowing two bees to pass each other back to back. The combs are precisely perpendicular to the cave floor, made possible by the sensory apparatus in the joints of the bees registering gravitational pull. The perfect geometric design of the cells which has elicited the awed admiration of artists, philosophers and scientists alike, is actually a function of the crystalline structure of the wax assuming hexagonal shape when heated by the bees to temperatures of 37-40 degrees C.

How did the bees learn to do this? Goethe would surely call it a "primordial phenomenon" (*Urphänomen*) reflecting irreducible laws of cosmic forces shaping the material world. Tautz would call it fixed rules of evolution. The inherent geometry of the cells is modified statically and dynamically by the bees, using propolis (i.e. sap scraped from plants) to manipulate the cell design as needed. Beekeepers and scientists have been puzzled by how the bees starting the construction of a comb from opposite sides can meet in the middle of the comb with a perfect seam. Jacqueline Freeman proposes an answer, reporting that she has observed bee construction crews hanging from the top bars of the hive, measuring the perfect arch of the comb-to-be with their bodies, just as the architect of a house might measure out the foundations with the measure of his feet.

Conclusion

What I have intended with this brief description of the complex and mysterious world of honeybees is to instill an appreciation of their ecological role beyond the economics of pollination. Certain themes emerge:
~ The honey bee superorganism represents a preeminent example of a complex adaptive system. As defined by John H. Holland, an adaptive organism features mutually responsive actors.
~ Decentralized controls arising from

competitive/cooperative individual decisions in a self-organizing and self-regulating homeostatic system, exhibit emergent phenomenological characteristics, including communication, specialization, spatial and temporal organization, and reproduction.

~ In biological, adaptive systems "the whole is more than the sum of its parts," and the whole shapes the actions of its parts. In the bee superorganism homeostasis occurs on the level of the individual bee and on the level of the whole colony.

~ Collective activities achieving balance of the whole include comb construction, climate control and hygiene. The socio-physiology of the colony shapes the characteristics of individual bees. Temperature regulation is effected through heating and cooling, optimal cell design and distribution of cell space for brood, feed storage and heating, in response to external temperature levels and to internal communication (dance). Faced with new tasks or emergencies, super-organisms respond by increasing the activity level of current workers, reassigning workers to new tasks, or recruiting new workers.

~Presence of honeybees is absolutely essential for biodiversity, and without honeybees sustainable management of renewable resources is not possible. As Albert Einstein is supposed to have said, "if the bees disappear, we have three or four years to live: no bees, no flowering plants; no plants, no agriculture; no agriculture, no food." Supporting the honeybees supports human existence.[8]

~As biodynamic beekeeper, Günther Hauck, urges we must shift beekeeping from a functional, profit-oriented, mechanical approach to Spiritual, organic practices on the basis of reverence and awe for life's inherent mystery.[9]

~ Holistic organic biology provides the framework to unlock the secret life of bees with modern physical and molecular methods. However, beyond the language of science, we must learn, as Jaqueline Freeman urges us to

grasp the unitary intelligence of bees by listening to the bees themselves with empathy, love, intuition, and imagination.

~Fundamentally, best management practices need to shift in the direction of allowing the bees to exist in keeping with their own nature, or (to speak with Aristotle) express their *telos* or "indwelling purpose."

A handful of simple but far-reaching recommendations sum up the biodynamic approach to beekeeping:

~Stop mobile beekeeping and urge industrial-scale orchardists to plow up some of their acreage to provide the plant-based diet bees require to maintain themselves year-round. Urge farmers and gardeners to create bee sanctuaries on their own acreage, however large or small.

~Stop the practice of replacing queens with artificially raised females and ensure genetic diversity by allowing bees to mate naturally in the wild.

~Stop using pesticides such as neo-nicotinoids to control Varroa mites and other pathogens. Instead, strengthen bee immunity by cautious applications of the remedies employed by the bees themselves, such as formic and oxalic acids, both (as Rudolf Steiner understood) essential to all life processes.

~Stop feeding bees with high-fructose corn syrup and other synthetic nutritional substitutes. Instead, feed the bees first by allowing them to nourish themselves from their own honey and pollen stores before removing any of these for profit.

~Stop using wax foundations and conventional box hives, and instead allow bees to build combs from their own bodies, and provide housing in structures that imitate natural bee caves such as top bar hives and Warre hives, tree stumps, straw skeps, or hives made from mud and manure.

~Ask the bees: Look what they do naturally and as much as possible emulate their practices!

Notes

[1] Paper presented at Huxley College for the Environment, WWU, and at Workshop on Beekeeping, S&S Homestead Farm, May 30, 2015.

[2] Wines, Michael 2015 "A Sharp Spike in Honeybee Deaths Deepens a Worrisome Trend, "*New York Times* (May 13.

[3] Gillam, Carey. 2015. "U.S. Honeybee Losses Soar Over Last Year, USDA Finds," *Reuters* (May 14).

[4] Tautz, Jürgen 2007. *Phänomen Honigbiene*. Heidelberg, Germany. English translation: David C. Sandemann 2008. *The Buzz about Bees: Biology of a Superorganism*. Berlin, Germany.

[5] Steiner, Rudolf 1923. *Über das Wesen der Bienen*. Dornach, Switzerland.

[6] Aristotle, 4th century B.C. τὰ μετὰ τὰ φυσικά *(Metaphysics)*, Book VIII, 1045a: 8–10. Fore further discussion, see Sehmsdorf, Henning 2016. "The Spirituality of the Soil: The Idea of Teleology from Aristotle to Rudolf Steiner," below.

[7] Freeman, Jacqueline 2014. *The Song of Increase: Returning to Our Sacred Partnership With Honeybees*. Boulder, Co.

[8] Even if this apocryphal statement is not evidenced in any of Einstein's published writings, it is still true. See Palmer, Brian 2015. (https://www.nrdc.org/onearth/would-world-without-bees-be-world-without-us). Retrieved May 18, 2021.

[9] Hauck, Günther 2002. *Toward Saving the Honeybee*. Junction City, OR.

Part V. Building Things

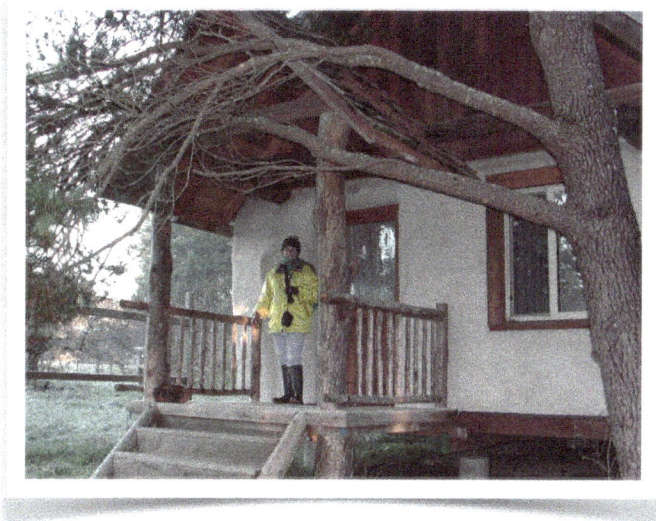

A Straw-Bale House for a Sustainable Farm

Jan Scilipoti, 2001[1]

Henning Sehmsdorf and his wife Elizabeth operate S&S Homestead — a small, organic, sustainable farm, providing vegetables and meat to our community. In Fall 1998, Henning and I started discussing his idea for a straw-bale bunkhouse to accommodate long-term interns and apprentices on the farm. Straw-bale, with its ties to sustenance and to well-managed agriculture, seemed a perfect complement to the ideals expressed by their farming methods.

Henning, a former professor, wanted to combine the building of the bunkhouse with education, by having the farm interns help design and build it as part of an internship, and by hosting a series of weekend workshops. As will happen in life, our plans changed as the project evolved. First, the call for workshop participants went largely unanswered. Then, our lead builder left for a full-time construction job.

We decided to cancel the workshops and focus on the building instead, reasoning that we had enough to do without arranging workshops in addition to building.

Our building team consisted of myself, Henning, intern Bill McKee and apprentice Brian Huntington. When our builder left, Bill's fiancé Elizabeth and Henning's son Johann offered to help. With them on the team, we decided that a lead builder would not be necessary. In hindsight, the project schedule

165

was affected dramatically without an experienced builder on site.

Still, my memories are punctuated with sweet moments of unexpected volunteers donating their skills and time, of delightful discussions on all sorts of topics, of working together as a team, and of delicious meals made from foods grown on the farm. Overall, it was an experience that strengthened our understanding of straw-bale construction, job-site management, and interpersonal relationships. The success of the building itself is reinforced by those who come to visit. We often hear that "it looks like a little house nestled in the Alps!" As a German immigrant who has devoted his life to small-scale farming in the Northwest, Henning finds this most appropriate.

The Design

In March 2000, we began meeting weekly to plan the project. We discussed and came to decisions about:
~ Research (regarding materials, methods, and local availability of supplies);
~ Budget (for materials, tools, and labor);
~ Design (site planning, heating and power, foundation, floors, walls, exterior cladding, insulation, roofing, doors/ windows, interior finishing, "beauty elements;"
~ Construction Drawings;
~ Construction Schedule;
~ Permits;
~ Materials Procurement and Storage.

To design the bunkhouse, we identified four basic functions. The floor plan should allow for tea preparation, study, lecturing, relaxing, sleeping, storage, and yoga. Meals, bathroom, shower, and laundry facilities would be available at other buildings on the farm. As drinking water would be brought in containers, plumbing would not be necessary. A "humanure" composting station would be built for convenience near the bunkhouse. Power use would be as minimal as the owner/builder code

allows. A wood stove would provide heat.

From sketches, we progressed to a plan for a rectangular, four hundred square foot (thirty-seven square meter) straw-bale infill structure with a gable roof, loft and covered front porch, following the code for an owner/builder project in San Juan County. No structural

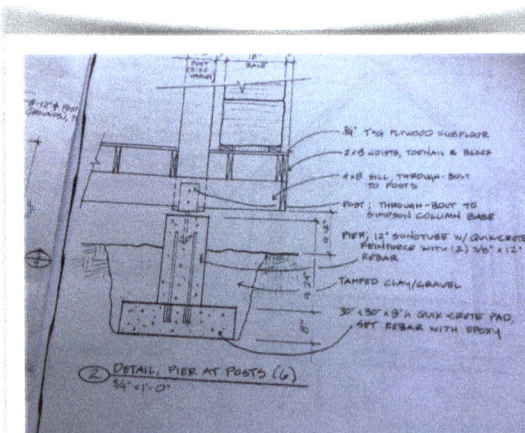

inspections were required, other than for limited safety concerns. This allowed us to use peeled logs from the site for the posts and beams, and earthen plaster mixed from clay from our fields and farm-produced straw to cover the bales without any lath or chicken wire.

In terms of detailing, we wanted to:

~ Keep it simple for our unskilled team;
~ Limit the impact on the builders and the environment;
~ Consider material use to limit toxicity;
~ Avoid straps under the earthen plaster, and compress the bales with the roof assembly instead;
~ Design simple, esthetically satisfying detailing at windows and transitions between materials.

Building Components: Foundation & Deck

In the Pacific Northwest, we experience frequent rain and wet conditions. To protect bales from ground moisture and splashing, they sit on a suspended deck. In addition to preventing wicking from the ground, the elevated deck also greatly reduces cement use (highly embodied energy and very expensive on the island) by limiting it to nine piers with footers. The entire structure hangs from three sills that are bolted to six posts. The posts are bolted to nine concrete piers, with footers below. The three deck sills support 2x8" floor joists at 16" o.c. with R-19 fiberglass insulation.

Post & Beam Structure

Six posts and four beams (made from rounds harvested and peeled on site) support the deck, walls and roof. We decided to leave these rounds exposed on the interior to enjoy the wood. We held the bales back about three to four inches from the posts to leave plenty of room to plaster behind them. The exposed round posts and beams are one of our most successful elements!

Beautiful! They look especially nice against the plaster walls. We also decided to extend the beams from the interior across the top of the

bales to the exterior. We wanted to see the beams supporting the rafters on both the exterior and interior. This required building the roof bearing assembly and the dimensional lumber gable ends around the beams.

Bale Walls

Our bales and roof bearing assembly — using straw bales produced on the farm — went up easily in a day. Instead of planning a work party, or to enlist help from friends, we decided — given the small size of our structure — to do the bales with our team of five. It was a wonderful decision. The raising displayed none of the 'bale frenzy' I have encountered at other job sites. Everyone had meaningful work to do, and the bales were installed with care, without a feeling of haste.

We tried temporary corner braces for the first time, and were very impressed with the resulting corners: tight and well-aligned. We recommend corner braces! We made partial bales by tying the two pieces off while the original twine was still intact. Even though we had a few tangled strings, the method worked well. We used a weed whacker with a heavy duty nylon to trim the bales. This created a great supply of chopped straw for our earthen plasters. After shearing the bales, the contours were easily seen, and we pounded the bales into place with a sledge before plastering.

Earthen Plasters

To keep the walls as permeable and natural as possible, we chose to use earthen plaster (no cement). Instead of wire lath or chicken wire, we used an "adhesion coat" clay slip worked well into the face of the bales. This provided

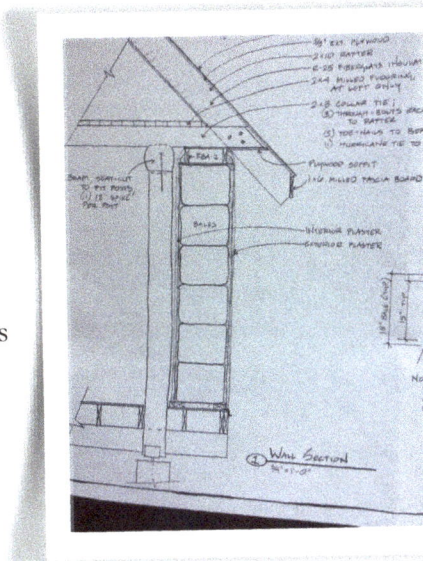

excellent keying for the first coat of earthen plaster. Through testing, the first coat became a wet, thick plaster comprised of clay slip, cow manure, coarsely chopped straw, and a small amount of sand. The second earthen coat was not as thick, and was made of the same ingredients, except that the straw was chopped more finely and there was more sand.

We tried a new method (for us) of creating slip for the earthen plaster, and were very pleased with the results. Previously, we had mixed clay with water in big "pits" that we made with bales and tarps. For this project, we tried to reduce the physical labor associated with the pit method. We got clay from a builder on the island who loaded it into our pickup to get it off his site. We stored this under tarps to keep it dry. We sifted the dry clay-based soil through a half-inch screen. and put this into water in a barrel. By letting this sit for a few days, an excellent slip was formed with very little effort. When we were putting a lot of plaster on the walls, we often mixed it in the barrel to increase the speed of slip production.

We used fresh cow manure from the farm in the earthen plasters, and found that it mixed in easily. We now prefer it to horse manure, which requires sifting to break up pieces. A local pit provided all the sand for the project.

To chop the straw, we used a chipper/shredder inside the bale walls to contain the flying straw.

Dusty and loud, but effective. We stored the chipped straw in bags under the deck until it was needed.

We mixed all plasters in an electric cement mixer and loved it! The mixer became our favorite tool. It was quiet, easy to operate, and gave us consistent, well-blended plasters. Having previously mixed by hand (or rather, by foot) and with a gas-powered machine, the electric method wins by far. On plastering days, one person at the machine could prepare slip, dig and trundle all the materials to the mixer, operate the machine, and provide enough plaster for three plasterers. With this method, we could finish a coat in two to two and a half days.

Over the earthen coat on both the exterior and interior, we used a lime-sand finish plaster. For the exterior, the reason was to provide weather protection for the earthen coats. For the interior, we wanted to use the traditional Italian fresco technique to color the walls, which entailed applying earthen pigments into fresh lime plaster. The integral technique looked especially nice with the organic curves of the bales.

A learning experience: we were plastering in the Fall, so we completed the earthen coats and one lime coat on the exterior before starting on the interior. The exterior coats dried successfully into a smooth, dense, hard finish. Then, weeks later, we started on the interior coats. Unfortunately, we had not yet installed

the wood stove. As a result, the interior earthen coats were very slow to dry. Mold was starting to form in small areas because of the moisture. We continued with the plasters anyway to stay with the construction schedule and finished the interior lime coat at about the time we installed the wood stove.

We kept a fire going as much as possible to help with the drying process. As the interior plaster dried, we found out that the earthen and lime plasters on the exterior were as permeable as we had hoped. So permeable, in fact, that moisture from the drying interior plasters moved to the exterior, and poppies of lime plaster bubbled open from the inside. That was months ago, and since then the exterior plaster has dried again, and seems as hard and firmly attached as it had been initially. In the Spring, we scraped off the loose bits of exterior lime plaster, and smoothed a fresh coat over it. Since we had been planning to do this anyway for additional weather protection, we felt grateful for the lesson that did not compromise the walls or the work plans. Next time, we will be sure to install the wood stove before the interior plaster goes up, and to provide plenty of ventilation for the moisture-laden air to exit the building through something other than the bales.

Windows & Doors

Based on experience with both wood and vinyl windows, Henning chose the latter for this project for their lower cost and maintenance. We ordered "almond" colored instead of the standard white, and are very happy with them esthetically. By ordering sliders instead of casements, we have a vertical line down each window opening that enhances the look of the building.

We built simple square bucks out of 2x10 lumber to house the windows. They were simple to build and install. Even with temporary bracing, however, we found it a

challenge to keep the bucks square, level, and plumb while installing the bales around them. Along one of the short walls, we have both the door and a window. To stabilize this wall, we bolted the top of the door and window buck to the roof bearing assembly. This was very successful.

We were not completely satisfied with the detailing of how the plaster met the buck on the exterior. Having decided that we wanted to inset the windows a little (three inches) on the exterior, and maintain a large sill on the interior, we found it difficult to provide nailers for exterior trim that allowed for the thickness of the plaster. We came up with an acceptable arrangement, but feel that we should improve this detail in future structures.

The two skylights we installed (one operable above the sleeping loft, and the other fixed

169

above the living area) were well worth the cost and effort. The light coming through them is magical.

Roof

Options for the roof raised much discussion. How to use a minimum of lumber, get it up quickly to protect the bales, and still provide the required R-25 insulation value? A given: we would have 18-24" eaves for protection against the rains. We settled on metal roofing, because of precedent on the farm, and our hope to install a catchment system in the future.

To keep installation as simple as possible, we decided on a gable roof. Planning for usable head-space in the loft, we planned a four/ twelve pitch. In hindsight, I would lower the pitch of the roof. Although it creates a generous loft, for a structure of this size and given its intended use, the loft is almost too generous. And in terms of installation, the four/twelve pitch created a steep, high structure that required heavy, larger rafter/ collar tie units. This tested the strength and height skills of our inexperienced crew. Reducing the pitch would also reduce the size of the gables, saving on lumber. A tip from a beginning carpenter: don't cut your rafter tails until after they are installed!

Although the manufacturer of the metal roofing did not recommend installing skylights, we found a local craftsman who had done it successfully in our area. We hired him (the only "subcontractor" on the site!) to install our skylights, and were very glad we did. We had a few major storms without any leaking.

Finishing Out

When scheduling your project, allow plenty of time for the trimming out! We were amazed at how long it took — although it was pleasant and satisfying to clean up all the little details around the gable ends, the roof bearing assembly, windows, the plank flooring sealed with a vegetable-based red finish.

We recommend checking carefully into options for weather proofing the stove pipe where it exits the roof. We tried an (expensive) rubber boot that supposedly was made for a four/ twelve pitch. Sure didn't seem to fit. And aesthetically, we are glad that the stove pipe is off to the side, and not at the front.

The rounds were one of our favorite elements. In addition to post and beams, Brian used peeled rounds to build a ladder to the loft and railings and stairs for the front porch. These look beautiful with the bales, and were so satisfying to work with. We enjoyed every aspect of this material, from peeling the logs to using hand tools for the fittings and notches.

We discussed the introduction of "beauty elements" from the beginning. Primary among these was the fresco technique on the interior walls. Behind the wood stove, Jan created a pattern of curving blocks, and filled each with a different earthen pigment. The other three walls were done with an ochre pigment. This gave the plaster a patina and softness that complements the bales.

Another handcrafted element was the installation of ceramic tiles painted by

Henning's mother in Germany. They grace the wood stove hearth, and are quite special.

We also cut three alcoves on the inside walls by cutting a few inches back into the bales. We cut shaped shelves out of plywood for the alcove, and firmly attached them to the walls with bamboo supports. We then plastered over the alcove and bamboo, leaving the plywood exposed. Jan designed a ceramic tile mosaic for each of these alcoves, and installed them after the plaster was complete. Using the same technique, we installed single-shelf bookshelves in two of the walls.

Finally, Jan created a six-inch wide stained glass panel installed in a framed space above one of the windows, which lights up the inside of the house when the sun is shining. Elizabeth sewed delicate, gauzy curtains for the windows and door.

Afterthoughts

These brief notes don't give a full sense of all the special situations, timely ideas, and hard work that together resulted in this sweet little straw bale house. They don't adequately describe the growth of a team of people working together, getting to know one another, and building friendships. They sure don't describe the big bowls of greens, tomatoes, and peas, fresh and full of life from the garden! I have a sense that when you make the decision to build with bale and with people you care for, these precious things will follow. May your building give you as much as this has given me.

Postscript

For some twenty years, the little bale house stood just east of the orchard and served as a dormitory for interns and apprentices, a place for them to study and for Henning to lecture, a venue for meetings and long conversations. Farm-stay guests spent weeks there in summers and winters. During all of this time, the roof never leaked, the plaster did not crack, and countless people enjoyed the peaceful, cozy and beautiful ambience of the place. Then, in spring of 2018, after a family had visited for an extended weekend, the house suddenly burst into flames and burned completely to the ground. The insurance company suspected electric failure; perhaps mice had chewed through the wires beneath the floor. Monetarily, the farm was generously paid for the damage. However, money could not compensate for the staggering loss of the building that had grown rich in experiences and memories. It will not be forgotten!

Note

---

[1] An earlier version of this essay was published *The Last Straw: The International Journal of Straw Bale and Natural Building*, No. 34 (Summer 2001), 33-34. The house was included in the Greenbuilders Registry. https://sbregistry.greenbuilder.com/search.straw?RID=12.

Solar-Powered Micro-Irrigation and Energy Production

Henning Sehmsdorf, 2004; updated 2021[1]

Irrigation System

Access to water for household use, animals, orchard and crop irrigation is a limiting factor on our small-scale family farm. In 2004, the only well on the farm pumped no more than one and a half gallons per minute, and the water was shared with twenty neighbors belonging to the Sunrise Beach Association, which owned an easement on our well.

Instead of drilling a second well to meet increasing agricultural water needs, the farm sought a competitive EQUIP (Environmental Quality Incentive Program) grant from NRCS (National Resources Conservation Service), a USDA (U.S. Department of Agriculture)

organization, to fund building a rain catchment system to collect water from two barn roofs and store it in a pond. Technical assistance in applying for the grant and implementing the project was provided by Tom Slocum, NRCS District Engineer, and Steve Nissley, NRCS District Conservationist. The solar pumping system at the pond was installed by Eric Youngren of Rainshadow Solar, Orcas, Island, WA.

The run-off from the two roofs measuring approximately 5,000 square feet is captured in water troughs in which pipes have been installed to funnel the overflow to a thousand-gallon cistern. The cistern overflows into open swales that carry the water more than eight-hundred feet to the existing pond whose holding capacity was expanded to accommodate approximately seven hundred-fifty gallons. The grassy swales filter the water.

At the pond, a four by eight foot building was erected to house a solar pumping station together with a solar shower for use by farm interns. The south-sloping roof supports two solar modules (Shell SM-110) to power a Dankoff solar twelve PV water pump, a sand media filter, and a twenty-gallon pressure tank. Valves at the pumping station control the flow of the water to three sites:
~ The cistern at the main barn, where a shallow well pump pressurizes the water for irrigation in two adjacent gardens and two greenhouses; flow to the cistern from the pond is regulated by a float valve in the cistern which

opens and closes as the water level in the cistern rises and falls;
~ The orchard, where an electronic timer controls alternation between three underground drip circuits providing water to trees and berry shrubs;
~ The two-acre CSA field next to the pond, where hundred-fifty foot long and five feet wide crop beds are supplied by valve-controlled underground drip lines.

The system came on line in July, 2004, during a three-month summer drought, saving ca. forty thousand gallons of well water per month. It was also noted that the plants responded favorably — as measured by increased vigor and productivity — to irrigation with soft rainwater from the pond, instead of hard groundwater from a well.

A major change in our water system arose in 2006, when membership in the Sunrise Beach

Association increased to a point where San Juan County mandated replacement of small storage tanks at individual residences with a large collection tank, and required that the water be chlorinated as prescribed by law for class-A water systems. Not wanting to drink chlorinated water ourselves, or give it to our farm animals, or use it for irrigation, we seceded from the shared well and drilled two additional wells, which, however, did not produce more than one gallon per minute each. This water supply proved barely adequate for household use, and without the rain water impounded in the pond, farming at S&S Homestead would have become unfeasible.

A decisive improvement offered itself in 2016. We were able to purchase a neighboring five-acre parcel that contained an untapped well producing fifteen gallons per minute. We developed the well, built a well house, and laid water lines to connect to the water system established on the farm. In spite of a plentiful supply of fresh and good quality well water, however, we continue to rely on water from the catchment pond for farm use. No one in the islands knows the quantity of underground water and how the aquifers are replenished. If the drought now bedeviling the West Coast becomes permanent due to climate change, water catchment systems will become ever more essential if farming is to continue in our part of the country.

Energy Production[2]

In late 2011, the final piece of the farm's plan to produce everything it needs was implemented: self-sustaining energy. Working with Opalco's MORE (Member Owned Renewable Energy) Program and Whidbey Sun & Wind, the farm installed a seventy-four panel photovoltaic system with a nameplate capacity of sixteen Kw.

The MORE Program collects voluntary donations from co-op members on their monthly electric bills to support producers like S&S Homestead with annual incentives based on the total amount of energy they produce. In addition, there are State and Federal tax incentives, rebates and credits, all of which helped to defray the cash investment of about $86,000 within ten years.

In planning this project, the farm set out to reduce its energy consumption by half and produce as much electricity as possible. For example, the home's electric floor heating system was replaced with an efficient wood stove fueled by waste lumber and timber from the farm. Incandescent bulbs were replaced with CFLs, and a solar water-preheater system was installed on the roof. The two combustion-driven cars of the farmers were replaced by electric vehicles. Since May of 2012, the farm has been able to meet its energy needs with the PV system. The net surplus in energy production is stored on Opalco's grid, banked as a credit and paid out each year in April. The

calculated annual financial return is about nine and a half per cent, which is better than the average stock market or any other investment today. But to consider only the financial benefits is to miss the point of renewable energy. The CO_2 sequestration readings on each of the three inverters show that the S&S Homestead system has sequestered more than three hundred metric tons of carbon during the last ten years, which demonstrates that the benefits and urgency of renewable energy are deeper than the financial equation.

Notes

[1] See https://sshomestead.org/wp-content/uploads/Solar%20Micro-Irrigation.pdf. Retrieved July 15, 2021.

[2] Based on "More Local Power: S&S Homestead Farm on Lopez Island" (September 18, 2012). *Energy Efficiency & Conservation, Membership Programs, Opalco.* (https://www.opalco.com/4873/2012/09/). Retrieved July 15, 2021.

Building a Barn
Henning Sehmsdorf, 2021

A wise old farmer once said to me: "A house won't feed a barn, but the barn will feed your house," and right he was. In the middle of the 1990s, after growing much of our food on our initial ten acres on Lopez for a generation, my wife and I decided to leave our secure teaching positions in Seattle. Intending to make our living on the island, we realized that we needed to expand farming operations in new directions, which required a barn. With a footprint twice the size of our home, the initial building cost of the barn also was twice that of the house.

Farm Center
Surrounded by a smaller barn, turkey run, greenhouse, vegetable garden, orchard in the middle of which we placed a wood shop, the family home, garages and sauna, the Main Barn lies squarely at the center of S&S Homestead Farm. It is also the center of most of our farm-related work. In the southwest corner of the building, we house our chickens next to the feed room, where we bin our grain. In the northwest corner, we inserted a covered area open to the outside for storing small machines such as a push mower, pressure washer, wire spindle for fence construction, and the like.

The entire east wing of the barn is occupied by a processing kitchen, where for many years we not only baked bread and preserved fruits and vegetables, but butchered cows, pigs, sheep and poultry, until the mobile unit built by the Lopez Community Land Trust made it unnecessary to process the slaughtered animals on the farm. In 2010, when we decided to add dairy cattle to farm production, we converted the kitchen and the adjacent entry and storage areas into a state-licensed cheese making facility.[1]

In the central mow of the barn we store winter food staples, such as potatoes, squashes, onions and fruit in bins constructed from recycled pallets covered with hardware cloth to keep mice and rats at bay. We also use the open space of the mow for occasional community dinners, dances, workshops and lectures, and we use the kitchen as a farm-to-school classroom.

In the loft running the entire length of the second story of the barn, we annually store a thousand square bales (twenty-five tons) of hay to feed our cattle and sheep during the winter. The bales are lifted into the barn with an electrified ladder through two gated windows, one on each end of the barn loft. The bales are stacked in such a way that they can be accessed from the inside mow and dropped into a truck bed at feeding time. The edge of the open mow is secured with removable gates in the form of an x-shaped cross. The cross is a heraldic symbol traditionally identified with St. Andrew, who is said to have been martyred during the Christianization of Britain, and the cross is often found in historic barns in Northern Europe.

Two important uses of the gambrel barn roof derive from its steep slope. One use of the roof is as a collection surface for rain water impounded each year in a pond holding an estimated seven hundred-fifty thousand gallons used for field and garden irrigation. The other use of the roof is as the surface on which we mounted seventy-four photovoltaic panels producing sixteen Kw of solar energy to supply most of the electricity used on the farm.[2]

In 1994, when I contemplated building a barn, I traveled through the Skagit Valley looking at historical and contemporary barns dotting the landscape. It struck me that the upward gesture of 19th- and early 20th-century wooden barns expressed a traditional attitude toward farming. Modern barns are strictly functional buildings constructed from minimal wooden frames clad with corrugated sheet metal for both roofs and walls. Here huge tractors assemble piles of grain, hay or other products in multi-use, undefined open spaces that also house farm machinery and supplies. The architectural design of these barns is akin to that of airplane hangars and other industrial buildings, reflecting the modern view of farming as an industrial process.

I was attracted to the gambrel roofs of historical barns, whose sloping design with each side becoming steeper halfway down, is both practical in that it allows for more headroom for hay storage, and — in my view — symbolizes the placement of the farm between earth and sky. The roofline of these lofty structures reminds me of the Gothic arches of medieval cathedrals and secular buildings, a style that has thrived in Europe since the twelvth century, and has roots in early Hindu and Islamic architecture. Some of the barns in the Skagit Valley have been lovingly restored, presumably for their beauty, expressing a connection to traditional ways of farming, while others are collapsing from neglect, because they are no longer seen as functionally useful structures.

French abbot Sugerius (1081-1151) wrote that the vertical, upward energy of Gothic rooflines, windows and flying buttresses expressed a longing for transcendence, "transferring that which is material to that which is immaterial."[3] This view rings true to biodynamic farmers who consider agronomic practice as promoting the inner development of the farmer by guiding "the Spiritual in the human being to the Spiritual in the universe."[4]

When I step into the barn and look up at the ceiling thirty feet above, soaring to a center point like two hands extended in prayer, I get a very different feeling from what I sense when I enter the squat, earth-bound cavern of a modern barn. The loft in the gambrel-roofed barn reminds me of Pope Gregory's (540-604 A.D.) saying about sacred imagery: "In it the illiterate read."[5] As a cultural counterpoint, consider Grant Wood's painting "American Gothic" (1930). This iconic image of modernist art has been read as a celebration of the sacred heartland,[6] but also as a critique of the oppressive confines of traditional American culture.[7] It is indicative of shifting perceptions

of pre-industrial farming that "American Gothic" has been widely parodied in Broadway shows, marketing campaigns, and pornography. During the eight-year run of *Desperate Housewives*, it regularly appeared in the popular opening title of the show.

Barn Economy

S&S Homestead Farm practices an economy of stewardship. Agronomic practices are designed in "imitation of nature"[8] to produce food, feed, fertility, livestock, water and energy in ways that build rather than diminish ecological resources. The barn has a central role to play in creating natural and social farm capital instead of extracting financial capital in the form of profit. On average, the annual processing of dairy, eggs, fruits, vegetables, meats, and bread in the barn kitchen, the storage of food staples and hay in the barn mow and loft, and the production of energy and water on the roof, amount to nearly half

of total farm production.[9] The barn indeed feeds the house and the whole farm!

The kitchen and the barn not only constitute the production hub of the farm, but also its educational and social center. Five years after building the barn, S&S Homestead was declared a "Demonstration Farm" by WSU's Center for Sustaining Agriculture, with focus on workshops and demonstrations in the kitchen. Here we taught short courses for the public on cheese making, bread baking, food preservation, fermentation, and seed saving. Students in the Ecological Food Production

class practiced culinary skills. They grew and harvested vegetables, grains and fruits in the farm gardens, orchards, and fields, and processed them in the barn and in the kitchen, before taking the produce to the school cafeteria. The methods taught were always non-industrial, simple and the kind that could be used in anybody's kitchen. For example, students learned to grow beans, then to dry and shell them using their feet. They learned to grow grain, harvest and thresh it, grind it in a table-top stone-burr mill, and bake the flour into long-fermented, flavorful bread in a wood-burning outdoor oven.

A vivid memory involves two students who asked us to teach them cheese making. They came every Saturday, bringing their homework to do while the curds ripened, until we had taught them everything we could about making soft and hard cheeses. One of them is now the chef de cuisine in a successful Italian restaurant, who notes on the menu that the mozzarella is hand-stretched, as he learned on the farm.

It is in the processing kitchen that we regularly gathered for dinner and long conversations with interns and apprentices about food, farming, and sustainable living; and it is here we received visitors, such as Congressman Rick Larsen, or the legislative assistants of Senators Maria Cantwell and Patty Murray, all of them

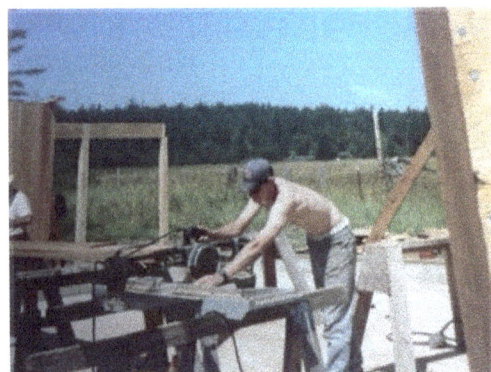

interested in farm-to-school programs and the role of the federal government in financing such efforts. An anecdote from one of Rick Larsen's visits is that how, when tasting a slice of farm-produced sausage, the congressman looked around at the students and asked: "Isn't anyone here vegetarian?" to which the students responded as if with one voice: "Not any more!" Rick Larsen later returned to tell us that his legislative efforts to fund school-farm programs had been successful inasmuch as Congress had passed the legislation; however, as he put it, "They created the

bucket, but left it empty," meaning that the intended funds went to finance war in Iraq instead.

Barn Construction

At the time we erected the main barn, my construction experience was limited to building pole structures from peeled logs and branches, and helping an experienced carpenter assemble our cedar home in the middle 1970s. During the intervening twenty years, I had improved my skill levels by working with neighbors on their projects and building a greenhouse and some sheds on the farm. However, I was not a proficient carpenter and therefore hired a professional builder who was able to implement my design drawings for the barn, with my help. Occasionally, we enlarged the team by hiring help to lift huge beams into place, pour the cement foundation, and install wiring and waterlines.

I also engaged my eighteen-year old son, Johann, who had just graduated from high school. For the entire summer, we got up at dawn, and after an early breakfast, worked on the barn project from six am until dinner at noon, after which we took a rest, and then spent the afternoon on farm work. Also on our team was my eighty-nine year old father, a retired lawyer visiting from Germany, who would sit on a chair watching us, and giving advice.

Working on the barn with Johann proved a journey of discovery for both of us. His teenage brain couldn't yet process how to make precise cuts on the radial arm saw; no matter how hard he tried, the planks always ended up a fraction of an inch too long or short. Together we came to the realization that before starting college studies, it would probably benefit him to undertake a hands-on apprenticeship first.

His grandfather eagerly found him a placement with a cabinet maker in Germany. Johann spent nearly three years learning the trade by attending a school for wood technology a couple of days every week and the rest working in his master's shop. He returned with journeyman's papers and such well-honed and precise skills that he had no trouble finding part-time work in a shop in Seattle, earning enough to finance study for a science degree in forest ecology at the University of Washington. Our favorite shared memory from the summer when we built the barn is from the day I was standing on top of an extension ladder cladding the rafters with cedar planks. When suddenly the ladder slid out from under me, I found myself hanging from the shaft of the hammer I had instinctively hooked over a rafter. Johann thought it riotously funny to see his father hanging from the beam by a hammer, but he quickly retrieved the ladder and helped me down. Later I told him that what had flashed through my mind when hanging there, was what my high school gym teacher said to me when I asked why we were practicing the difficult giant swing on the high bar. He said: "You will remember this exercise when you can still climb a tree in your sixties and seventies." He was right!

Pole Barn

Three years after building the main barn, we erected another structure we call the Pole Barn, where we initially housed farm machinery and pens for pigs and sheep, and later built a milk parlor to supply the cheese room in the main barn. The main barn had been inspired by an idea about biodynamic farming expressed in formal architectural drawings and manifested in dimensional lumber. The pole barn was improvised in response to immediate needs as they arose, and to the serendipitous availability of recycled and farm-produced building materials. As the local power company was replacing wooden poles along our road with steel poles, it struck me that these would make excellent supports for the utility barn I was thinking about. The power company gifted me with twenty heavily creosoted fir poles, of

which I set sixteen in four-foot deep, plastic-lined holes. I used the front loader on my tractor — an exercise undertaken with some trepidation lest the heavy timber tip backwards and crush me — to delineate a rectangular building measuring thirty by sixty feet. I tied the poles together with heavy planking salvaged

we peeled and shaped with broad axes. The improvised, cumulative design of the barn, its squat shape and the sheet metal roof notwithstanding, the blend of ancient, weathered lumber and of new wood shaped by hand in organic forms, lends to the finished barn a kind of wild and unconventional beauty.

from a hundred-year old, collapsed barn, and erected the rafters for a regular, two-sided roof on top of these ties. My brother came from California to help me install the twenty-foot long sheets of metal roofing.

Then I clad the narrow front and back sides with thin planking of salvaged lumber, but left the broad sides open. Inside I built feeding stanchions for cattle and sheep, and a pen to house weaner pigs until they could live outside. A year later, realizing that the ground inside the barn was too soft in winter to store farm machinery, I poured a concrete floor encasing the poles. I did the same in the cattle stalls because the heavy animals needed a solid floor to stand on, and in the pig pen because pigs dug up the soil underfoot, but not in the pen of the sheep, because they are light enough not to pug the ground.

To protect us from rain, I added a deep shed roof over a cement slab poured on the west side of the barn, where we could prepare feed, clean the animals, butcher, press apple cider, and even stack compost.

Another year later, we added a hayloft over one third of the barn's footprint — the maximum allowed by county regulations for this type of building. The massive poles and planks that support the loft holding one ton of hay were harvested from trees in our own forest, which

Notes

[1] See Sehmsdorf, Henning 2021. "Ecological Livestock Raising," and "Blood Into Milk: Micro-Dairying on a Biodynamic Farm," below.

[2] See Sehmsdorf, Henning 2004 (2021). "Solar-Powered Micro-Irrigation and Energy Production," below.

[3] @presbyformed 2016. "Gothic Cathedrals & Medieval Symbolism." https://presbyformed.com/2016/09/07/gothic-cathedrals-medieval-symbolism/. Retrieved October 6, 2021.

[4] See Sehmsdorf, Henning 2016. "The Spirituality of the Soil: The Idea of Teleology from Aristotle to Rudolf Steiner," below.

[5] @presbyformed, op.cit.

[6] Fineman, Mia 2005. "The Most Famous Farm Couple in the World: Why American Gothic still fascinates." https://slate.com/culture/2005/06/the-most-famous-farm-couple-in-the-world.html. RetrievedOctober 6, 2021.

[7] https://en.wikipedia.org/wiki/American_Gothic.

[8] See Sehmsdorf, Henning 2021. "Farming for Health: The Economics of Stewardship," below; published as "Small-Scale, Self-Sufficient Farming for Health," *Lilipoh (Life, Liberty, and the Pursuit of Happiness)*, Summer, 2021: 36-57.

[9] See S&S Homestead Farm Production 2016. https://docs.google.com/spreadsheets/d/1Na6dxEs-3AfhDFISvr7Jnoh5eRsGdoy5Qi5r0TfglC0/pubhtml. Retrieved October 7, 2021.

Part VI. Economics

THE VIABLE FAMILY FARM

Elizabeth Simpson, 2000[1]

This morning, my husband and I banded our last-born calf, Hunding, turning him painlessly from a bull calf into a steer. We will slaughter him in the field, without trauma, after he has fed on grass for eighteen months. He and his siblings will feed us and our customers on beef that is tender and flavorful, but has less fat than a skinless chicken breast and is free of chemicals, hormones, or antibiotics.

This afternoon, I chopped up some of our stored apples and fed them to Loveday, our beautiful Jersey cow, who is due to give birth in April. At that time, her calf will stay with her, and she will still provide us with enough milk, butter, cheese, yogurt, and cream to supply our customers, our animals, and us.

On the way back from grooming Loveday, I helped my husband pen our chickens after their day of free-range foraging. He collected their eggs, and then fed the sheep and cows hay and vegetables produced from our own fields. Tomorrow, I will pick vegetables from our one-quarter acre garden, which supplies twenty-five people year round. In the summer we raise a few pigs on vegetables and local grain, and our orchard bears fruit, berries, and flowers.

The received wisdom is that family farms are a thing of the past. But this farm has fed the family for thirty years, and in the last seven years, my husband and I have gone from two full-time salaries to one half-time salary with no change in living standards. My half-time teaching salary pays the utility bills, buys what the farm cannot produce — toilet paper and cleaning supplies, books, magazine subscriptions, gasoline and clothing. I bake our bread, and we make our own pasta. We use no machines but a small tractor, a hay rake and a mower. The farm is debt free.

We grimace when we read that family farms have gone the way of the horse-drawn plow, and that only corporate farms can survive economically As a matter of fact, one of our neighbors farms eighty acres using two teams of handsome Belgians — real horsepower — and he does well economically.

The first fiction in that claim is that, economically speaking, family farms were ever completely self-sufficient. Except in places where people homesteaded in isolation, there was always an outside income to provide — as mine does — cash for shoes, clothing, a kitchen stove, fencing materials. A farmer would sell butter and eggs. His daughter would teach school in town. His sons would take winter jobs in a local sawmill. But the main support of the family came from the farm. It still can. And the farm can feed its non-farming neighbors.

The other fiction is that corporate farming is economically viable. Currently, over thirty per cent of corporate farming profits come from government subsidies. Our taxes. Passage of the Daschle-Harkin Bill would provide $73.5 billion, over ten years, in addition to $98.5 billion to maintain existing programs.

By supporting corporate farming, our taxes also contribute to the abuse of factory farm animals, the genetic modification of crops, and a dead sea the size of New Jersey in the Gulf of Mexico, due to runoff from large-scale pig and corn farms into the Mississippi River. We pay for the massive climate changing gases emitted by such farms. We also pay for the arsenic, hormones, and antibiotics fed to animals in order to increase their appetites, productivity and resistance to disease while they await slaughter in muddy feedlots. We pay taxes that fatten the pocketbooks of corporate farmers and their shareholders, and poison ourselves.

Wouldn't it be better to support small, organic, sustainable farms like ours? We follow a few simple rules to keep our farm environmentally sound and economically viable:

~ We keep it small. My husband and I manage the animals, vegetables, greenhouse, and orchard by our selves. When we can no longer do that, we will cut back the size of our operation.
~ We feed ourselves first, and sell the excess to our community. Even our modest production nets several thousand dollars per year.
~ We incur no debt: we save money for water systems and outbuildings before we install or build them.
~ We maintain a closed system: The cows, sheep and chickens are bred and raised here, and gain natural immunities from living in one place. Our piglets come from a neighbor. We do not import replacement animals, feeds, or sources of fertility. The animals fertilize the pastures, and our compost builds rich topsoil in the garden. I preserve fruits and vegetables in summer so that we have a generous supply in winter.

Our farm is self-sufficient and sustainable. Because it is organic and complex — soil organisms, vegetables, flowers, fruit and animals all thrive together and feed each other — it is free of pests and disease; it sustains people, wildlife, farm animals, birds, insects, and the microorganisms that help plants make food from sunshine.

We live in a place that we have made beautiful. We have the best food in the world, perfect health, robust, peaceful animals, and the satisfaction of providing delicious, healthy food to our friends and neighbors. In this time when foods that are nutritionally deficient, poisoned with herbicides and pesticides, travel, on average, 1,400 miles from field to market; when BSE, salmonella, and E-coli threaten people's health, isn't it time to realize that the family farm is a viable prospect?

Note

1 Written as an opinion piece for *Newsweek Magazine*. Original water color by Tom Hoffman, Lopez Island artist.

Home Food Security
Henning Sehmsdorf, 2004[1]

If you are a farmer on Lopez Island — or anywhere in San Juan County — and you produce barley, for instance, you can sell it at the Cargill grain elevator on Highway 20; or if you produce beef or pork, you can sell the animals at the Marysville auction or to a number of feedlots further south; or if you produce milk you can make a contract with Darigold to take it off your hands: these are all examples of producing for commodity markets. For a small producer, commodity markets are problematic, however: the producer has little or no control over prices, and profit margins are typically very small.

An alternative often recommended to the small producer is local niche markets. For instance, in San Juan County, you can use the USDA-approved mobile slaughter unit — developed by the Lopez Community Land Trust and run by the Island Grown Farmers' Cooperative — and either butcher on a custom basis, or have your animals processed by the butcher, and sell your meats and sausage out of your farm freezer, or at local food stores.

Or, if you grow vegetables, fruit, flowers, or herbs, you can market these directly through a CSA — an arrangement by which customers pay a lump sum to the farmer at the beginning of the season in exchange for a weekly basket of produce; or you can sell your produce to local restaurants and food stores, or at the Farmers' Market; or you can process your calendula flowers, for instance, into skin cream and market it through the Internet; or process your berries into jams and sell those through various outlets.

At S&S Homestead Farm, over the years we have tried a number of these approaches, from commodity to niche marketing. We have come to the conclusion that what works best for us is not to concentrate on a particular commodity or niche, but to produce a lot of different things to meet our own food needs first, and sell the rest to our immediate community. I want to tell you about the economics of our experience, or about the bottom line of homestead food self-sufficiency.

On our fifty-acre farm, we produce beef, pork, lamb, chicken and eggs, milk and other dairy products, vegetables, and fruits. We also produce the feed for our animals — green forage, hay and grain — and the needed fertilizers in the form of cover crops, composts and compost teas. Everything we produce is grown biodynamically. What does that have to with economics? Quite a bit, actually, because foodstuffs produced this way are nutritionally whole. This means that we don't need to buy vitamins or other nutritional supplements — for which Americans spend more than $5 billion each year. It also means better health for us and for our animals, saving us thousands of dollars every year in medical and veterinary bills. It means that our plants are healthy, and we don't need to spend money on pesticides.

By USDA standards, S&S Homestead is not a commercial farm, because the total annual economic value of our production is less than $50,000. In a typical year we produce about $15,000 in vegetables and fruits, $12,000 in beef, pork and lamb, and $3,500 in dairy, for a total of a little more than $30,000. In 2002, fifty-eight per cent of that total represented cash sales. The remaining forty-two per cent was consumed by the farm household and four interns (see chart of Economic Data, 2001-2002, below). The chart shows that after deducting fixed production costs — depreciation for buildings, machinery, fences and water systems — and variable (or direct) costs — supplies, utilities, taxes, insurance, and the cost of our internship program — the farm was left with about twenty-four per cent in net profit, or about $7,500.

How do these data throw light on home food security? According to the U.S. Department of Labor Statistics, and the Washington Association of Churches' minimum standard for an average two-person household in San Juan County, food, transportation, housing, health care and entertainment altogether represent seventy-five per cent of typical household spending.

Because S&S Homestead produces its own food and sells more than half of it, the household nets a surplus of more than $15,000 in the food category. Because we live where we work — and rarely leave the island — our household has minimal transportation costs, about one third of the county minimum standard, and one seventh of

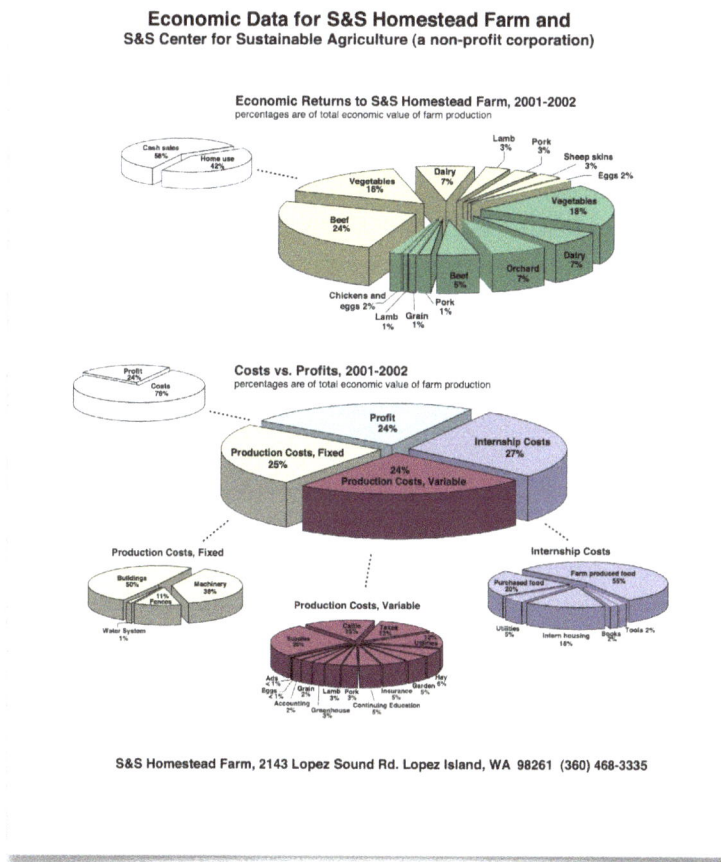

Economic Data for S&S Homestead Farm and
S&S Center for Sustainable Agriculture (a non-profit corporation)

Economic Returns to S&S Homestead Farm, 2001-2002
percentages are of total economic value of farm production

Costs vs. Profits, 2001-2002
percentages are of total economic value of farm production

S&S Homestead Farm, 2143 Lopez Sound Rd. Lopez Island, WA 98261 (360) 468-3335 the national average. Similarly, because we live on the farm, and used our own labor to build our house long ago — and have stayed in place — our housing costs are a fraction of the county and national average.

The comparison of health care costs is particularly instructive. Last month, the New York Times reported that for the first time in history Americans spend more on health care than on food. Forty-five per cent of health care is paid for by public spending, the rest by personal insurance or out-of-pocket, about $5,984 per household nationally, and an average of $3,156 per household in San Juan County. By contrast, our household spends an average of $1,200 for medical and dental checkups and minor medical consultations per year. We spend zero for drugs, supplements or other health aids, largely due — we believe — to the food we eat and the exercise we get in growing it. Nature deficiency, which has become a major source of physical and mental ailments in Western society, is not a concern for farmers who live and work in a symbiotic relationship with the soil, plants, animals, wind, weather, and sun.

Finally: Entertainment. Interestingly, our household budgets about one tenth of the national average for entertainment and vacations. Could it be that life and work on a small farm is so interesting and fulfilling that less commercial recreation is needed?

These statistics tell us that after paying for food, transportation, housing, health care, and entertainment — constituting seventy-five per cent of typical household spending — our farm household shows a surplus that is the same as the figure calculated by the U.S. Department of Labor for the remaining twenty-five per cent of household spending for utilities, supplies, clothing, personal care products, education, charity, tobacco, insurance, and pension. In other words, the farmers on

S&S Homestead do indeed make an adequate living.

This answers the question often posed to us: Can small farmers succeed economically? A second question often asked is: Can our model of home-based food self-sufficiency be replicated? Of course it can, and there are many ways of doing it, not necessarily the way we have done it.

For many years, my wife and I had two professional salaries to help pay for the purchase of the land and development of farm infrastructure. This is not unusual. We don't know many, if any, farmers on Lopez Island who did not bring outside income to establish themselves on the land. Typically, they brought savings or earnings from a previous enterprise, or an inheritance. This was true of most agriculture in the U.S. until the growth of large-scale industrial farming after WWII. Traditionally, family farms have always relied on outside income to provide additional cash to pay for goods not produced on the farm. The land, livestock and buildings were typically inherited from parents who lived and died on the farm — much as the Amish still do today. But the main livelihood came from the farm's feeding and housing the family. In our case, for twenty-five years outside jobs supported the farm that produced much of our food. For the last ten years the farm has supported itself.

To feed itself, a two-person family or household on Lopez would not need a complex, integrated farm like ours, nor would it need nearly as much land. For several years now, we have grown more than $15,000 worth of vegetables annually on less than a quarter acre, consumed half and sold the rest. According to the USDA, in 2003, the average gross income for large-scale American farms per acre was $463. As our example shows, the productivity of a highly diversified, intensively managed small farm producing a

much greater variety of food, is vastly greater.
Home food security is economically realistic.
What is required is for the farmer to stay put
and do the work.

Note
_____

[1] Presented at Navigating Our Future: Food Forum, Friday Harbor (Winter, 2004); WSU Livestock Advisors (Winter, 2004); Snohomish Conservation District Small Farm Expo (Spring and Fall, 2004); King County Small Farm Expo (Spring 2004, 2005), Whidbey Conservation District Small Farm Workshop (Spring 2005), and WSU Snohomish Extension: Self-Reliant Homes and Farms (Spring 2005). Original water color by Anne Whirledge.

Marketing A Philosophy[1]

Ariel S. Agenbroad & Karen Faunce, 2005

In 1970, Henning Sehmsdorf designed a fifty-year holistic plan for his farm on Lopez Island in Washington State. Thirty five years into that plan, S&S Homestead is debt-free and almost entirely self-sufficient, providing food for the family, interns and livestock, generating fertility for soil, crops and pastures, and drawing income through custom slaughter of livestock, and a community supported agriculture subscription service (CSA).

Several interns each year come to S&S Homestead for the opportunity to learn biodynamic farming from Sehmsdorf and his wife, Elizabeth Simpson, who are also actively involved in research and education, garnering grants from such entities as the USDA Sustainable Agriculture Research and Education (SARE) program, and National Resource Conservation Service (NRCS), pioneering farm to school programs in their area, and sharing their experiences and expertise through educational outreach both on and off the farm.

S&S Homestead is farmed biodynamically, an agricultural method based on a series of

lectures given by Austrian philosopher Rudolf Steiner in the early twentieth century. Biodynamics seeks to actively work with the health-giving forces of nature. Henning Sehmsdorf explains, "The whole idea behind the biodynamic farm is that it's a closed organism, with a kind of individuality where you produce your own feeds and you produce your own health." Emphasis is placed on integrating crops and livestock, recycling nutrients, maintaining soil, and promoting the health and well being of crops, animals, and humans. This holistic conceptualization leads to a series of management practices that address the environmental, social and financial aspects of the farm.[2]

Sehmsdorf grew up near Dresden, Germany, amid the destruction and aftermath of World War II. He lived in a household with fourteen other family members. Of his early experiences in farming, Sehmsdorf recalls, "Since I was born during the war and there was no food, we survived partly by going to neighboring farms as kids and gleaning." Later, in school, he also participated in "harvest vacations," where groups of schoolchildren worked the fields in exchange for food. Sehmsdorf says, "I always wanted to be a farmer, and my father, who was a lawyer, pointed out that there was no way that I could be a farmer in Europe, unless I just wanted to be a farmhand. There was no way I could get ahold of any land. Land was much too expensive. So, in a sense, doing what I'm doing now is really an old dream."

Entwined with the dream to farm was the desire for an education. He recalls, "I came over in 1956, when I was just 19, basically because in postwar Germany there was no way I could go to school and make a living. So I worked in a meat factory for a year and learned to hate how animals were treated and meats were produced, and basically said to myself, if I'm going to eat decently I have to do a lot of this myself, but it took me ten years to get there, because I had to get a PhD first."

Sehmsdorf earned his doctorate at the University of Chicago and took a teaching position at the University of Washington in Seattle, and again, his thoughts turned to farming. "In 1967…I came out here and I spent the first summer driving around looking for land to grow food for the family I didn't have. You know, I had no wife and no children yet. But I knew that somehow this was going to be connected. I couldn't afford to buy 200 or 2,000 acres, it was going to be a small piece. I suddenly found these islands," Henning remembers. Before the construction of dams that diverted water to larger farms east of the mountains, Lopez Island was an agricultural community with a population of 7,000, and a net exporter of dairy, grain, fruit, meat and fish. But when Sehmsdorf arrived the population had dwindled to 500, and "the attitude was that farming was dead."

Henning Sehmsdorf is now the steward of forty acres on Lopez Island. He bought ten in 1970, has since purchased five more and leases approximately twenty-five acres from neighboring landowners. For three decades they have built their farm here, and in that time they have also seen a gradual resurgence in agricultural production on the island. "This is the only county in the State of Washington where agriculture has been increasing over the last fifteen to twenty years. But the increase has all been in small-scale farming," says Henning.

He began very small-scale. "We started out with just a chainsaw and an axe, basically." He and his wife, Elizabeth Simpson (both teachers at the University of Washington: he taught comparative literature, she English and American Studies), worked full time. Over the course of twenty-five years they raised two children and slowly built the infrastructure of the farm. He says, "We would just come up here on weekends and summers and grow most of our vegetables. In the summer we had rabbits and chickens in that quarter acre which is now our orchard. We also ran a cow and a calf as part of a neighbor's herd, in exchange for helping him in the summer bring in several thousand bales of hay from his and our fields."

In 1994, when the youngest child turned eighteen, Sehmsdorf left the University and the couple moved to the farm full time. In ten years they have completed the construction of their home, outbuildings, intern housing, barns, a rainwater cistern, ponds, fences and paddocks. They have established a direct market approach for their farm and its in-demand products. They have become a model of self-sufficiency. Elizabeth still teaches part time at the local high school, but Henning claims, "At this point we would not need her income any more. In other words, the farm pays for everything. The farm provides us with food, shelter, all our needs, and we can pay for everything from transportation to health care, taxes, all from the farm income."

Direct Marketing Strategies & Profitability
S&S Homestead's marketing strategy is simple.

Sehmsdorf explains, "We answer the phone. That's pretty much it! In other words, we don't advertise, we don't really do anything like that. Basically we respond to people expressing a wish to eat the food we grow here. And what we produce is beef, pork, lamb, eggs, dairy products, vegetables, and fruit." Simpson adds, "What Henning is trying to do is to redefine economics, in different terms." Sehmsdorf elaborates, "OK. We have a fifty-year holistic management plan. And we review it every year and I do this with my wife and with the interns. A holistic plan means that economic goals are integrated with life values. As a matter of fact, we have defined profit as a tool, not a goal. You know, we have to have a certain amount of money to be able to operate, so we aim for that, but we don't aim for maximization of profit. We have certain ways we measure whether we're succeeding or not. Financial viability is one of them. Another one is environmental soundness. And I also like to think that we are concerned about questions of social responsibility and justice."

Though now it seems a natural extension of their farm philosophy, direct marketing was not their first approach. In fact, Sehmsdorf says, "When we first started out, when I first produced anything at all, I thought I was going to sell to commodity markets." However, he soon realized that the cost of producing the grain was more than what he could get from the grain elevators. "I very quickly caught on to the fact that if I was going to make any money at all, it would have to be sold locally on a custom basis. And that has proven to be very, very, true," says Sehmsdorf.[3] Direct marketing requires a certain ability to work with people, of which Sehmsdorf was well aware: "I think it takes a certain personality. You have to be interested in people and not just plants or animals." S&S Homestead is unique in that it not only markets its products directly, but the farm itself has become a marketable product, generating income through the internship program, educational outreach and grant funded research projects.

Vegetable Production

The climate on Lopez Island is temperate, with the average last frost date falling around April 15. However, in order to provide CSA customers with produce year round, season extension techniques are used. A greenhouse and hoop house keep tender plants protected and producing, while cold frames and row covers warm field crops. Several crops are grown which can be continually harvested throughout the winter season, such as kale and salad greens under cover. Increasingly, the farm is seeking sustainable, socially responsible sources for biodynamic vegetable and grain seed. The farm grows all its own seedlings in the greenhouse, potted in compost. This saves money and ensures that outside disease is not introduced into the system. Rainwater nourishes the plants, further reducing costs. Because of the direct way Sehmsdorf's produce is sold, on-farm, there are no transportation costs and the products are sold at their fresh-picked best.

CSA

Sehmsdorf and Simpson began their Community Supported Agriculture (CSA) subscription service ten years ago, after moving to the farm full time. Henning remembers, "People were coming to us saying, 'Would you sell us stuff out of your garden?…We love your produce.' And we said, 'OK.'" Henning and Elizabeth developed a questionnaire listing what they were currently growing and what they thought they could produce. Then they distributed the form to all their neighbors, asking them to indicate what they would be interested in buying. They received twenty responses. At first they tried to fill individual orders based on the questionnaires. Sehmsdorf explains, "We said, 'OK. We should grow x number of this and x number of that…' and it was very complicated." They found that planning for and accommodating these specific orders was nearly impossible, especially when customers changed their minds about what they thought they wanted. He continues, "When we filled people's trays and they saw what was on another tray they would say 'Why didn't I get some of this?' and I'd say, 'Right here, it says you don't want radishes, you wanted something else…' Then they'd say 'Oh, I didn't know, I want radishes too!'"

The system needed rethinking after the first season, Sehmsdorf recalls, "We said, 'OK. We're just going to grow stuff. And we're going to tell people they're going to get a basket of good food. And we're going to give them recipes and teach them how to cook the food.'" It took four years to create a working, successful CSA.

During this time, they developed the concept of FLOSS, which they use to explain what they believe their customers really want, and what they, as a farm, can provide. FLOSS is an acronym for Fresh, Local, Organic, Seasonal and Sustainable. By implementing principles of biodynamic farming and by selling only to their neighbors on the island, Sehmsdorf can be sure that his products are adhering to the standards of FLOSS and keeping his customers happy. And these customers have remained loyal for ten years.[4]

S&S Homestead's CSA brings in about $10,000 a year in sales, plus home consumption. It consists of two twenty-week seasons. The first runs from April to August, then the second begins and runs through the winter, with a break around Christmas. They sell shares for each season. Quarter, half, and whole shares were offered, and in 2003 were priced at $135, $270, and $540, respectively. The fresh-picked shares are picked up at the farm on Saturday mornings. A typical order in April or May might consist of lettuces, spinach, chard, kale, mustards, peas, corn salad, salad mix, herbs and sprouts. During the winter months customers

may receive potatoes, onions, and cool weather greens. Today, they produce enough for about twenty CSA shares, including enough for the family and interns in a garden the size of a postage stamp.

Sehmsdorf explains the economics: "And you think about this, two and a half thousand square feet in bed space, $10,000, and that's a twentieth of an acre. Right? So that's $200,000 per acre. Of course, I couldn't manage a whole acre of vegetables by myself, plus the rest of the farm; but still, I don't know how that sounds to you, but to me it sounds like economic viability. It works because we don't have to buy much of anything. Our inputs are mainly diesel, and total fuel costs last year for both farm machinery and personal transportation were $400. No other chemicals of any kind."

Livestock Management

The high quality of Sehmsdorf's livestock is due to selective crossbreeding and sustainable practices.[5] The cattle and sheep are entirely grass fed, either by pasture during warmer months and homegrown hay in the winter. Two acres per cow is allotted. A rotational grazing system ensures that the animals continually have access to fresh grass and the areas left behind are naturally fertilized. Moving the sheep regularly to fresh pasture also eliminates internal parasites.

Sehmsdorf's theory as to why his animals are so healthy stems from the concept of the biodynamic system: "I believe we've actually developed place-specific immunities, because the animals eat the plant matter that is produced from the farm composts and manures, and so it goes around and around, and so our animals are healthy without ever being inoculated and we have zero vet bills. I think that's a good record."

Because he produces his own inputs,

Sehmsdorf can charge less for superior products, a fact his customers enjoy. And by direct marketing his beef, lamb, pork, poultry and eggs, customers visiting the farm can see what they're getting: fresh, healthy food that is produced sustainably. Garden soils are double-dug and enriched with composts and biodynamic preparations. "We're very concerned about selling not at the high end premium rate, but at a rate that people can pay, and we also make allowances for people who have no money."

Custom Slaughter: Lamb, Poultry, Beef and Pork

When customers began to notice the sheep, chickens, pigs and Simmental cattle Sehmsdorf was raising for the family, they naturally began to inquire about adding meat to their orders. "People would come to the farm, they'd see the animals, and somehow we would talk about meat and they would say, 'Oh, I want some of that,'" says Sehmsdorf. Even though there was now a market for their meat, the logistics of filling that demand were problematic. When feeding the family, Sehmsdorf would do the butchering himself, and it could take him hours to butcher their yearly steer. Processing enough meat to satisfy his customers required a different approach.

He spearheaded a project at the local land trust to establish a USDA-certified mobile slaughter unit on the island and surrounding areas, now run by the Island Grown Farmers' Co-op, which also manages a processing facility near Bow where the meat is packaged. The unit now processes $400,000 worth of product yearly, including Sehmsdorf's animals. In 2004, Sehmsdorf had the mobile processing unit slaughter seven head of cattle, at a cost of $35 a head to kill. He explains, "The way I do it is that I sell the meat at $2 per pound to my customers, and then the customer pays for the processing. Because it's really custom slaughter, so they're buying the animal but what they pay

is determined by the hanging weight. The average is 750 pounds hanging weight."

Growing all his own feeds, managing the livestock biodynamically and selling the animals directly give Sehmsdorf an economic advantage over conventional feedlot operations. Per cow, yearly expenses total $25, and only because he must hire a neighbor to bale hay. Sehmsdorf says, "That's my total out of pocket expense. Everything else is provided by the farm, right? 'Course, it involves my labor, not included, but $25 per cow. And each animal will net me $1500. So that's over $10,000 that I get for these animals. Now no one can tell me that's not good economics!"

The Whole Farm as a Marketable Product

S&S Homestead's distinctive, well-ordered brand of self-sufficiency naturally draws curiosity. "I tell you, it's a steady stream of people coming here. And I know what this says: that they are hungry. They are culturally, physically, and Spiritually hungry for a better life," Sehmsdorf reflects. An important tenet of the farm's philosophy involves education and outreach. "We consider ourselves an education farm, not a profit based production farm." Through an internship program, on and off-farm education and grant funded research projects, Sehmsdorf is able to satisfy those who desire to learn from his experience. These programs earn their keep, by bringing in tangible income or valuable research and information. Education is the cornerstone of the farm, and the area in which Sehmsdorf would most like to direct more of the farm's resources and energy in the future.

Intern Program

Over the past ten years, S&S Homestead has been host to two dozen interns and apprentices from around the globe through their registered nonprofit organization, S&S Center for Sustainable Agriculture. Potential students find the program through the internet or cooperating universities. Internships are available for three to six months, and apprenticeships can last up to two years. Interns receive room and board, live in quality accommodations and are well fed by farm products. However, Henning states, "We're not making any money from the interns; I mean, they don't increase our production that much…but the contribution that interns make is that they leave behind substantial collections of data. So that's the real return, I would say." Intern reports have provided the farm with valuable studies on everything from soil nutrients and livestock production to water systems, resource management and pasture health. Interns can bring opportunities for new enterprises as well by contributing their fresh ideas, labor and energy to any number of projects.

Educational Outreach

Educational outreach occurs on and off-farm. On farm, an intern has spearheaded a program in horticultural therapy, providing special needs children with alternative education opportunities. The goal for the program is to access educational funding and become a self-supporting permanent program.

Another program involves growing produce for local schools through a farm-to-school project.[6] The unique element of this program is the involvement of students. "The kids come out here and help me grow the stuff, and then the product goes to the school cafeteria, where the students wash and prep the greens for the salad bar," Sehmsdorf explains. "And we teach a class, a high school class called Ecological Food Production, where the kids do the growing and they keep track of field data, growth rates and any disease."[7] This year, for the first time, the class is supported by a SARE (Sustainable Agriculture Research and Education) grant worth $7,500 to build a deer fence, a hoop house for winter vegetables, and pay partial support for a technical advisor, Dr. Carol Miles.

Dr. Miles, a WSU plant systems specialist, together with the students, carried out heirloom bean trials on the farm.[8] A high school senior, Tasha Wilson, is using the trial data as the basis of her senior project required for graduation.[9] Sehmsdorf hopes that the success of this project will lead to a permanent school curriculum in environmental and nutritional health. Henning has only recently realized the profitability of sharing his expertise with an audience. He had sought money for educational outreach in the past, but when the King County Small Farm Expo invited him to give a presentation on the economics of homestead food self-sufficiency, he said, "You know, I've always done this pro bono, but do you have a fee structure? And they say, we'll pay you $250 plus expenses…and they're going to pay me! I'm amazed, but hey, I never asked before, and they do have funds for this." In three weeks, Sehmsdorf made $1000 in fees speaking at various small farm workshops in the area.

Grant-Funded Research

Sehmsdorf is always looking for new ways to collaborate with academics, to explore current research, and to find funding. He and Elizabeth are open and articulate when it comes to describing their farm and aspirations, and have successfully pursued grant monies to fund various farm projects. In 2001-2002, S&S Homestead received a $2,000 grant from SARE to demonstrate the feasibility of low-tech and low-cost barley production in a two-acre field where the cattle winter, to prevent groundwater pollution from nutrient run-off, while at the same time strengthening farm self-sufficiency in animal feeds.

With seeds provided by Washington State University wheat breeder Steven Jones, S&S Homestead is currently experimenting with growing and preserving the valuable genetics of heirloom wheat while growing enough grain to satisfy the food needs of the farm household. In 2004-2005, a third SARE grant ($7,500) is supporting replicated field trials comparing farm-produced biodynamic soil stimulants with lime applications to balance soil pH, increase available NPK, micronutrients and soil organic matter in small-scale forage and hay production.[10]

The same year Sehmsdorf received a $6,000 grant in cost share funds from NRCS (National Conservation and Resource Service) to research and develop a solar-powered irrigation system that collects rainwater off two barn roofs, and stores the water in a seven hundred-fifty thousand gallon pond. From the pond the water is returned to irrigate the orchard and vegetable production sites during the typical summer drought, thus minimizing demand on limited groundwater resources, while at the same time benefiting plant health through irrigation with soft rainwater instead of hard groundwater. The grant also includes funds to build covered composting sites, protect pond water from fecal pollution by the cattle, and plant shelter belts to prevent soil erosion. While these grants are typically small, they benefit both the production side of the farm and its educational outreach programs by focusing energy on finding solutions to specific problems, and by bringing research expertise from the land grant university to the farm.

During the last few years S&S Homestead has benefited enormously from collaboration with university and extension agents and researchers bringing their know-how in engineering, soil science, microbiology, plant and forage systems, and agricultural economics. It has also been possible to write modest support for interns into these grants, so that students pursuing graduate degrees in various fields have opportunities to integrate their research interests with on-farm training. This year, the farm is hosting three interns pursuing advanced degrees in soil science, nutritional science and agricultural economics.

On Producing Locally

"Think local is my advice. I know that there's a lot of other advice out there, like 'sell through the internet,' or that there are a lot of niche markets lined up. They're all real and important, and I'm not denigrating that. I say think local because I think that if we are going to improve food availability to the average person, we need to have more people who grow locally for local populations."

On Financing the Small Farm Dream

"Most interns don't want to hear about economics. They just want to know 'What's the magic bullet? How do I make lots of money?' — but they don't want to know about economics, really. But let's think about this again. Why is it that you think you can start from scratch with no money and go to the bank and take out a mortgage and then grow enough strawberries to pay for that mortgage, and do it in a sustainable way? At first, keep your job, save your pennies, make your down payment, earn some equity, and then you make the switch. I don't know a single farmer on this island who has built his land base and infrastructure without outside income, another job, an inheritance or investment. I wonder whether I could have paid for what I have here through my CSA or through my milk production or through meat production alone. I don't think I could have. What I did instead was to keep my job and grow the food I wanted for myself and my family. And in the process I developed the infrastructure, and when the time was right, I made the transition and now this is supporting us entirely."[11]

S&S Center for Sustainable Agriculture and Homestead Farm has applied for further funds from SARE to help support the planned transition to an institutionalized teaching farm. S&S Homestead is governed primarily by the farm philosophy involving all the systems as a whole. Quality products are produced through the excellent soil fertility, animal health and year round vegetable production, ensuring the best products for the farm and its customers.

Soil Fertility

Soil fertility is the basis of the healthy systems at S&S Homestead. Feeding the soil organisms feeds the pastures, the fields and the gardens, and in turn the animals and humans who are also part of the system. Perpetual fertility means that Sehmsdorf continues to market excellent products to his customers, as well. Over ten tons of compost produced on the farm each year is comprised of garden and kitchen waste and animal manures. The compost is inoculated with a series of biodynamic preparations, intended to enhance the metabolic processes in fermenting the compost-forming bacteria. Sehmsdorf explains: "If you nourish the microorganisms in the soil, through these methods, according to biodynamic thinking, they're supposed to be able to supply everything needed. We've been doing it for thirty-four years and we do not see any deficiencies, nor do we see any disease. And the productivity is really good, both in terms of vegetables and fruit and animal protein."

Personal Goals, Philosophy and the Future of the Farm

Sehmsdorf and Simpson seem to be well on their way to fulfilling their fifty-year plan. They have established a market for their products on their terms, and feel confident that they are providing their neighbors with the best possible product. Occasional challenges arise, which Sehmsdorf approaches with introspection and a sense of humor. "People have said to us, 'We buy your food because it's cheap.' Which, to us, is the wrong motivation. We want them to buy our food because it's better for them, and because it supports the local economy and the physical and social environment we all live in, but, you know, people are funny!" However, affordability remains an important aspect of his philosophy. "We're very concerned about

selling not at the high end premium rate, but at a rate that people can pay, and we also make allowances for people who have no money. We take payment on a sliding scale. And we give long-term credit. And people are remarkably loyal about that. And because we're making a good return, why make more? Everyone benefits from that."

The current channels of marketing are working so well, Sehmsdorf must evaluate the possibility of expansion. "So, if you ask me, do I have any ideas for new expanding markets, yes, but only if we go to the next step." The next step would be the dream Sehmsdorf has for S&S Homestead. "I want to turn this into an institution that is linked to WSU's small farms program, to the island school, and to the whole community through our local land trust."

Twenty-five of their leased acres are part of a larger hundred twenty-five acre property owned by a neighbor. Sehmsdorf dreams of acquiring the entire acreage for community-based education and food production. However, land prices on the island now fetch between ten and twenty thousand dollars an acre. "I would love it…if someone with deep pockets could buy the land, the Lopez Community Land Trust could hold it in trust, and it could really become a major training site and educational site, right here. You know, I see all kinds of possibilities."[12]

Henning Sehmsdorf inherited his optimism and hope from his mother, whom he deeply admired. He keeps a woodcut she carved and gave to him after the Dresden bombings. "It's a quote by Martin Luther which says, 'And if tomorrow the world were to perish I would still plant my little apple tree today.' And it shows a man with a boy and they are planting a tree under birds flying. This has become my life's inspiration. I mean, you might say in a way I'm doing what I'm doing because of that."[13]

Notes

1 Published by Rural Roots (Colette DePhelps, Executive Director) and the University of Idaho in *Northwest Direct Farmer Case Studies*, funded through an Initiative for Future Agriculture and Food Systems grant by the USDA. Edited for brevity and reprinted by permission. For the full text, including tables and figures, see https://sshomestead.org/wp-content/uploads/Marketing_a_Philosophy_on_SS_Homestead.pdf. Retrieved July 8, 2022.

2 See Diver, Steve 1999. "Biodynamic Farming and Compost Preparations." National Center for Appropriate Technology. www. attra.ncat.org. Retrieved July 8, 2022.

3 See Sehmsdorf, Henning 2004. "Home Food Security," below.

4 For a discussion of FLOSS in the context of a local, seasonally based food system, see Simpson, Elizabeth and Henning Sehmsdorf 2021. *Eating Locally and Seasonally: A Community Food Book for Lopez Island (And All Those Who Want to Eat Well)*. Lopez Island, WA.

5 See Sehmsdorf, Henning 2008. "Ecological Livestock Raising," below.

6 On the public discussion of farm to school programs, see Simpson, Elizabeth and Henning Sehmsdorf 2002. "Nutrition – Body and Soul," keynote at the First National Conference on Farm-to-Cafeteria, Seattle, WA. Published in "Community Food Security News" (Spring 2003) https://foodsecurity.org/CFSCSpring2003.pdf. Retrieved June 22, 2022; and "Kids Can Make a Difference" (Spring 2003, vol. 8, no. 2.) https://kidscanmakeadifference.org/. Retrieved June 22, 2022.

7 https://lopezislandsd.ss19.sharpschool.com/our_schools/lopez_elementary/l_i_f_e_garden_program. See also Sather, David 2018-2019. "LIFE on Lopez," *Washington Principal*, 28-31. https://cdn5-ss19.sharpschool.com/UserFiles/Servers/Server_176833/File/Unique/Garden/life_on_lopezfall18.pdf. Retrieved June 22, 2021.

8 See Sehmsdorf, Henning 2004. "Class on a Lopez Island Farm" (Western SARE Report on S&S Homestead Farm grants in support of a class for sustainable agriculture, published by Western Sustainable Agriculture Research & Education). https://sshomestead.org/wp-content/uploads/FarmSchool_txt.pdf. Retrieved June 22, 2022.

9 Wilson, Tasha 2005. "Seeds of Life," below.

10 For the trial results, see Sehmsdorf, Henning 2008. "On-Farm Research: Biodynamic Forage Production," below; see also Lynn Carpenter-Boggs, Jennifer R. Reeve & Henning Sehmsdorf, 2011. "Sustainable Agriculture: A Case Study of a Small Lopez Island Farm," below.

11 For a broader discussion of small-farm economics, see Sehmsdorf, Henning 2021. "Farming for Health: The Economics of Stewardship," below.

12 On S&S Homestead Farm's plans beyond the farmers' retirement, see "Future Farm Project," https://sshomestead.org/future-plans/. Retrieved June 22, 2022. See also Sehmsdorf, Henning 2021. "Retiring on the Commons," below.

13 See Sehmsdorf, Henning 2023. *Trauma & Blessings: Biography of a Prussian Immigrant*. Lopez Island, forthcoming.

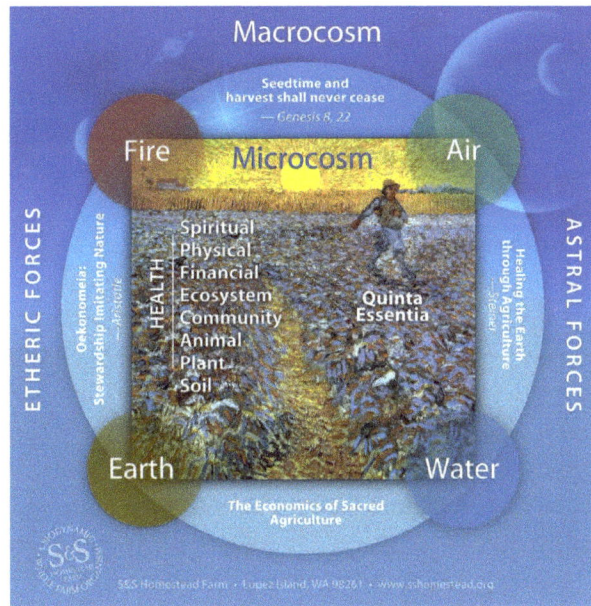

Farming for Health: The Economics of Stewardship

Henning Sehmsdorf, 2021[1]

Whole-Farm Organism

When I started farming on Lopez Island in 1970, I began with a fifty-year farm plan that considered the "whole" under management — people, resource base and money — by relating it to overarching quality of life goals, intended forms of production and a vision of what the farm should look like in the future. Over the years, the farm has become highly diversified: we grow practically all our food, most of the animal feed, all the fertility, much of the electric energy we use, all the water from several wells and from a rain catchment system, and a goodly share of the wood products we need. Self-sufficiency has been at the core of our farm vision from the very beginning. Sales of farm surplus have provided enough cash flow to meet necessary expenses. The farm is debt-free, and over the years increasing market values of the land, buildings and

infrastructure have contributed to the solid capitalization of the farm enterprise.

My topic will be the economics of a small, homestead-scale farm. Ours is a biodynamic farm, and I want to explain how its economics are defined in the context of biodynamics. Fundamental to biodynamic economics is the idea that while the farm organism is bounded in that it provides for most needed inputs independent of the market, it is open to the cosmos. The farm is conceived as a microcosm that exists in the larger, macrocosmic context. The cosmic perspective on biodynamic agriculture means that farming expresses what twentieth century Protestant theologian Paul Tillich called "ultimate concern."[2] In other words, for biodynamic farmers, agriculture is an inherently Spiritual, sacred task, and economics have a larger significance than the material bottom line.

If you look at the farm goals depicted in the poster above, you see that they represent a continuum summarized under the term "Health." We don't farm for money, but for health in the comprehensive sense of that term. Of course, we realize that in order to be successful, the farm has to be financially viable. However, you will notice that the top production goal is Spiritual health, and the second goal is physical health, the two being closely related. The third goal is financial health, which means mostly that we have no debt and can pay our bills. Fourth, the farm is committed to ecosystem health: that is, we are conscious of our place in the ecology of our natural environment, meaning that we aim to implement the Aristotelian vision of agricultural enterprise as an imitation of nature.[3] Fifth, we are committed to community health and to supporting a resilient local and seasonally based food system. Last, but not least, we prioritize soil, plant and animal health as the very foundation on which the farm depends.

Displayed in the poster are the four elements or essences conceived by pre-Socratic natural philosophers as fire, air, earth and water which constitute the building blocks of every ecosystem. Fire is the metaphor for the source of all energy, the sun. It is understood as the divine transformative force that gives shape and purpose (telos) to air, earth and water in natural phenomena. Water, for example, manifests as a liquid, solid, vapor or boiling substance depending on the presence of fire. Needless to say, this concept of the elements differs radically from that of materialistic science, as exemplified by Mendeleev's Periodic Table of Elements (1869), which describes matter in terms of a hundred eighteen chemical elements organized on the basis of atomic number, electron configurations, and recurring chemical properties.

However, Aristotelian cosmology is strikingly akin to modern chaos theory, which posits that the cosmos has a kind of deterministic "dark realm that conditions the formation and shapes of the galaxies, their interactions, and everything that's going on within them."[4]

Notice that the poster depicts a fifth element or quintessence (quinta essentia) in the person of the farmer. The term "quintessential" literally means that something is of the fifth essence. It points to the quintessential role of the human in shaping natural ecosystems, specifically in regard to agriculture. Today the human role in shaping planetary ecology is recognized under the term anthropocene.

Prior to the beginning of agriculture about eight to ten thousand years ago, ecosystems were naturally in balance, whether a river system or a desert, a forest, a prairie or a mountain range, ever evolving according to the laws of nature. For example, during the early development of the Cascade volcanic arc (from 275,000 to 35,000 years ago) successive eruptions gave shape to the stable range of rocky peaks, plant and animal life that includes Mount Adams and Mount Saint Helens. However, when Saint Helens blew its top in 1980, it not only altered the structure of that volcanic dome, but destroyed much plant, animal and human life as well as countless homes, roads, bridges, and railroads in the region. The result was a temporary imbalance in the ecosystem, as well as in the human habitation and socio-economic structures reliant on that stability. What had organized itself naturally over eons was temporarily disrupted, but over time it reorganized and corrected itself to achieve a new balance. That's what natural ecosystems do.[5] When a river floods in a storm, there is a temporary disruption of the drainage system, but it corrects itself, perhaps by carving a new channel for the river. When lightning starts a forest fire, there is a temporary disruption of tree and animal life, but the forest ecology rebalances. Not only do new trees take the place of the ones that burned, but the regular fires occurring whenever lightning strikes clear out the underbrush, and the ashes fertilize and support new plant and animal life. Natural ecosystems, at whatever scale, are self-organizing and self-correcting. They are also self-healing.

But when human beings come into the picture and take a stick and scratch the soil and put in seed, things change. What happens is that human interference potentially upsets the ecological balance and prevents nature's ability to self-correct and heal. This fact is amply demonstrated by the world-wide, massive loss of topsoil, the loss of soil fertility, and the pervasive pollution of groundwater and oceans due to industrialized, chemical and mechanical, agriculture. But, of course, the human presence also offers opportunities to heal the imbalance agriculture has caused, as governments, corporations, producers and consumers are gradually realizing. That's why the human is the fifth — quintessential — element, in maintaining natural ecosystems on the planet, as much as on the farm.

Economics as Stewardship

Aristotle's coinage of the term *oikonomia* connects agricultural enterprise with stewardship of natural systems. Some 2,500 years ago, the ancient philosopher combined two words, *oikos*, which means "household," and *nemein*, which means "to steward," or "take care of." In Aristotle's definition, "natural" economics means the stewardship of the farm household in imitation of "the way nature

takes care of her offspring." So, if you are an economist in the Aristotelian sense, you're a steward; in other words, you are concerned about the health of the next generation whether plant, animal or human.

Aristotle contrasted *oikonomia* with another term he also coined, *chrematistika,* from *chremata,* which means coins, in other words, money. He said, if your enterprise is focused on making money, then that's not economics but *chrematistics,* and therefore "not natural."

Aristotle's Economics

'oekonomia'
builds **natural** & **social** capital

'chrematistica'
depeletes **natural** & **social** capital to build **financial** capital

It does not take care of the young, nor provide for the next generation. Instead it extracts capital from the enterprise. By contrast, Aristotelian economics returns the value of farm production to the farm organism and the community. Extracting capital from the farm as profit depletes natural and social capital in order to build financial capital. This is of course the reason why modern, conventional agriculture is failing, both socially and ecologically. That's why our rural communities are destroyed, why we have lost our soils, why the groundwater is poisoned, why the air is polluted, and why our public health system is failing: the food that is produced by conventional agriculture depletes the natural and social resources on which our personal health depends.

In the largest sense, in the personal as well as in the ecological sense, in the Spiritual as well as in the material sense, we are vitally dependent on responding to the task of economic stewardship as described by Aristotle. However, what actually happened in the history of agricultural economics is that the two terms were reversed in the interest of profit and power, substituting the term economics for chrematistics. It fell to Rudolf Steiner, the founder of biodynamics, to remind us of the true meaning of economics, to remind us of our obligation to be stewards of the earth. This meaning of stewardship becomes the economic principle by which biodynamic farms are organized.

It is pertinent to note that Judeo-Christian religion, which for millennia practiced scriptural exegesis to privilege "hostile domination of nature," has lately come to the recognition that conventional agriculture poses the "largest threat to biodiversity and ecosystem function of any single human activity,"[6] and that the solution to the ecological crisis is "principally moral and theological rather than technological."[7] Hence the poster above is topped by the quote from Genesis 8:22 that "Seedtime and harvest shall never cease," provided that humanity, the quintessential element, will keep the covenant to be stewards of God's creation.

Ecosystem Services

Steiner's concept of ecosystem services rests on what he meant when he called on farmers to be "healers of the earth." When I first came to the property in 1970, I spent two summers just camping on the land, feeling it and measuring and mapping it topographically. I noticed that there was almost no bird life, and it became clear to me that the reason there were no birds was that there was no surface water. So one of the first things I did was to dig a pond, which eventually became the basis of our water catchment system, irrigating the whole farm.

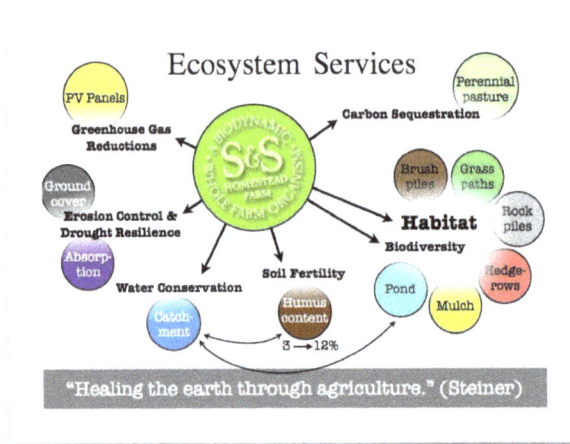

Ecosystem Services

"Healing the earth through agriculture." (Steiner)

Besides supplying the water needed for crop production, the pond restored the ecological balance of the land. That balance had been disturbed a century earlier when the land was first carved out of verdant forest by European settlers, who removed the trees that had supplied the ground with sufficient moisture to support a wide range of wildlife. Records kept over fifty years document the wild animal species that have returned to the farm because of that pond: not only otters and deer, but also herons, mallards, mergansers, newts, snakes and frogs, and all kinds of bees and pollinators, because there once again is water. So digging the pond was a *quintessentially* ecological decision that had all kinds of economic implications for the farm organism.

The second thing I noticed when I first arrived, was the absence of earthworms in the soil. The fields had been hayed for decades without animal inputs. The shallow soil was dry, compact and brittle, unable to absorb the heavy winter rains that washed off the rocky island carrying sediment into the ocean. There were mostly large clumps of native quack grass, interspersed with imported, naturalized perennial rye, canary reed grass, and meadow foxtail, leaving much bare soil in between. There were very few legumes, broad-leaf plants, wild flowers or other edibles.

In collaboration with a neighbor, I brought rotationally grazed ruminants to the fields, initially beef cattle, later dairy cows, sheep, pigs, chickens and turkeys. In the winter, we fed the animals good hay from the best fields on the farm, using the same rotational practices, and thereby reseeded the pastures with remarkable results. After a few years of animal droppings being worked into the soil by hoofs, snouts, beaks and claws, and enlivened by biodynamic preparations, the soil organic matter and the range of forages increased dramatically. Over the course of half a century, the humus content of the soil grew from three to twelve per cent. Earthworms now abound, the capacity of the soil to hold air and water has grown concomitantly, and the range of forage plants has expanded from an average of five to forty.[8]

Because of their health and productivity, our pastures and hayfields have become perennial. This means that they do not need to be tilled and reseeded every few years, as in conventional agriculture, thereby sequestering carbon in the soil. We have also surrounded our fields with hedgerows to provide additional habitat for wild animals and shelter the fields from punishing winds.

In our vegetable gardens and fruit orchards, we jettisoned mechanical tillage after we observed how destructive rototillers were to earthworms and soil fungi. Instead, we established triple-dug beds to create permanent three-foot organic soil horizons. We fed the soil with cover crops, composts, fermented nettle and comfrey teas, and with biodynamic preparations. The preparations put *fire*, i.e. cosmic energy, into the soil. We also surrounded the beds with grass paths as habitat for soil organisms. The humus content in the garden beds currently is at fifteen percent. Plant disease is practically non-existent because the life in the soil is in balance.

In our forest, we gradually removed dead trees and flammable understory. We turned the woody debris into biochar by burning it in home-made kilns constructed from recycled oil tanks. The pyrolytic transformation of the wood by high heat sequestered the carbon in the char instead of sending it into the atmosphere as climate changing gases. The biochar is crushed and applied to composts, pastures and vegetable plots, where it provides favorable environments for soil organisms.

In late 2011, a final piece of our original farm plan was implemented to provide a source of self-sustaining energy. Working with the local power company, we installed a seventy-four panel photovoltaic system on two barn roofs with a nameplate capacity of sixteen kWh. About nine months into the first year of production, the system had already produced 15,860 Wh.[9] In planning this project, our goal was to reduce farm energy consumption by half and produce as much electricity as possible sustainably. We replaced the home's electric floor heating system with an efficient wood stove fueled from the farm's woodlot. We installed a roof-top solar water pre-heater at the house, and at the pumping station at the pond installed PV panels to energize the farm irrigation system.

The cost of the barn roof installations was recovered over ten years from incentives paid by the power company, and state and federal governments, providing an annual financial return of about nine and a half per cent. Equally important, within the decade since installation, the carbon sequestration readings at the inverters show savings of nearly three hundred tons. Harvesting solar energy by means of PV panels forestalls the production of climate changing gases at some distant, fossil fuel-driven power plant. As Bill McKibben pointed out, this is the level of CO2 savings households in general need to make to save us from impending climate disaster.[10]

Measuring the value of ecosystem services is easy when it comes to technical installations like a PV system: the inverter will do it for the consumer who can then monetize his carbon credits, if he chooses. It is much harder to assess the benefits of sequestering carbon in perennial pasture or soil organic matter, even more difficult to ascertain the immeasurable benefits of sustaining wildlife with a pond or hedgerows. There is little, if any, public support for small farms in this regard in the U.S.

By contrast, in Norway, farmers are paid a public salary no matter what they produce. They are considered "culture workers" (*kulturarbeidere*) because of the significant contribution their farms make to the natural and socio-cultural matrix and quality of life in rural districts. A farmer may be raising strawberries, or making cheese, or whatever, but the profits from those enterprises come on top of the farmer's base salary. They are paid from the public purse (*jordbruksoppgjør* = Agricultural Settlement) for keeping the land open, protecting the water, and providing jobs, which means that the populations of rural districts and towns are kept intact. People are not forced to migrate to Oslo or other urban centers to find work, with all the negative consequences of that migration. The public taxes itself to incentivize farmers to become stewards providing ecosystem services for the common good.

On biodynamic farms ecological stewardship is the principle of self-organization. No one biodynamic farm is like any other; they are all highly individualized organisms. But every such farm organizes itself around the *quintessential* task to "heal the earth." If the farm veers off in one direction or another, it corrects itself, provided the farmer pays attention to the underlying principle of stewardship.

What we have been teaching to apprentices, interns, students and adults coming for workshops and farm tours, is that "responsibility"means what it says: "the ability to respond." It means that the farmer is able to observe the farm from the perspective of stewardship and act accordingly. Responsibility in the biodynamic context does not mean

keeping this law or that regulation, or abiding by this schedule or that convention. It means taking *quintessential* responsibility for a defined ecosystem. The farmer is fully aware that as humans we have disturbed the natural environment in disastrous ways and now must heal the disturbance and return to sustainable balance. "Disease" means being "not at ease." We get back to "ease," back into balance, by practicing the economics of stewardship as envisioned by the ancient philosopher.

The return to balance, which today is often labeled "regenerative agriculture," began with Rudolf Steiner's *Agriculture Course*, a series of lectures Steiner gave in Eastern Germany (now Poland) in 1924. After a century of chemical farming, agricultural soils in that region were depleted. Seeds and crops were failing, rates of animal reproduction becoming deficient, and people falling ill. In addressing these problems, Steiner pointed out two things: One was that farmers had depleted the soils by spreading chemicals toxic to soil organisms. More importantly, farmers had forgotten what biological life is. As agricultural producers increasingly confused farming with input-centered industrial processes, they forgot that they were involved in organic life processes. Steiner's initiative was to restore agricultural health by strengthening the biological life in the soil in keeping with cosmic rhythms. The term *biodynamic* derives from *bios,* meaning "life" and *dynamis,* meaning "rhythmic force." Basically, biodynamics is about the rhythms of biological life, which is what farmers had forgotten when they turned to chemical farming, said Steiner.

Whenever I attend an agricultural conference, I invariably hear: "How do I control disease, how do I control weeds, and how do I put fertility back into my fields?" The answer to these questions is found in what we have discussed. It is complex, but also very simple. Once the farmer is committed to the *quintessential* idea that he is primarily a steward imitating nature, and organizes his farm around that principle, everything else falls into place. We have very little if any disease on the farm, we have few weed problems., and we have good fertility in our fields and gardens. If you measure the NPK ratios in our soils, these are not particularly high. But you don't need high NPK levels if you have a living soil that is rich in organic life. The organisms in the soil — of which there are as many in a teaspoon of soil as there are people in the U.S. — harvest the fire, the solar energies, and convert them into carbohydrates, into sugars and all kinds of nutrients the plants need from the soil. We don't fertilize our plants at all. Instead, we feed our soils. We don't put any fertilizer on our broccoli. Instead, we make sure that the soil the broccoli is planted in is rich in life. And that takes care of everything. It takes care of the nutrient needs of the broccoli, and it takes care of establishing the *stasis*, the balance needed to protect the broccoli from disease. You don't have to cure the broccoli by spraying it with some chemical, because it isn't sick. Because it is whole. That is what "wholeness" and "holistic" mean: both of these words mean "health." That is why we farm for health, not for money. Money is a tool that we have to have because we have to pay bills, taxes and insurance, but it is not the goal, just a means to an end.

Farm Capitalization

In 1967, I got a teaching position at the University of Washington. Having been born on the Baltic Sea, the cooler growing conditions of the Evergreen State appealed to me more than the drier and hotter climate of California and other Southern states. I drove up and down I-5 looking for a piece of land on which to grow food for a future family. Seeing that the corridor between Vancouver, B.C. and Vancouver, Washington was developing into a continuous commercial strip, I found my way to Lopez Island. A banker at Fanny Mae (Federal National Mortgage Association) urged me to finance my farming dream by "keeping my day job," rather than taking out a longterm mortgage to pay for the land and infrastructure. Good advice, which we have followed ever since! A typical thirty year mortgage at the then current rate of over six per cent would have cost me three times the original loan. It took me three years of saving twenty-five per cent of my annual salary at the University to buy the original ten acres in 1970 without incurring debt.

Agricultural economists often recommend to beginning farmers to lease land rather than to invest cash, on the argument that land "very rarely pays for itself in farm-generated cash flow."[11] I learned differently from a friend who, prior to moving to Lopez Island, had bought and resold several small farms in states further east to earn enough of a stake to buy land here. He did, but still needed a mortgage to finance the purchase of forty acres on the island, and he continues to pay that mortgage today. Would he have been better off financially if he had delayed ownership and used his earnings as a professional orchardist to buy his current farm for cash? In other words, "kept his day job," paid cash for his land and enjoyed the dramatic increases in value due to the proximity of the San Juan Islands to major urban centers on the near mainland?

Over the years we bought up neighboring acreage and houses at ever increasing price levels, always paying cash from current income and savings, so that today we farm land that has multiplied in value many times over, without having to pay rent or lease fees. In the mid-seventies, we built our home for a cost that today equals its annual rental value, and have lived here for nearly fifty years without paying rent. We refurbished the houses we bought from former neighbors for use as rentals or apprentice housing. After retirement, we sold them to two families in exchange for labor, thus enabling us to stay on the farm for the rest of our lives. Similarly we built solid barns, outbuildings and infrastructure, mostly with our own hands, which return great value to the farm every year, without indebtedness.

Labor

While the family was growing its own food from 1970-1994, it never occurred to any of us to count the cost of our labor. Without giving it much thought, we ignored the fundamental principle of conventional economics that every choice has an opportunity cost. The idea behind opportunity cost is that the expense of an item is the lost opportunity to do or consume something else. In short, opportunity cost is the value of the next best alternative. The next best alternative for my wife and me would have been to work additional hours at the university, for instance, by teaching during the summer quarter, rather than working on the farm. With the extra money earned, we could have afforded the best food available in the market. Financially our gain would have been greater. However, the cost to the family's health, the health of our farm and to the ecosystem would have far outweighed any monetary advantage. Once we became aware of the conventional definition of farm profitability, we chose to ignore the opportunity costs riding on our investments in the land, buildings and infrastructure, and in our labor. The non-cash values produced by the farm were more important to us than the conventional measures of farm profitability.

The question of labor cost took on new significance when we started farm teaching programs in sustainable agriculture in 1999, the year Henning was appointed adjunct professor at WSU Center for Sustaining Agriculture & Natural Resources, and S&S Homestead was designated a WSU Demonstration Farm. In 2002, S&S Center for Sustainable Agriculture

incorporated as a non-profit teaching institution under the State of Washington. The appointment at WSU did not carry any salary, but it allowed us to apply for on-farm research and workshop grants and for tuition support for international and minority interns to earn academic credit at WSU for their internships. In 2004, the farm received a USDA grant to fund a year-long high school class on "Principles of Ecological Food Production." The class evolved into a permanent Farm-to-School curriculum that brought students of all grades to the farm several times a week. In 2012 Henning became a Mentor Farmer under the apprenticeship program offered by the Biodynamic Farming Association. In 2014, the farm was licensed as a state-approved milk processing and cheese making facility.

In teaching farm economics, we foregrounded the difference between biodynamic and conventional approaches. In articulating how to think about labor and its cost to the farm, we relied on our own experience as well as on the views of agrarian writers such as Gene Logsdon, John Ikerd and Jan Douwe van der Ploeg.

Logsdon, who homesteaded a remnant of his grandfather's thousand-acre farm lost to bank foreclosure, was a well-known critic of conventional farm economics. In defining "pastoral economics," he described the labor of the homesteader as "profit, not a cost as it is in industrial accounting." Because the homesteader doesn't have to pay for his own labor, the value of his work is an income to his farm.[12]

Ikerd, Professor Emeritus of agricultural economics at the University of Missouri, contrasted Aristotle's notion of the "economics of happiness" (*eudaemonia*) with the conventional concept of "economic wealth (as) a pursuit of individual, hedonistic, or sensory pleasure." Ikerd's definition of small-farm success links labor to service in the pursuit of "righteous living." Farmers who labor to "make a decent living while caring for the land and caring for other people, not only are building sustainable agriculture for the future, they are opening the doors to happiness."[13] This view of labor echoes the biblical notion that working the soil to serve its needs, is a form of worship.[14]

Douwe van der Ploeg, Professor of Transition Studies at Wageningen University, defined labor in the context of "new peasants" (in Peru, Italy and The Netherlands) struggling for resource autonomy and sustainability in the age of globalization. In this context, internalizing nature, independence from commodity markets, intensively skilled production, and life quality take precedence over labor costs. Labor is valued as the means by which to "get ahead," which is to achieve independence and social, rather than financial, wealth.[15] This view of labor is in close agreement with the biodynamic view.

"But you can't take that to the bank!" some of our apprentices would shout in frustration. Looking for transactional compensation for their work, they wanted to learn how to earn a living by entrepreneurial farming in a capitalist society, where success is measured by profits earned. How to provide fair labor compensation in a non-profit enterprise focused on health? Part of the challenge was to make clear that for the duration of their training, interns and apprentices shared farm ownership in the sense that the farm supported them with food, housing, instruction, guidance, land, machinery, tools, and infrastructure. It was the experience of the farm owners that trainees took as much of the farmers' time as they gave to the farm. They also cost the farm heavily in wear and tear on machinery and other infrastructure.

Every intern and apprentice was required to do assigned readings and keep a daily journal

answering three question: What did I do? What did I learn? And what does it mean in the context of the whole farm organism and the world beyond the farm? Academically inclined trainees had the option to spend half the day studying a prescribed curriculum and carry out farm-based research resulting in a (preferably publishable) paper. Entrepreneurially inclined trainees were given opportunities to pursue market-based activities of their own choice — such as running a vegetable CSA or raising pigs — and for their own risk and profit. The financial value of room, board, and the physical infrastructure made available to the trainees (plus a cash allowance to cover expenses for utilities), amounted to the equivalent of the average minimum wage in the U.S. However, this compensation (except for the allowance for utilities) was not paid in cash, but rather in the cash-equivalent value of the services and goods received. In order to make these values visible, we developed farm budgets that reflected both cash and non-cash value flows, showing labor and compensation as both income and expense. The apprentices could now see the values of incomes (and expenses) they created through their work, even though they could not "take it to the bank."

Associative Economics and Price

Another seminal idea of Rudolf Steiner's is "associative economics," which he distinguished from competitive economics, and focused on the notion of "true price." How do biodynamic farmers determine the price of their products? Steiner held that all workers should receive sufficient remuneration for any commodity to meet their needs until they have produced another such commodity. In other words, price should be a reflection of social need and responsibility. By contrast, the conventional notion of price proceeds from the assumption that price is the reflection of supply and demand. There is no mention of social need or responsibility. It's simply a question of what's available and what will the market bear. What price will suppliers accept and what is the remuneration food purveyors are willing to pay farmers?

As a result of this market-based principle, farmers today are squeezed two ways: they have little or no control over the price of the goods they sell to the wholesale market; nor do they control the price of supplies they buy from the market. Most small food producers today don't enjoy the affluence of earlier generations of farmers, who depended mostly on community-based, local markets and on their own self-sufficiency. Until WWI, farmers in the U.S. and in Europe were largely self-sufficient. They produced the food their families ate, the fodder for their animals and the fertility of their fields as a matter of course. They didn't import their food, feed, or fertility from some other place; they produced it on their own, which meant that they had control over their finances in a way most farmers today can't even imagine.

Biodynamic farmers think about price in terms of their role as stewards rather than as competitors. All workers should receive sufficient remuneration for a commodity to enable them to produce another such commodity. If you sell your apples or your beef to your community, you should receive enough money that you can produce another apple or another cow, and do so sustainably. That should determine the price. Actually, this makes common sense, but the competitive market doesn't work that way. There is a whole lot of practical idealism working in biodynamic economics. It's all very concrete and down to earth.

Steiner argued that in a global economy meeting one another's needs, and the needs of the environment, should be managed by coordination of producers, distributors, and consumers through industry and consumer associations, rather than by the "invisible hand" of capitalist markets, or by socialist governments.[16] While Steiner's idea of an "altruistic stakeholder-managed economy"[17] has not been realized on a global scale, the idea of associative economics has inspired significant initiatives such as community land trusts,[18] community supported agriculture (CSA),[19] social finance,[20] and local currencies.[21]

For the biodynamic farmer, associative economics works best in the context of personal relationships, that is, in community. In commodity markets, pricing is pretty much out of the hands of the producer, but if growers sell directly to neighbors through a CSA, farm stand, farmers' markets, and food hubs, they have the choice between setting the price at the level of what the market will bear, or at a level of social and ecological responsibility. For example, the pandemic has thinned the ranks of mechanics, construction workers and repair providers on our island. Some of them have responded to market opportunity by doubling or even tripling the prices charged for their services. Likewise, livestock replacements such as chicks — on which small-scale egg producers must rely because most chickens by now have had the brooding instinct bred out of them — during the pandemic increased the price ten-fold in response to growing demand for home-based food security.

This development corroborates the startling assertion by agricultural economists that local food production is rapidly becoming an unaffordable "lifestyle choice." Their analysis ignores the impacts of market-based food systems on human health and ecological survival.[22] On Lopez Island, the upward pressure on price has been exacerbated during the pandemic by the influx of financially advantaged investors, who compete for land, housing, services and food with local consumers.

A complaint that constantly arises in producing and selling organic produce is that such food is too expensive for anyone but the well-to-do. Rudolf Steiner made the case that nutritional value depends on the "life force" in the food. By contrast, the conventional market is based on the assumption that what American consumers want is convenience, low price, and availability of foods in all seasons. In other countries, for example, in Scandinavia, people budget nearly thirty per cent of income for food. Therefore the quality of their diets and health is much higher. Ideally, biodynamic producers charge prices that provide nutritionally whole ("enlivened") food for everyone in the community, while practicing ecological stewardship and maintaining the financial viability of the farm.

Farm Budgets

Our annual farm budget projections articulate the practice of *oikonomeia* (household stewardship) by making the economic values flowing through the farm visible in monetary terms. Over the years, we have learned to distinguish on-farm cash and cash-equivalent value flows. Everything that we produce on the farm is assigned a monetary value so that we can tell what a product would be worth in today's market-place, if we were to sell it there. However, we sell only a limited portion of what we produce, and consume the larger share at home in the form of food, feed, fertility, water, energy, wood products, land, infrastructure, housing and, not least, labor. Whatever is not sold in the market still has monetary value to the farm as essential resources. If you don't have to pay cash for those resources, they constitute cash-equivalent farm income. If you don't buy your sausage because you produce it yourself, it represents monetary value that is determined by comparing prices in the local market for organic sausage. If you don't pay for electricity or housing because you produce them at home, the monetary value of those resources constitutes cash-equivalent farm income. Likewise, as these products are consumed on the farm, they constitute non-cash farm expenses, while monetary outlays for taxes, insurance, or supplies constitute cash expenses. The purpose of our farm budget is to estimate, plan, control, and adjust cash and non-cash value flows as necessary to achieve a balanced farm economy.

It follows that the farm budget distinguishes between three distinct levels of income and expense: Cash incomes and expenses, Cash-Equivalent of Consumables, and Cash Equivalent of Lease Value.

Based on past experience, the budget for 2017 projected a total production value of $436,670. Cash income was estimated at forty-three per cent of total. Cash-equivalent value of products consumed on the farm during the year

Cash Income	Cash Expenses
Working Capital $40,000	Working Capital 40,000
Farm Sales $34,500	Farm Development $28,000
Educational Programs $45,500	Amortization $7,000
Rentals & Farm Stays $46,500	Supplies & Services $30,000
Social Services $24,000	Repairs $30,000
Cash-Equivalent Consumables	Machinery & Tools $12,480
Food $37,200	Labor Hired $29,000
Feed $18,000	Insurance $4,650
Fertility $9,850	Taxes & Licenses $4,670
Livestock Replacements $5,470	Travel, Accounting, Communication $4,700
Wood Products $6,560	**Cash-Equivalent Consumables**
Water $2,460	Farm Prod. Consumed $82,000
Electricity $2,460	In-Kind Labor $93,000
Cash-Equivalent Lease Value	**Cash-Equivalent Lease Value**
Infrastructure $42,000	Infrastructure $42,000
Housing $24,000	Housing $24,000
Land $4,670	Land $4,670
Total Income $436,670	**Total Expenses** $436,670

S&S Homestead Budget 2017

was projected at forty-one per cent. Cash-equivalent of the lease value of infrastructure was assessed at sixteen per cent.

Income

~ Projected cash income for 2017 included *working capital* (nine per cent), which is the farmers' original investment rolled over from year to year to provide farm cash flow without a bank loan; *farm sales* (eight per cent): CSA, custom meat, farmers' market and farm gate sales; *educational income* (ten per cent): farm-to-school programs, farm tours, and workshops; *rentals and farm stays* (eleven per cent)*; and *social services* (five percent): elder care.

~ Projected cash-equivalent income from products to be consumed on the farm (*cash-equivalent consumables*) included: f*ood* for ten to twelve persons (nine per cent): meat; eggs; dairy: milk, cheese, butter, yogurt; fresh & processed vegetables; staples: potatoes, root crops, squash, onions, garlic; fresh & processed fruit; processed grain (wheat & rye)

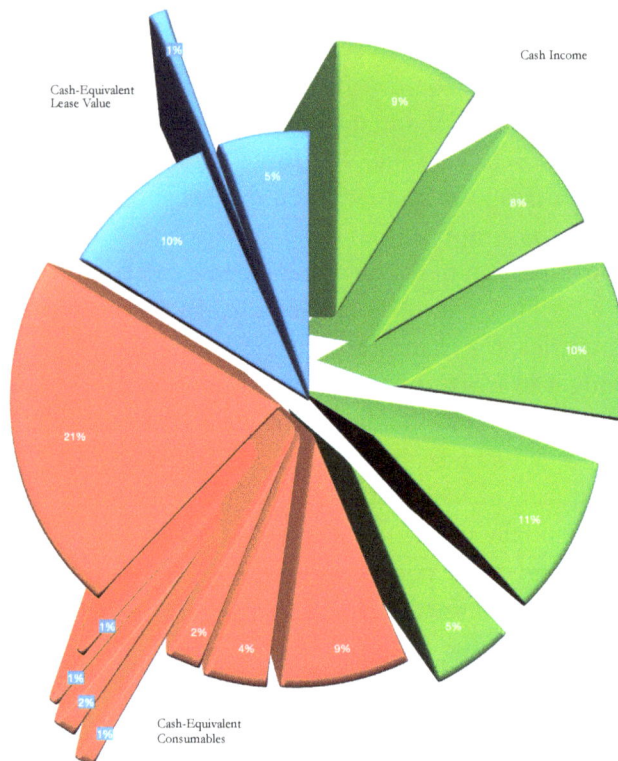

Income Projections by Percentages

for bread; *feed* (four percent): forages, hay & grain for one beef bull, six cows, six yearlings & six calves; one dairy bull, two cows, two yearlings, two calves; one ram & two dozen sheep; three pigs; three dozen chickens; *fertility* (two per cent): manures, composts, mulch, compost teas and BD preparations; *livestock replacements* (one per cent): calves, lambs, chickens; w*ood products* (two per cent): lumber, firewood, wood chips and sawdust; *water* (one per cent): output of water catchment system & three wells; *electricity* (one per cent): output of Pv system, water pre-heater, solar pumps; and i*n-kind labor* (twenty-one per cent): two farmers and five trainees.

~ Projected cash-equivalent lease value included: *infrastructure* (ten per cent): two barns, three greenhouses, woodshop, dairy facility and processing kitchen, water and PV systems, fencing, driveways, machinery and tools; *land* (one per cent); and *housing*

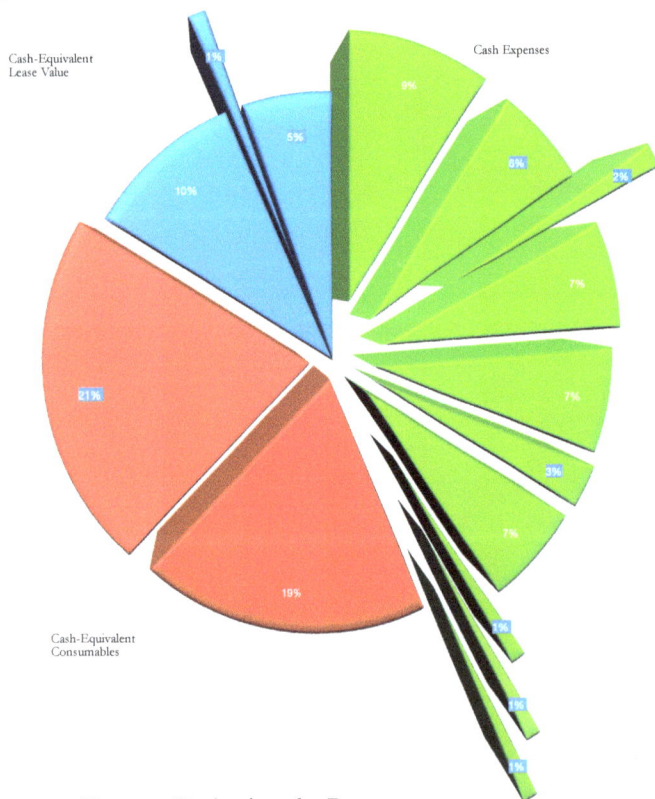

Expense Projections by Percentages

214

(five per cent): owner and trainee housing.

Expenses:
~ Projected cash expenses for 2017 included: *working capital* (nine per cent) set-aside for 2018; *farm development set-aside* (six per cent); *amortization set-aside* (two per cent); *supplies & services* (seven per cent); *repairs* (seven per cent); *machinery & tools* (three per cent); *labor hired* (seven per cent): machine work, butchering, etc; *insurance* (one per cent); *taxes & licenses* (one per cent); *travel, communication, accounting* (one per cent).
~ Projected cash-equivalent expenses of consumables included: *farm production consumed* (nineteen per cent): food, feed, fertility, livestock replacements, wood products, water, electricity; *in-kind labor* (twenty-one per cent): two farmers, five trainees.
~ Projected cash-equivalent lease value included: *Infrastructure* (ten per cent): two barns, three greenhouses, garages & storage buildings, dairy & processing kitchen, wood shop, water catchment system & three wells, PV system, water pre-heater & solar pumps, fencing, driveways, machinery (truck, two tractors, haying, cultivation & seeder equipment) & tools; *land* (one per cent): cropland, pastures, building sites, drives; *housing* (five per cent): owners' home, trainee house; two farm-stay houses).

Conclusion

We wrote our last complete farm budget in 2017, the year before we retired. We continue living on the farm and grow our food with the help of two young families now residing on the farm permanently, exchanging their labor for equity in land and housing. Retirement meant that we gave up our educational, social service and rental programs, while continuing farm production on a reduced scale to meet our own needs, with minimal sales to the market.

The purpose of the farm budget was to plan for financial balance, which is an important aspect of overall farm health. The budget identifies farm resources and their economic functions. By assigning a monetary value to every product, including those not sold on the market but consumed on the farm itself, the budget provides an objective measure of farm productivity.

By USDA standards, S&S Homestead was barely a commercial farm because the total annual value of marketed food products was less than $50,000. This did not include income from educational programs, rentals, farm stays, and social services, which was four times greater than the sale of foods. In other words, much of the financial viability of the farm was based on non-food production. However, even the modest amount of meat, eggs, dairy products, vegetables, fruit and staples provided by S&S Homestead Farm made a substantial contribution to the food security of about fifteen households (representing about sixty end consumers). This means that if there were about fifty Lopez farms of the same scope as ours, local producers could provide much of the food consumed on the island. Nor would it take as much land as we steward on our farm. For years, we provided a twelve-member CSA with fresh vegetables from a quarter-acre garden and with berries and tree fruits from an orchard the same size. The production of meat, dairy products and grain, of course, required more land in pasture, hayfields and crop fields.

The budget also makes clear that the economic viability of the farm does not primarily rely on cash sales, whether of food, educational programs or social services. Fully forty per cent of farm production is consumed at home on the farm. The budget recognizes that while the production for on-farm use of food, feed, fertility, animal replacements, wood products, water, and electricity does not result in increased sales, it has cash-equivalent value to the economy of the farm. Accordingly, this value has been expressed in the budget as farm income. Similarly, the budget recognizes that the acreage on which we farm, the housing we live in, and the infrastructure without which the farm could not function, such as machines, tools, barns, greenhouses, shops and other service buildings, driveways, irrigation pond and water catchment systems, wells, PV system and other systems to capture solar energy for water heating, for example, all have major economic value captured in the budget by counting as income the equivalent of what it would cost to lease the land, buildings and required infrastructure. Because the farmers from the start took the banker's advice to "keep our day jobs" and avoided going into debt while capitalizing the farm, the budget has never been encumbered by liabilities from mortgages or other loans. Instead the farm has benefited from available structures and resources at no financial cost other than routine maintenance and repair, which benefits are budgeted as farm income.

Also reflected in the budget is the ecologically based principle of not extracting profit from the enterprise, and instead investing annual surpluses to build farm capital. In 2017, the projected margin of surplus was estimated to be six per cent of the total budget. This surplus was to be set aside in an account for "Farm Development," from which we finance infrastructure improvements, such as the PV system installed in 2012. The system produces most of the energy the farm needs, while keeping an average of thirty tons of carbon dioxide annually from spewing into the atmosphere at a distant power plant. It usually takes several years to accumulate enough funds in the "Farm Development" account to finance major projects without having to resort to a bank loan.

The most unconventional feature of the budget is the classification of labor as farm income; both the farmers and the farm trainees contribute their work without cash compensation. Instead, the budget recognizes labor as value flowing through the farm, for which all workers are compensated in kind with food, housing, learning, entrepreneurial opportunity, and a high quality of life.

Our decision to "retire" on the farm in 2018, the same year Elizabeth retired from her "day job" as a public school teacher, had major budgetary consequences for the farm. Retirement meant giving up the educational programs, rentals, farm stays, and social services. We shrank farm production by eliminating the beef herd and cutting the dairy herd and sheep flock in half, thereby reducing the need for hay production. We converted the housing built for interns, apprentices, and farm-stay guests into homes for two families who are earning equity through their labor on the farm.

Essentially we have reduced the farm to the scale where it can provide for our food and other needs with the help of younger workers. In turn their needs for land and housing are met by sharing in the capital gains made over the half century of our tenure on the homestead. The continued goal is to "farm for health" to the benefit of the farmers, the community and the ecosystem. We continue to cherish the hope that some day soon the community will feed itself from food grown on the island.

As quoted in *Hands at Work* (2009) and again in *Bounty: Farmers, Food & Community* (2016): "Our dream is the community will feed itself. The only question people will ask about their food is which of their neighbors' farms it came from. We believe we're the future."[23]

Notes

_____

[1] The views on biodynamic economics articulated here were first presented in the article "The Economics of the Small-Scale, Self-Sufficient Farm" in *Stella Natura,* 2013, and presented at the 2014 Biodynamic Conference. Updated with economic data from 2017. A version of this essay was published in *Lilipoh: The Spirit of Life*, Summer 2021, 36-57.

[2] Tillich, Paul 1953-63. *Systematic Theology.* Chicago, Il.

[3] For a detailed discussion of Aristotle's derivation of the term economics (οικοσυστημα) from the noun "(nature's) household" and the verb "to steward," see the essay "The Spirituality of the Soil: The Idea of Teleology from Aristotle to Steiner," below.

[4] Sheldrake, Rupert et. al. 2001. *Chaos, Creativity and Cosmic Consciousness.* Rochester, VT, 17.

[5] See Jantsch, Erich. *The Self-Organizing Universe: Scientific and Human Implications of the Emerging Paradigm of Evolution.* New York, NY. Jantsch views the unifying paradigm of self-organization as guiding all interactions of micro-structures and ecosystems within the entire biosphere and the macrocosm.

[6] U.N. Millennium Ecosystem Assessment 2005, quoted in Davis, Ellen F. 2014. *Scripture, Culture, and Agriculture: An Agrarian Reading of the Bible*, Cambridge, UK, 2.

[7] Op.cit, 9.

[8] Reeve, Jennifer R., et. al. 2011. "Sustainable Agriculture, a Case Study of a Small Lopez Island Farm,"*Agricultural Systems*, 104: 572-579.

[9] "More Local Power: S&S Homestead Farm on Lopez Island.2012. Posted in: *Energy Efficiency & Conservation, Membership Programs, Opalco.* (https://www.opalco.com/4873/2012/09/. Retrieved July 15, 2021.

[10] McKibben, Bill 2012. "Global Warming's Terrifying New Math," *Rolling Stone. https://www.rollingstone.com/politics/politics-news/global-warmings-terrifying-new-math-188550/.* Retrieved July 8, 2022.

[11] Ekarius, Carol 1999. *Small-Scale Livestock Farming: A Grass-Based Approach for Health, Sustainability, and Profit.* North Adams, Mass, 147.

[12] Logsdon, Gene 1995. *The Contrary Farmer*, White River Junction, VT, 27.

[13] Ikerd, John 2008. *Small Farms Are Real Farms: Sustaining People Through Agriculture*, Austin, TX, 134-136.

[14] Davis, op.cit, 28ff.

[15] van der Ploeg, Jan Douwe 2008. *The New Peasantries: Struggles for Autonomy and Sustainability in an Era of Empire and Globalization*, London & Sterling, VA, 35, 43-45, 114.

[16] Steiner, Rudolf 1993. *Economics: The World as One Economy.* Chartham, UK.

[17] See Lamb, Gary 2010. *Associative Economics: Spiritual Activity for the Common Good.* Ghent, NY, 145. See also Karp, Robert 2007. "Toward an Associative Economy in the Sustainable Food and Farming Movement," *New Spirit Ventures. https://www.biodynamics.com/content/toward-associative-economy-sustainable-food-and-farming-movement-robert-karp.* Retrieved July 8, 2022.

[18] In the U.S. there are over 225 community land trusts which are nonprofit organizations governed by a board of CLT residents, community residents and public representatives that provide lasting community assets and shared equity homeownership opportunities for families and communities.

[19] According to USDA data, 7,398 farms in the US sold products directly to consumers through CSA, accounting for 7% of the $3 billion in direct-to-consumer sales by farms.

[20] RSF Social Finance, for instance, provides funding to social enterprises in the U.S. and Canada that are working to create long-term social and ecological benefit. https://rsfsocialfinance.org. Retrieved July 8, 2022.

[21] Community currencies in most U.S. states play a role in better valuation of environmental resources and providing an incentive for more sustainable behavior.

[22] Blank, Stephen 1999. "The End of the American Farm?," *The Futurist*, 22-27. See also Blank 1998. *The End of Agriculture in the American Portfolio*, Westport, CT.

[23] Graville, Iris (story) & Summer Moon Scriver (photography) 2008. *Hands at Work - Portraits and Profiles of People Who Work with Their Hands*, Flemington, NJ, 81; Graville, Iris (profiles), Robert Harrison, Steve Horn & Summer Moon Scriver (photography); Kim Bast (recipes) 2016 . *Bounty — Lopez Island Farmers, Food, and Community*, Lopez Island.

Retiring on the Commons
Henning Sehmsdorf, 2021[1]

Generally, the idea of the commons is defined as "the cultural and natural resources accessible to all members of a society, including natural materials such as air, water, and a habitable earth."[2] The idea of the commons, as it is applied today, is most closely related to the counteracting of the social inequities arising from exploitive private capital by returning value to those who have contributed to the commons in one way or another. This essay discusses the concept of the commons as it applies to the following question: how can landowning or non-land owning farmers who pass their farms to succeeding generations obtain and receive security in retirement?

The owners and farmers of S&S Homestead Farm — my wife now in her seventies, and I in my eighties — are currently pondering the means by which we might retire without abandoning our lifelong commitment to the commons. The conventional thing to do would be to sell the farm — its market value now probably more than a couple million dollars — and "go and play shuffleboard in Florida." Another option would be to optimize our investments here, rent out the houses and hire help to maintain our life on the "farm estate" without the burden of working the farm in its entirety. Neither of these two options is attractive to us, however, because in either case, the farm would cease to exist as a self-supporting, ever evolving organism providing a livelihood for people, animals and plants to its fullest potential. Without like-minded successors to work and live on the farm, it would decline into a consumable resource to be used up in support of our retirement.

Over the last decade, a new (or rather, an old) perspective on intergenerational farm succession has emerged in public discourse, namely the idea of the commons. In *Utopia* (1516), Thomas More suggested the legal establishment of "commons" as a way to help feudal farmers economically disadvantaged by the conversion of land (which traditionally was available for communal use) into private land for the exclusive use by wealthy landowners.[3] More's ideal society hinged on the abolishment of private property. In his utopian vision, farmers did not own the land they worked;

instead they were "cultivators who come in succession to live there." More's conceptualized utopia was not met with success, however. On the contrary, between 1604 and 1914, the British Parliament passed 5,200 individual enclosure acts creating private property rights to nearly seven million acres of agricultural land previously considered held in common.[4]

As cultural anthropologist Sonya Salamon points out, privileging private over common ownership of agricultural land in the U.K. had far-reaching consequences beyond the British Isles.[5] She found that many of the English-surnamed, large-scale farms in the American Midwest were established by immigrants from England who had lost their holdings under the enclosure laws, but came to the U.S. to establish privately held farms which perpetuated the same ethos of private land ownership. By contrast, German-surnamed, small and midsized farms in the Midwest were frequently established by religious communities, among them the Amish, who emigrated from Europe in search for land. While succession of the former group of farms typically was by sale within or outside the family, succession in the latter group was by inheritance within the family and within the community.

Ownership of the land in so-called "covenant communities" tended to be stable and inter-generationally permanent. Historically, the idea of the commons was frequently tied to various economic arrangements by which "commoners" support the elderly in retirement. In *Agrarian Justice* (1797), Thomas Paine, rather than promoting the abolishment of private property, supported retirement payments to be made to farming populations traditionally living on the commons to compensate for the "loss of his or her natural inheritance by the introduction of the system of landed (private) property."[6]

In pre-industrial agrarian Europe, support of the elderly from privately or commonly held land — which was to be passed on to the succeeding generation — usually took the form of traditional means of compensation, which eventually became part of public law. Before industrialization, support of the elderly was encoded in informal tradition; when the traditional ways of life were being replaced by industrialization, support of the elderly was encoded in the law. In Norway, for example, the term *føderåd* (means of livelihood) referred to contractual entitlements in the form of food, on-farm housing, and other services received by retiring landowning or tenant farmers from the new farm owners.[7]

In Greece until quite recently, the *gerontomoiri* (old-folks' share or provision) included not only shelter and food, but also daily care such as washing the clothes of the elderly.[8] Similarly, in Germany, the *Altenteil* (old folks' share), as defined by law in 1900, spelled out in legal terms the benefits and obligations incumbent on a property owner to provide for the entitled *Altenteiler* (old folks shareholders).[9] Indeed, that legal custom is still current, as exemplified by my brother-in-law having inherited the family farm in northern Germany. The terms were such that he would inherit the farm at the time of his marriage; his parents would move to an attractive villa (owned by the farm) on the edge of the village, where they were to be supplied with a pension that included food from the farm and certain services.[10] The same inheritance terms applied to my nephew (my sister's oldest son), who claimed the farm at the time of his own marriage. My sister and brother-in-law, then in their early sixties, moved into the same villa on similar terms, repeating the succession custom that had been practiced ever since the farm was established by his family in the 1700s.

In the United States, similar practices survived in immigrant communities in the nineteenth century and later. For example, as long as

Norwegian settlers lived in their own communities separated from the American mainstream, the native custom of contractual care of elder farmers passing their land on to the next generation continued to prevail.[11] The custom is still in use among the Amish, where parents retire as early as in their middle forties or fifties and transfer control of the farm to their adult children, in exchange for being cared for the rest of their lives. It might be said that to the Amish, even today, their idea of ownership overlaps with that of the commons, and their inter-generational succession traditions and practices are less focused on preserving private property rights than they are on ensuring the continuity of shared values and community identity.[12]

In a talk he gave at the Caux Forum for Human Security, near Montreux, Switzerland in 2011, David Bollier — American activist, policy strategist, writer and Senior Fellow at the Norman Lear Center at the USC Annenberg School for Communication — described the commons as "a self-organized system by which communities manage resources (both depletable and replenishable) with minimal or no reliance on the Market or State," but predicated on the obligation that "the wealth that we inherit or create together must pass on, undiminished or enhanced, to our children."[13] The communal inheritance practices among the Amish involving secure retirement for the elders, leaving the farm to their own children or other community members, accomplish that goal.

In 2016, the Lopez Island Land Trust established a new organization in our county, Lopez Island Farm Trust (LIFT), designed specifically to hold farm land in San Juan County in trust as a commons in order to build and secure a thriving local food system and rural culture in the county, forever.[14] The owners of S&S Homestead Farm have proposed gifting the bulk of the farm (two

thirds of the acreage and all the infrastructure) to the community commons through LIFT, while securing their own retirement, and that of succeeding generations, in perpetuity. We consider ourselves "cultivators" in the spirit of Thomas More, "who have come to the land in succession to live (here)," and "economists" in the spirit of the ancient philosopher Aristotle, who coined the term by combining the noun *oikos* (meaning household) with the verb *nemein* (meaning to steward).[15] As Aristotelian economists we have endeavored to steward the land "in imitation of nature as a mother provides for her children,"[16] mindful of not only our needs, but also of those of the community and future generations.

Teresa Opheim, Senior Fellow at Renewing the Countryside, an organization dedicated to strengthening rural communities by assisting in environmental and economic planning, has asked aspiring and established farmers in many places how important land ownership was to their enterprises and to agriculture in general. Some argued that ownership from the start was important "so that a farmer can put up buildings and infrastructure to support production and build equity in an appreciating long-term asset."[17] Others warned against locking up dollars in land ownership that could be invested more profitably in machinery and other current costs. We were impressed by the sentiment linking land ownership to ideas first introduced by Thomas Jefferson and promoted today by agrarian writers such as Wendell Berry, Aldo Leopold, Fred Kirschenmann, and Barry Lopez: "It is important for many small farmers to own small farms. It is what our democracy is based on, and it is the underlying foundation of our republic. It develops character … (and) makes it more difficult to pack up and leave when the going gets tough."[18]

For us, one of the means by which to pass ownership and stewardship responsibility for the land to the next generation, while securing

our own retirement, has been to settle two
young families on the farm by offering them
equity for housing and about one-third
of the farm acreage in exchange for labor. The
contracts with the two young families
(including five school-age children) will be
fulfilled in eight and nine years, respectively, at
which time the tenure of the present farm
owners will most likely have run its natural
course. It remains to be seen how they and the
Lopez Island Farm Trust will hold the
remaining two-thirds of the farm as a
community asset in the future.

Notes

---

[1] A version of this essay was published in *Biodynamics*, Fall 2021/Winter 2022, 21-24.

[2] N.a. 2022. "Commons."https://en.wikipedia.org/wiki/Commons. Retrieved June 22, 2022.

[3] More, Thomas 2006. *Utopia*. San Diego, CA, 59.

[4] Carroll, William C. 1994 "'The Nursery of Beggary': Enclosure, Vagrancy, and Sedition in the Tudor-Stuart Period," *Enclosure Acts: Sexuality, Property, and Culture in Early Modern England*, eds. Richard Burt and John Michael Archer, Ithaca & London, 34–47.

[5] Salamon, Sonya 1992. *Prairie Patrimony: Family, Farming, and Community in the Midwest*. Chapel Hill, N.C., *passim*.

[6] Paine, Thomas 2007. *Common Sense for the Modern Era*. Eds. Begler, Elsie and R. F. King. San Diego, CA, 215.

[7] N.a. "*Føderåd*." The Norwegian Tax Administration. https://www.skatteetaten.no/en/rettskilder/type/handboker/skatte-abc/2020/føderad/. Retrieved June 22, 2022.

[8] N.a. "*Γεροντομοίρι*." Slang Greek. https://www.slang.gr/definition/28189-gerontomoiri. Retrieved July 15, 2021. See also Evandrou, Maria and Jane Talking 2005. "Demographic Changes in Europe: Implications for Future Family Support for Older People," in *Aging Without Children: European and Asian Perspectives on Elderly Access to Support Networks*, eds. Kreager, Philip and E. Schröder-Butterfill. Oxford, UK.

[9] N.a. "*Altenteil*." https://de.wikipedia.org/wiki/Altenteil#Rechtsgrundlagen. Retrieved June 22, 2022.

[10] N.a. "*Auszugshaus*." https://de.wikipedia.org/wiki/Auszugshaus. Retrieved June 22, 2022.

[11] N.a. "*Føderåd* or *kår*," *Norway-Heritage*. http://www.norwayheritage.com/snitz/topic.asp?TOPIC_ID=5109. Retrieved May 12, 2021. See Sandy's post (15/06/2011): "My ancestors settled in Wisconsin in 1850. The oldest son to immigrate entered into contracts to purchase 120 acres 1853. In 1861 he sold the contract to buy the land to his brother, Elias, for $300. Elias borrowed the $300 from his parents. I have found a contract between Elias and his parents stating the 'condition of their (Elias, his heirs, executors, and administrators) obligation...for and in consideration of a competent sum to him (Elias) in hand paid by (the parents),' basically that Elias will maintain his parents with 'meat, drink, clothes, and all other things, necessary and convenient.'"

[12] Wesner, Erik J. 2015. "Old Age in Amish America." https://amishamerica.com/old-age/. Retrieved June 22, 2022. Among the Amish, the residence provided to the retiring farmers under the terms of the inheritance contract, is referred to as *Daudihaus* (grandpa house), corresponding to the German *Auszugshaus* (retirement house).

[13] Bollier, David 2011. "Commons — Short and Sweet." http://bollier.org/cp\commons-short-and-sweet. Retrieved July 15, 2021.

[14] N.a. 2016. "Lopez Island Farm Trust (LIFT)." Lopez Community Land Trust. https://www.lopezclt.org/lopez-island-farm-trust-lift/. Retrieved June 22, 2022.

[15] Aristotle, 4th century B.C., *Πολιτικά* (Politics*)*, I, iii, 1258a, 34 -1258b, 8.

[16] Ibid.

[17] Teresa Opheim. "How Important Is It for Beginning Farmers to Own Farmland?" Sustainable Farming Association. (https://www.sfa-mn.org/how-important-is-it-for-beginning-farmers-to-own-farmland/). Retrieved May 17, 2021.

[18] Ibid.

Part VII. World View & Holistic Science

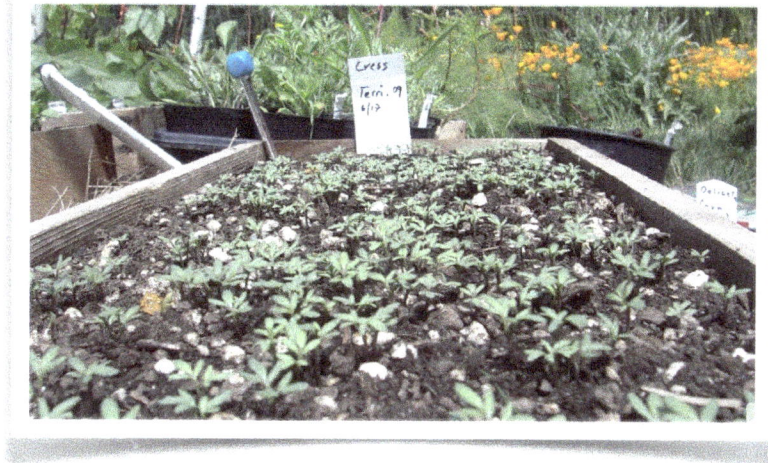

Biodynamic Farming and Goethe's Concept of Plant Morphology

Kelley Palmer-McCarty, 2009

Beginning to Understand Biodynamic Farming

I had never farmed so intensively until I set foot on the soil of S&S Homestead Farm on Lopez Island, Washington, this June. My previous experience was limited to a few hours a week for a high school agriculture course four years ago, and a job on a lettuce farm for the past two summers where I was primarily involved in harvesting and processing. Despite my lack of hands-on experience, I had been exposed to many discussions centered on different farming practices. These discussions taught me — in bits and pieces — the differences between conventional and organic farming.

A simple answer identifies conventional agriculture with the use of pesticides and synthetic chemicals to coerce the plants to grow, grow, grow! By contrast, organic agriculture identifies with the prohibition of such chemicals, turning instead to more "natural" remedies, but is often based on the same focus on economic production, although by definition organic farming's aim is to "promote and enhance biodiversity,

biological cycles, and soil biological activity."[1] You may notice today that while a growing fraction of our country's main cash crops are being grown "organically" or without chemicals, they are still produced in an intensive manner such as mono-cropping that does little to encourage biodiversity or soil activity. Biodynamics, however, was an expansion of my ideas of organic farming that I thought I had a sense of, but could never quite explain to anyone who asked.

It took three months working on S&S Homestead to even begin to be able to describe it. Now I always start by saying, "Let me give you an example." I have heard this example numerous times from both Henning Sehmsdorf and Elizabeth Simpson, the two incredible people who have owned and managed the farm for over forty years. I heard this first four years ago when I was in high school in the agriculture class that went to the farm several times each week, and I heard it again this year, but it is an example that made an impression on me I will never forget and I find it very handy in introducing strangers to biodynamics.

Say there is a plant that is infested with aphids. The conventional farmer would see the bugs and immediately spray them with a pesticide. The farmer would repeat this application whenever the aphids appeared again, spraying more and more as the insects developed immunities to the pesticide. An organic farmer would see the aphids and find an appropriate natural, organic remedy for the pests that could be applied immediately. A biodynamic farmer, however, would first ask the question: "Why are the aphids here?" The farmer would look for the root cause of the infestation because to the biodynamic farmer aphids indicate that the plant is stressed — that it needs something. By the farmer's healing the plant by amending the soil conditions, the aphids would most likely diminish or disappear because the plant was no longer weakened.

These responses demonstrate the drastically different ways of thinking behind the different methods of farming. On one hand you have a quick fix for a problem with insufficient thought to the possible consequences for soil or human health, while on the other hand you have a carefully considered long-term solution that aims to create an environment that is self-sustaining and healthy, but also a solution that may not be effective right away.

Yet this only scratches the surface. Being able to ask, "Why?" is only one facet of biodynamic farming. Biodynamics is also about awakening our senses to things and ideas in this world that we have forgotten how to see. Biodynamics is a life stance, a unique worldview that is not as common or easy to accept as it used to be. It is about reopening our senses to the forces that permeate the universe. It is about harnessing these forces to create an environment that can sustain and provide for itself.

The beginning of the biodynamic movement was in response to some very troubling trends showing up on farms in Germany in the early 1920s. Strains of crops that traditionally could be grown on the same plot of land year after year began to weaken and stop producing. Farmers were forced to turn to new strains, but these were also weak and had to be replaced with new strains frequently. Animal fertility was decreasing, and certain diseases were becoming commonplace. The farmers turned to Dr. Rudolf Steiner for the answers, a leading name in the world of anthroposophy. One of the most fundamental texts for biodynamics is the *Agriculture Course*, a series of eight lectures Steiner gave at a workshop in 1923, two years before his death.[2]

Before proceeding any farther into the details of biodynamics, it will be useful to explain the basis of anthroposophy or what Steiner termed *Geisteswissenschaft,* usually translated as Spiritual science.

It can be argued that in translation, the meaning of the term has changed from what Steiner originally intended. During a discussion with Henning one afternoon, we clarified that *Geist* can mean many different things: *Geist* is reason and conscious thought; it is an inner tendency; it is a life force; it is a soul, a ghost, or Spirit. *Wissenschaft* literally means "knowledge-ship," or "systematic inquiry in academic disciplines such as the social sciences, humanities, logic, mathematics and physics."[3] Altogether, *Geisteswissenschaft* comprises the "totality of all sciences whose objects of study are the different aspects of culture and of intellectual and Spiritual life."[4] In the U.S., we consider this the humanities. In contrast to other branches of scientific inquiry, the humanities (*Geisteswissenschaften*) examine all aspects of human existence from the physical to the Spiritual. Steiner considers *Geisteswissenschaft* to be a necessary complement to natural science, which is more narrowly based on "knowledge depending on mathematical proofs and exact measurements (*exakte Wissenschaft*)."[5] In Steiner's view, Spirit (*Geist*) should be studied with the same rigor as matter is in natural science, and many of the findings of *Geisteswissenschaft* are based on and supported by natural science.

As I have come to understand it, biodynamic farming and the thinking and philosophy that inform it require an open mind. It requires the ability to see all the different influences and forces at play in the world, and to work with these to sustain ourselves and our environment.

In the preface to Steiner's *Agriculture Course*, Pfeiffer explains that a sick plant is not "diseased in a primary sense" because the health of the plant is a reflection of the health of its environment, especially the soil.[6] Over time the life forces have decreased in the crops we grow because we have lost the traditional ways of farming that kept us connected to the rhythms and forces that influence the organisms we care for. In modern reductionist science, these forces and influences have fallen by the wayside, and Steiner explains that because of this, the life forces in our food are diminished. In turn, so are we.

In Steiner's first lecture he explains that one reason our agriculture has suffered is because "modern Spiritual life" has become destructive and narrow-minded. He explains that "opposing forces in our time are all too numerous, preventing one from calling forth a proper understanding."[7] The lack of communication between branches of modern science — the disconnect between, for example, economics and soil biology — has wreaked havoc on the quality of the products now sold in our markets.

The forces Steiner speaks of are found on earth but come from the cosmos. He describes the instinct behind traditional farming as an inherent human response to these forces over many generations, until recently. Traditional farmers formed instinctive relationships with the rhythms of the seasons, sun and moon, planets and stars. Steiner describes the forces that work through these mediums. For example, the "silicious force" working through silica (i.e. quartz) is influenced by the outer planets, Mars, Jupiter, and Saturn. It plays a large role in plant shape and development. Complementing this force is the limestone process that is influenced by the inner planets, Mercury and Venus, and the Moon. The limestone process influences "everything that contributes to the sequence of generation after generation in the plants," channeling directly from the cosmos to the plants.[8]

Beginning in the third lecture, Steiner describes a set of biodynamic preparations that channel and enhance these forces to benefit the soil and plants. The purpose of these preparations is to reconnect the plant world with the cosmic influences modern agriculture has come to ignore. Steiner's goal was to spread the use of the preparations far and wide, to begin "the earth's healing."[9]

It will take a lifetime of reflection for me to understand these forces, but for the time being, it is enough to realize that over the course of three short months my mind has been opened to a new, more engaged way of viewing my surroundings. Henning told us many times that the most important and valuable product on a biodynamic farm is the farmer. I can sense that in myself. Although I gave many hours of labor on the farm and helped maintain its health, the real fruit of my labor was the change that occurred in me. When I left, I came away with transformative knowledge, and that is the most valuable knowledge, and most valuable gift, I could have received.

Response-Ability: Developing the Ability to Respond

When I first arrived on the farm, I wasn't sure what to expect. I entered into this internship knowing only a few things: I would do physical work four hours a day, five days a week, 8 am to noon, then study for four hours in the afternoon, and complete an experiment on plant morphology. What I was to read was mainly a mystery, and the work to be done unknown. I have known Henning and Elizabeth for about six years now, but under different circumstances. Elizabeth taught me both English and Spanish for two years in high school. She also taught me how to bake French bread and make raspberry jam as part of our two-hour, twice-a-week agriculture course. It was during those brief hours that I got my first taste of farming and first met Henning. It was then that I transplanted broccoli, stirred a preparation, and weeded a garden. Yet during that time I got only the simplest glimpse of all that is S&S Homestead Farm. With the first day of my internship began a re-working of my idea of the farm.

During the first month, fellow intern Colleen Conboy and I dutifully carried around our journals, frantically taking notes as Henning explained everything we were doing and why. At the end of every day I found myself thinking that I would never have learned this much in a classroom. I also found myself absorbing basic farming skills, such as how to tell good compost from "dead" soil, how to herd sheep effectively, and how to transplant seedlings without drying out their roots. But I was also beginning to absorb something subtle, beneath all this: the philosophy, the gentle driving force, behind every action Henning takes. I remember during one of those first weeks, Colleen and I were taking wheelbarrows across a large field swaying with tall grasses. There was a path down the middle of the field, and I followed it most of the way, but veered off as I neared the gate, not realizing there was a path near the fence I could have taken — if I had been patient. Henning pointed out that the paths were mowed so that the rest of the field could remain untouched. Once the grass is flattened, it becomes more difficult to harvest — good hay, wasted. I remember thinking: Even the placement of the paths is carefully considered! I must be more aware of the purposes behind everything here!

This deep attention to detail is not surprising, considering that Henning follows a fifty-year farm plan, and an annual work plan. It is remarkable and inspiring! To create a healthy farm organism the farmer must be deliberate about everything he does. He must observe every detail and use those observations to be as efficient as possible, as with the paths in the fields. The keen skill of taking into account the tiniest things, and not dismissing them as unimportant — as most people would — became the foundation of the work we did during our internship. We had to foster the ability to observe both the individual parts of the farm and the whole entity in order to understand how the parts interact and become the whole.

Henning put Colleen and me in charge of the two orchards on the farm. We were to

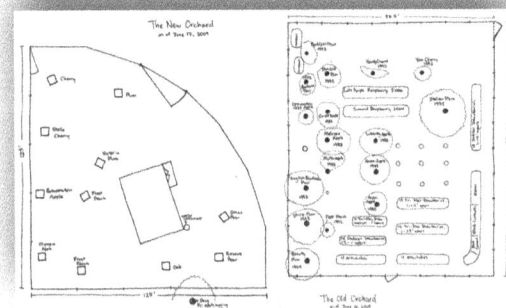

The Orchards on S&S Homestead

observe each each as an ecosystem and work toward understanding all of the components of the orchards, from soil ill health, weeds, and varieties of trees to pollinators, pests, and the effects of the weather. We learned to feel the soil to decide if we needed to water instead of following a pre-scheduled regimen. Throughout the course of the summer we created, shaped, reshaped, and added to a compost pile, monitoring temperature changes to determine the activity of the microorganisms in the pile We pruned, weeded, and tended to all the beds to keep them clean and uncrowded. We raked up rotten fruit and harvested each bush, plant, or tree as the crops ripened, discovering the subtle differences in size, color, and taste between different varieties of strawberries, currants, raspberries, plums, pears, and apples.

The orchards became our main responsibility on the farm. We began every morning in the old orchard and I found myself checking the health of all the plants as I would my pets, or my children. It was a good way for us to exercise our skills of observation and to recognize a problem and find responses and remedies.

For example, we discovered that as the strawberries ripened, they were being nibbled by birds. In response, we covered the strawberries with netting. Soon we noticed that the berries were still being nibbled, this time by insects and slugs. We sprayed diatomaceous earth, a white powder made of the fossilized remains of diatomic zooplankton skeletons that slice up the mucus membranes of slugs and other animals. When the heavy apple-laden branches were beginning to bend and a few broke under the weight of the fruit, we built temporary props to support the branches. Once the apples were harvested, we removed the props, and saved the wood for next year.

If we had tried to do this detailed work on the whole farm, we would have gotten lost. For beginning farmers, the entirety of the system would have been too much to take in, and at the end of three months I still find question after question popping into my head about why we're doing what we're doing, and what comes next. Developing keen observation skills to notice the smaller, subtler events on the farm (like a discolored leaf, or a larva on a stem) takes time and practice. To hold the entire farm system in your mind takes even more practice. Although I am much more confident in my ability to look at the farm ecosystem and notice the minute changes that occur daily, I have a long way to go to be able to observe the world around me the way Henning does.

Once I began learning the reasons behind every action Henning takes, every project he has completed, I saw all parts of the farm intimately connected. In a way I was not able to experience in high school, I began to see the farm as a whole, a living organism where all individual constituents complement each other (animals complementing plants and vice versa). Not only was I able to see the farm as a functioning organism at any given point in time, I began to visualize how all the processes and patterns work together through time.

This was the main philosophy I spoke of: the concept of the farm as a symbiosis of plants and animals balancing and supporting each other throughout time with little outside input. Steiner described this relationship as approaching the ideal of a "self-contained individuality."[10] For example, any plant scraps we collected were composted and later the finished compost was applied to the soil, which cycles back the nutrients the original plants had drawn from the soil.

Imagine a cow in a pasture. The cow eats and digests the grass, excreting what her body does not need. The partially digested grass fortified with the gastric juices, results from a composting action: the nutrients from the grass are cycled through the cow's stomach and returned to the soil. When the cow is slaughtered, Henning takes the offal and composts it with sawdust and straw for two seasons. The finished compost is applied to the soil, cycling back all the nutrients and forces contained within the cow's body.

At first I found myself picturing the farm organism with humans — us, the stewards — as separate components, almost like conductors acting from a distance. But with time, and after numerous conversations with Henning, I understood that I could not think of myself, or any human, as separate. Despite the brief three months I worked on the farm, I was just as much an integral part of the farm organism as the cows that live and die there. Humans are the consciousness behind the farm. We guide everything to fruition or to some intended use (apple trees for fruit, lettuce plants for salad, milking cows for butter, lambs for meat, carrots for roots, sunflowers for seeds).

There was one conversation in particular I remember clearly. We were told to imagine a forest with no human influence and picture the forces at work. There is the inherent life force coming from the plants and animals, and they all have their being in a delicate, self-ordering balance.

Now imagine the farm. The difference is immediately apparent. The consciousness we bring to the land creates boundaries, systems, and patterns not found in a forest devoid of human influence. Instead of the natural balance of inherent purposes found in plants and animals working together in harmony, the farm is a combination of plants and animals whose

purposes have been molded by our will and our goals.

Just consider the Tri-Star strawberry and the wild strawberry. Every few days we harvested the Tri-Star, pulling off large juicy red berries up to two inches long, with firm flesh, and bright shiny color. The plants themselves were large with broad leaves. Then every day as I walked home along the farm driveway I saw tiny wild strawberries at my feet. The plants were a tenth of the size of the Tri-Stars, and the fruits were mere red specks, perfect tiny globes, that easily squished between my fingers. Their taste was intense and deep, but the Tri-Stars have been bred over the years to grow larger and more abundant, albeit less flavorful, fruit. The wild strawberry is fulfilling its own purpose, while the Tri-Stars are fulfilling ours. I realized that every action I took impacted the plants and animals I dealt with.

So I began to see myself as one element in the large, complex organism that is the farm. Henning says that the farmer is the *quintessence* (fifth essence or element besides soil, air, water, and fire), and it is the farmer's task to listen to the farm and guide it toward the balance that exists in nature. Every morning I opened my eyes and observed, and asked why I was seeing what I did. I took on whatever tasks needed to be done that day, and took pleasure in the mystery of my days. One day I might move the sheep to new pasture, prune tomatoes, prepare an empty bed for the next crop, or buck hay. I am a tool of the farm and although we guide the farm toward our goals, the farm has its own voice. Through observation we pick up the signals that tell us what the farm needs, and we listen. If I were to identify one concept I will walk away with this summer, perhaps it is this. Although I can guide my future and my environment, my environment also guides me. I must remain aware of that so I can maintain a strong connection with my surroundings which in turn support my health.

Goethe's Plant Morphology

Today most of our world is informed by reductionist science, based on the idea that in order to understand a whole object, organism, or system, you must first understand its most minute parts. An early and influential example of this method is embodied in the system of biological classification developed by Swedish botanist Carl Linné (1707-1778). His method was to create a "hierarchy of nested groups within groups" on the basis of physical characteristics.[11] This system is called hierarchical ordering. What started as a simple "natural" classification system turned into a complex, infinitely changing taxonomical order whose details changed depending on whom you ask. The decision to split one group from another is arbitrary. As minute differences in the physiology of plants and animals became apparent, the splits and branches in the classification became more numerous. This system's goal is to classify every possible differentiation, no matter how small. Linné's approach to the natural world was adopted in modern chemistry focused on molecular, atomic, and subatomic interactions, and in other branches of science such as biology, zoology, and physics which operate in similar ways.

Modern science dissects our world and then tries to reassemble it from its individual parts. Steiner says in his lectures: "The men of to-day say and do many things in life and practice as though they were dealing only with narrow, limited objects, not with effects and influences from the whole Universe."[12] When we pick our world apart piece by piece, we lose the Spirit or life force, so when we try to reassemble everything, we don't end up with the integrated whole we started with. There is a passage in Johann Wolfgang von Goethe's *Faust,* where the devil mockingly describes the work of the scientist:

> *To know some living thing and describe it,*
> *He hastens to expel its Spirit;*
> *Now he holds the parts in his hand,*
> *But, alas! He lacks the Spirit band.[13]*

During a discussion with Henning, we tried to discover the roots of the modern tendency to dissect and divide everything we find. He asked, "What is motion?" According to the ancient philosopher Aristotle (384-322 B.C.), motion is movement toward a purpose. This could be the motion of an apple as it falls to the ground, or it could be the inward movement of the tree toward creating the apple. In comparison, according to Enlightenment philosopher René Descartes (1596-1650 A.D.), motion simply is the transposition of an object from point A to point B. Descartes dismissed Aristotle's concept of movement toward a purpose in exchange for simple physical motion through space. Which definition of movement do we usually turn to in today's world?

When Henning asked us that question, immediately my mind went to Descartes' definition. Most of us have lost touch with the concept of motion as movement toward a purpose. It seems to me that early on, we were trained to think like Descartes, and the modern system of studying the parts to understand the whole has become the template by which we construe reality. As a result, we ignore the purpose, the living force behind the movement and behind the individual parts.

We find in modern science that no matter how we dissect our world, we still have trouble reaching a truth we are satisfied with. The individual branches of science we learn about in school are inadequate to synthesize all of the information we gather, and we find ourselves no closer to answering such questions as, "What is life, how did it come to be, and what is its purpose?"

Scientists of the Romantic period undertook to heal the dichotomy between matter and Spirit. Samuel Coleridge described life in a way that encompasses the views of both Aristotle and Descartes. He borrowed two terms from Spinoza: *natura naturata* and *natura naturans*. In *Aids to Reflection* (1839) he describes *natura naturata* as "nature in the passive," that which has come into being. It amounts to an observation of facts, of events in nature that have already occurred or come into existence. Essentially it is the study of the past, looking at individual data observable in the present. These observations are registered through our senses, filtered through the experimental method.

Coleridge describes *natura naturans* as "the sum of the powers inferred as the sufficient cause of the forms...in nature in the active sense," meaning that *natura naturans* describes the process of coming into being in nature.[14] Emerging nature can be observed with the help of the senses, but to truly understand the phenomenon, one must use imagination to synthesize sense observations into a whole that can only be held in the mind. In attempting to understand both *natura naturata* (material being) and *natura naturans* (being in process of becoming) we approach Aristotle's concept of movement.

Johann Wolfgang von Goethe (1749-1832), the poet and natural scientist who inspired Coleridge, developed the practice of science based on the observation of *natura naturans* — the constant flux of nature coming into being. Goethe's science was based on the concept of unity in nature and he approached a given subject as a whole, undivided entity. In his words, "in organic being, first the form as a whole strikes us,

then its parts and their shape and combination."[15] Goethe carefully followed the life cycles of plants, made notes of patterns and rhythms in nature, and as a result of a lifetime of study and keen observation published *The Metamorphosis of Plants* (1797-1830).

Goethe's book-length essay describes in great detail the different changes a plant undergoes throughout its lifetime and how all the organs are transformations of the leaf prototype, what Andreas Suchantke, quoting Goethe, calls the "true Proteus."[16] Goethe notes that "everything is leaf, and through this simplicity the greatest diversity becomes possible."[17] His essay follows the leaf through all stages of a plant's growth, beginning with the emergence of the cotyledon and followed by the so-called first true leaves. As leaves unfurl along the stem, the shape becomes more defined and distinct, developing characteristics that make the plant unique. When the leaves reach the peak of their "expansion and elaboration," they begin to "transition to inflorescence" which is marked by a contraction of the subsequent leaves.[18] He continues with a description of the metamorphosis of the leaf into the various organs of the flower and fruit, following similar patterns of expansion and contraction until once again the plant approaches the seed from which it emerged.

For many years after Goethe wrote *The Metamorphosis of Plants*, no further studies used the methodology he had developed. The first to receive much notice were the experiments conducted by Jochen Bockemühl, whose findings were published in 1982 in his essay *Morphic Movements in the Vegetative Leaves of Higher Plants*.[19] The goal of Bockemühl's experiments was to explore the plant archetype by studying a plant's morphology in an attempt to understand the idea and purpose behind it. He decided to study the morphology of the leaf as a part of the plant representing the whole.

In Bockemühl's words, the leaf sequence "reflects in a special way, the developmental movement as a whole."[20] A leaf sequence is a composite of individual leaves that form on the stem of a plant from bottom to top, and therefore each is an individual phenomenon. Bockemühl's method is to show the fluid transformations from leaf to leaf in the imagination of the observer. What the observer perceives is what Bockemühl calls a "morphic movement."[21]

In his experiments, Bockemühl systematically harvested leaves and arranged them in a visual diagram showing the changes in the leaves up the stalk of the plant, and in the leaf itself through time. He arranged leaves graphically to help readers simulate in their minds the actual movement of the leaves through space and time. By looking at the physical changes in the leaves from one week to the next, and as the plant approaches the flowering stage, Bockemühl observed four formative activities taking place at different stages of the leaf sequence.

July 31, 2009
Garden Cress

Figure 1

Shooting is a growth in a certain direction; *dividing* is the multiplication of a certain event, such as shooting; *spreading* is the filling out of the leaf blade; and *elongation* is a stretching of the leaf usually originating from the base.[22] He studied the patterns of influence the four activities exerted on the plant's form through its development in order to conceptualize the plant's archetype. As summarized by Bockemühl's collaborator, Andreas Suchantke, "the visible body of the plant results from the activity of the archetypal plant."[23]

Bockemühl conceived of three regulatory principles to describe the interactions of formative activities: separation, interpenetration, and merging.

First, there is separation. In the expansion phase, the actions of elongation and spreading are usually separate from each other, and the other two activities, dividing and shooting, remain hidden. Second, there is interpenetration. Near the center of the movement, when elongation and spreading begin to work together, particular motifs unique to each species emerge. Third, there is merging. In the last portion of the movement the act of elongation and spreading merge into one action and shooting becomes more dominant. In this way the first and last leaves both possess form elements that make them polar opposites.

As part of my three-month curriculum-based internship on S&S Homestead Farm, I designed an independent project to recreate Bockemühl's experiment. I submitted the proposal to Fairhaven College of Western Washington University with professor John Bower as my advisor. I worked directly with my on-farm mentor, Dr. Henning Sehmsdorf, Adjunct Professor at Washington State University's Center for Sustaining Agriculture, and fellow intern, Colleen Conboy, a student in the Environmental Studies program at Seattle University. We spent much of the summer studying Goethe,

Suchantke, Bockemühl and others in preparation for our experiment in order to understand what we were seeking.

We began our trial on June 17 by seeding garden cress (*Lepidium sativum*) in a planting box. The potting soil we used was a mix of compost, garden soil, and perlite. We scattered the seeds evenly and covered them lightly with potting soil. They germinated on June 20, and we harvested our first plant for drawing, photographing, and drying on June 22. We thinned the cress as it grew. However, the plants grew very quickly and even with the thinning, they became leggy and reached the flower stage within only seven weeks. This may have been because we conducted our experiment in the hottest part of the summer and experienced a heat wave. For future experiments, it may be worthwhile to plant the seeds farther apart or in several wooden boxes or conduct the experiment during a cooler time of year.

Once the cotyledon was well formed, we began harvesting plants. We harvested twice a week, pressing either the whole plant or just the leaf sequence as the plants grew larger. Once a week I drew each leaf in the same manner as Bockemühl's illustrations, alternately picking leaves off the stalk as I moved vertically up the plant. I sketched the leaves so that the resulting record preserved their relative scale.

We stopped harvesting on July 31, after the seventh week, when all the plants had flowered and we could no longer easily harvest a complete leaf sequence without having to work around the flowers. We arranged our drawings as Bockemühl had done (figure 1). We hoped to discover similar patterns of formative activities in the cress morphology.

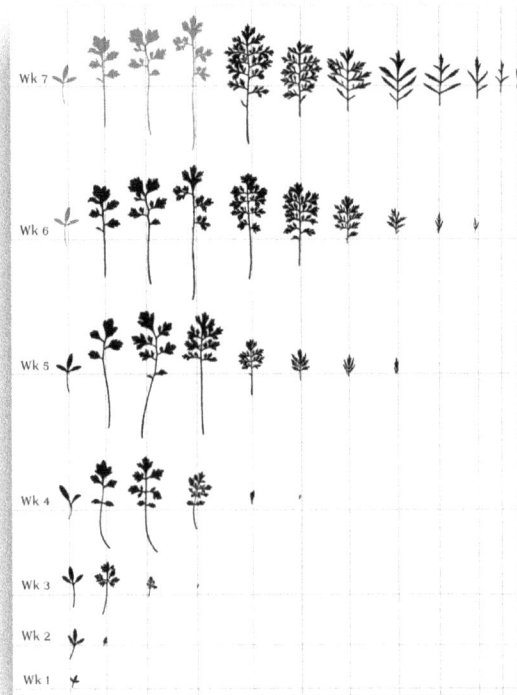

Figure 2

Once we finished collecting our leaf sequences, the question was whether we could sense the same patterns of morphology at work in our results as he had found. Let us look at the most mature leaf sequence gathered in week 7 (July 31).

During the separation phase, on the left-hand side of the oval (figure 1), there is obvious elongation occurring in the stems of the leaves, as well as some spreading and dividing occurring in the leaflets. Near the apex of the oval we see the most fully developed leaf, the one Bockemühl would consider closest to the archetype. Moving farther to the right the stem retracts as elongation is no longer the dominant force. The divisions lessen, and as we approach the flower the leaflets show signs of shooting, growing outward in slender, unified forms.

Next, let us examine all seven leaf sequences we laid out horizontally (figure 2), similar to

Bockemühl's diagram,[24] where he examined the morphology of Nipplewort (*Lapsana communis*).

Moving from left to right along the rows, each drawing represents a leaf as we move vertically up the plant, beginning with the cotyledon and ending with the highest leaf. Vertically from bottom to top in the columns, we can follow the development of the individual leaf through time. Therefore, it represents a spatial movement from left to right and a temporal movement from bottom to top. Once again, the general patterns of formative activities that Bockemühl described can be seen in each leaf sequence.

Figure 3

As we conducted this experiment, I found myself focusing less on the specific formative activities and more on the general motion of the plant. What fascinated me was the beautiful expansion and contraction of the leaves, as if breathing in — expanding — and breathing out — contracting (figure 3). As Colleen and I studied biodynamic farming and Goethe's plant morphology over the course of the summer, this particular rhythm became a common theme.

As I came to understand the concept of a plant's teleology or indwelling purpose,[25] I began to see leaf expansion and contraction in a new light. As a seed germinates, it moves up and out of its protective hull toward sunlight. With the first young growth the plant expands its leaves, expanding surface area to increase its ability to photosynthesize. Then there is a shift in the plant's life, when its energies are redirected. Now, instead of putting energy toward leaf expansion, the plant moves toward reproduction. The energies move toward the flower, then the seed or fruit. With this change we see the contraction of the leaf, the breathing out of the plant, directing its inherent energy toward immortality through reproduction.

In summary, we did find evidence in our leaf sequences of the same interactions and patterns of the formative activities that Bockemühl had observed. But beyond that, our biggest success in completing this experiment was to train our minds to sense the morphic movement — and purpose — of the plant. To understand *natura naturans*, you cannot just examine the diagrams that Bockemühl created, nor simply read his essay. You must take the time to conduct the experiment yourself. You must be the observer, the one to synthesize the observations and arrive at your own conclusions.

By witnessing the changes the plants go through with time, you can accurately (and imaginatively) sense morphological transformation as a fluid movement, connecting the individual images of leaf formation into a continuous stream in your mind. Then you may consider yourself one step closer to perceiving the archetypal plant.

Figure 4

Goethe invites us to practice what he calls "a delicate empiricism" where observers so deeply identify with the object observed that they arrive at an intuitive understanding of "true theory"as an act of the imagination.[26]

Figure 4 show the leaf sequences harvested between weeks 1 and 7, arranged in circular patterns to make visible expansion and contraction, and the four formative activities: elongation, spreading merging, and shooting.

Notes

[1] Iowa State U. 2009. "What is Organic Agriculture?" http://extension.agron.iastate.edu/organicag/whatis.html. Retrieved September 8, 2009.

[2] Pfeiffer, Ehrenfried 1958. "Preface." *Agriculture Course: The Birth of the Biodynamic Method.* Dornach, Switzerland, 5-16; Hindes, Daniel 2009. "Rudolf Steiner and Anthroposophy." http://www.rudolfsteinerweb.com/Rudolf_Steiner_and_Anthropsophy.php. Retrieved September 8, 2009.

[3] Kambartel, Friedrich & Jürgen Mittelstrass 1978. "Zum Normativen Fundament der Wissenschaft," *Syntese,* 37/3: 471-477.

[4] Sehmsdorf, Henning. Personal communication, June, 2009.

[5] Ibid.

[6] Pfeiffer, op.cit, 5.

[7] Steiner, Rudolf 1958. *Agriculture Course: The Birth of the Biodynamic Method.* Dornach, Switzerland, 18f.

[8] Ibid, 23-25.

[9] Ibid, 8.

[10] Ibid, 29.

[11] Blamire, John 1998. "Classification" *Science at a Distance.* http://www.brooklyn.cuny.edu/bc/ahp/CLAS/CLAS.Linn.html. Retrieved September 10, 2009.

[12] Steiner, op.cit, 20.

[13] Goethe, Johann Wolfgang von 1962. Faust: Eine Tragödie (Faust: Tragedy), I, 58 (lines 1936-1939):
Wer will was Lebendig erkennen und beschreiben,
Sucht erst den Geist herauszutreiben,
Dann hat er die Teile in der Hand,
Felt, leider! nur das geistige Band. (English trans. Henning Sehmsdorf)

[14] Shed, W.G.T. (ed.) 1884. *The Complete Works of Samuel T. Coleridge.* New York, II, 310.

[15] Quoted by Blunden, Andy 2010. *An Interdisciplinary Theory of Activity.* Boston, Mass, 28.

[16] Suchantke, Andreas 1995. "The Leaf: The True Proteus," *The Metamorphosis of Plants: Essays by Jochen Bockemühl and Andreas Suchantke,* Cape Town, South Africa, 8. In Greek mythology Proteus is a sea god known for his changeability. Proteus' name (from Greek πρῶτος meaning first) suggests that he was imagined as the primordial or the first born.

[17] Ibid.

[18] Goethe, Johann Wolfgang von 1963. *Schriften zur Botanik und Wissenschaftslehre* (Writings on Botany and the Theory of Science), München, Germany, 22.

[19] Bockemühl, Jochen 1995. "Morphic Movements in the Vegetative Leaves of Higher Plants" *The Metamorphosis of Plants: Essays by Jochen Bockemühl and Andreas Suchantke*, Cape Town, 21-46.

[20] Ibid, 21.

[21] Ibid, 22.

[22] Ibid, 23.

[23] Suchantke, op.cit, 10.

[24] Bockemühl, op. cit, 31.

[25] Compare Sehmsdorf, Henning 2014. "The Spirituality of the Soil: The Idea of Teleology from Aristotle to Rudolf Steiner," below.

[26] Goethe, Johann Wolfgang von 1963. *Maximen und Reflexionen* (Maxims and Reflections), München, Germany, 68.

Sky Phenomenon over S&S Homestead Farm 2021

Goethe's Color Theory on the Farm

Henning Sehmsdorf and Barry Lia, 2008[1]; updated Henning Sehmsdorf, 2021

"The desire for knowledge is first stimulated in us when remarkable phenomena attract our attention. For it to continue we must find a deeper sympathetic connection, which will lead us by degrees to a deeper acquaintance with the subject" (Goethe).[2]

Ever wonder why the sky is red in the morning and evening, but blue at noon? Goethe's color theory based on phenomenological observation explains it better than any other. Much of biodynamic farming is rooted in the phenomenological method exemplified in Goethe's practice of observation. The quote above speaks not only to his approach to understanding color, but to knowing anything deeply through personal experience. The sympathetic engagement he urges results in profound acquaintance with the things we study, and yields the kind of experiential knowledge that is essential to good farming.

Contextual Phenomenology

At an on-farm workshop in 2008, we asked participants to stand in front of a certain building and scan the sky and tree and roof lines of surrounding barns and outbuildings, using prisms held closely to the eye. What they saw was that the overcast sky appeared white and colorless, but the tree tops and the upper edge of the barn roof appeared blue. The upper roof line of the composting structure in front of the barn was orange, but its lower roof edge was green. From these observations, the participants were able to draw the following conclusions. When seen through a prism held closely to the eye:

~ Blue colors appear when a dark field (such as a tree or barn roof) is below a light field (such as the pale sky);

~ Orange colors appear when the reverse (the light grey composter roof beneath the darker wall of the barn);

~ Green colors appear when orange and blue overlap (at the lower edge of the composter roof).

~ Therefore, color is not primarily a function of prismatic refraction — which was the same in all instances observed by the workshop participants — but of context, meaning the juxtaposition of fields of relative darkness and light.

Reductionist Phenomenology

A hundred years before Goethe, Isaac Newton had demonstrated experimentally that clear white light was composed of the seven colors visible to the human eye (red, orange, yellow, green, blue, indigo, and violet). He thereby established the measurable spectrum that led to fundamental breakthroughs in visual perception, optics, physics, chemistry, and the study of color in nature, made possible by exact mathematical description of the angles of refraction.[3]

Newton's famous *experimentum crucis* (crucial experiment) showed that a beam of light coming from a small hole cut into the shuttered window and passing through a prism, separated into the seven spectrum colors. But any beam of colored light passing through a second prism did not subdivide further. Newton concluded that white light must be a mixture of colors, while the colored beams in his experiment were pure, that is, not a mixture.

A century later, when Goethe tried to reproduce Newton's experiment, he discovered to his astonishment that a prism held to the eye while looking out of a window, produced only two colors instead of seven: blue and orange. How to explain the difference? From his exhaustive survey of studies of color since the ancient Greeks, Goethe knew that two thousand years earlier, Aristotle had developed the first known color theory, suggesting that all colors came from interactions of white and black (lightness and darkness).[4] Had Newton overlooked something important?

In examining Newton's experiment, Goethe noticed that the English physicist had darkened a room with a window shade which he perforated with a small aperture to admit a narrow beam of light. The beam of light was gathered by a lens and then directed through a prism, which seemed to separate the light into seven colors that appeared on a white wall at about eight feet away. What Newton did

Newton's Drawing of the *Experimentum Crucis*

not investigate was whether the dark edge of the aperture itself had any bearing on the creation of color by the prism. Nor did he notice, or investigate, whether overlapping colors created additional colors, such as green or magenta, or whether the distance of the prism from the wall determined which colors overlapped and which did not.

Goethe's Phenomenological Experiments Re-Visited

As Goethe had, our workshop participants noticed that when, at noon, we directed the prism held to our eyes toward the open sky, i.e. without confining the light through a small aperture, the prism did not create the complete color spectrum as posited by Newton. Rather the two colors that appeared when a prism was used to refract

Figure 1

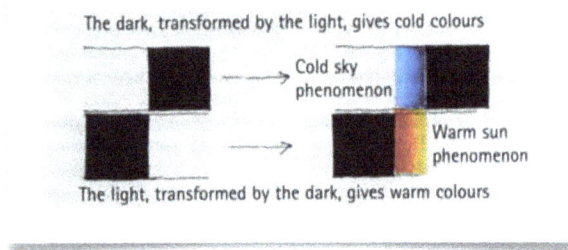

Figure 2

unconfined light were shades of orange-yellow and blue-indigo.

Looking at the sky without a prism in the late afternoon, we also noticed that the sky above appeared a shade of fading blue, but orange-yellow above the red sun setting in the west. Early the next morning, we observed the opposite: when a red sun rose in the east, it colored the horizon orange (figure 1).[5]

According to Goethe's theory, the blueness of the sky appears where the sun-filled atmosphere meets the darkness of outer space. The orange-yellow color appears where an area of fading light on the horizon lies above an area of greater brightness coming from the setting sun. It demonstrated to us that even without a prism, the human eye perceives colors at the edges between areas of relative darkness and light (see also the photo over the essay title above).

Danish author Lone Schmidt verified these sky phenomena by arranging two adjoining plates alternating between white and black fields seen through a prism. In the top plate, the white

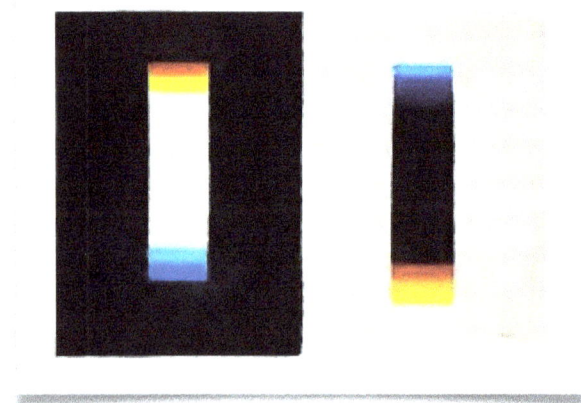

Left: figure 3; right: figure 4

Left: figures 5-6; right: figures 7-8

field (equivalent to the sun-filled atmosphere) followed by the black (outer space), when seen through a prism, was transformed into blue. In the bottom plate, the dark field (equivalent to the light-deficient atmosphere in the evening) followed by the light (representing the setting sun) produced red and yellow (figure 2).

Goethe tested his intuitions regarding the origin of color through a painstaking series of visual experiments, some of which we reproduced at the workshop.

Viewing a simple checkerboard or a design of curved lines through a prism confirmed that the color blue appeared above, and orange appeared below any dark edge. In a black rectangle placed on a white background, layers of blue appeared above, red and yellow below. In a white rectangle placed on a black background, the reverse happened (figures 3-4).

Another one of Goethe's experimental plates reproduced at the workshop featured white and black bars between rectangles (figures 5-6). Goethe discovered that at the narrow edge between white and black, the blue and orange colors overlapped and thereby created green in the case of the light bar, and magenta in the case of the dark bar (figures 7-8).

Turbid Medium (Trübes Mittel)

Unlike Newton, who focused his *experimentum crucis* on mathematical, *quantitative* description of the refractive angles of the color spectrum, Goethe focused his research on the *qualitative* conditions making perception of color by the human eye possible. He referred to these conditions summarily as the "archetypal phenomenon" (*Urphänomen)* of how humans experience color. Not all sentient beings see color; earthworms and bats, for example, surely don't, while others perceive color differently from humans. Bees, for instance, see ultraviolet light, i.e. short waves, while humans see long wave light. Bees therefore see black where humans see red. In the course of evolution, flowering plants have developed reflective four-fold patterns on their coronas, which mirror ultraviolet light visible to

Morning Sun, S&S Homestead Farm, 2008

bees but not to humans. This means that bees can see optical patterns and colors invisible to the human eye. In speedy flight (thirty km/hour), bees do not see color at all, that is, their color sense is shut off until they reach their target, at which point it is turned back on.[6]

Early in the morning on an August day, workshop participants were treated to a spectacular display of the sun rising above the trees on the eastern edge of the farm. What they saw with the naked eye confirmed the observations they had made looking at Goethe's experimental figures through a prism: an area of blue color appeared above the dark tree line, while orange color appeared below the white light of the sun.

But what was the medium through which workshop participants observed the atmospheric color phenomena, if not a prism? As defined in Newtonian optics, "a prism is a glass or other transparent object in prism form, especially one that is triangular with refracting surfaces at an acute angle with each other and that separates white light into a spectrum of colors."[7]

Goethe, by contrast, posited that color is created on the edges between dark and light fields, whether we look through a prism or at color in the sky with the naked eye. This observed fact is what he meant by the primordial phenomenon (*Urphänomen*).

To explain this continuity, he pointed out that whenever we look at color phenomena in the sky, we see it through the more or less turbid medium of the atmosphere. The air may seem crystal clear, but it is actually filled with dust debris from the earth or from outer space. When we observe the blue sky, a yellow-red sunrise or sunset, a multi-colored corona around the moon, or a rainbow, we see these phenomena through the atmosphere, i.e. the turbid envelope of gases between the earth and

outer space. For this turbid medium Goethe coined the term "turbid medium" (*Trübes Mittel*).[8]

To demonstrate the universality of this principle of color perception, Goethe used various media such as a semi-translucent glass egg or a glass jar filled with cloudy water or milk imitating turbid substances. Workshop participants found that holding the egg (or a glass of diluted milk) before luminous backgrounds, the egg appeared to be orange in color, rather than white. Held before dark backgrounds, the egg appeared blue, rather than black or grey. In other words, light seen through a turbid medium appears yellow-orange, and darkness seen through an illuminated medium appears blue. These observations confirmed the conclusion that the physical colors of the light spectrum created by a prism are a function of the contextual polarity of light and darkness in which they occur. The phenomena of the blue sky (during the day) and the orange horizon (in the morning or evening) result from the juxtaposition of darkness and light filtered through a semi-transparent medium such as the atmosphere.

The blueness or orange is not in the sky or the atmosphere as such, but arise from the interaction of light and dark perceived by a human observer (or an instrument such as a camera imitating the human eye). The interaction of polarities in creating the perception of colors is the *Urphänomen* (primordial or archetypal phenomenon) Goethe described in his "theory of color."[9]

Theory of Color (Farbenlehre)

The English translation of the term *Farbenlehre* as *Theory of Color* is actually misleading. Goethe would have agreed with Ludwig Wittgenstein who asserted that Goethe's "theory" was really not a theory at all, because he did not provide us with a mathematically based *experimentum crucis* to decide for or against his hypothesis.[10] What Goethe did instead was to provide an exhaustive sequence of observations of color under the most varied circumstances, from which he at long last distilled the archetypal concept of color perceived by the human eye as being created at the edge between dark and light fields.

Tender Empiricism (Zarte Empirie)

A better translation of the term *Farbenlehre* would actually be "Teachings About Color." What Goethe teaches us is how to know anything in the phenomenological world through a process of intimate observation involving all the senses: sight, hearing, smell, touch, taste, and intuition. He urges the observer to withhold conceptual abstractions, and to resist the urge to rush into premature explanatory hypotheses until sustained observation brings observers to a point where they intuitively identify with the object of observation, and "become the theory."[11] By "theory" Goethe means intuitive experiential knowledge.

For the biodynamic farmer, Goethe's color teachings and more generally his theory of knowledge emphasize the importance of the whole person in knowing anything on the farm: plants, animals, soils, fertility, weather, and energy. This contrasts with a reductionist approach that emphasizes mathematical measurements and instruments to the exclusion of human beings observing, feeling, intuiting, and interacting with the environment of which they are an inextractable part.

For example, a laboratory soil test to measure pH or NPK levels provides the farmer with important data. However, these data become meaningful only in the context of the farmer's intimate knowledge of the soil from observing animals, soil organisms, fruits, vegetables, weeds, rainfall or drought over time. An experienced, observant farmer will be able to intuit the liveliness (or deadness) of the soil at various times of the season.

On S&S Homestead, we have learned to take soil tests with a grain of salt. Laboratory recommendations for additional fertilizers are usually based on market-oriented production goals rather than on site-specific observations of soil, plant, or animal health, which is the overriding production goal of biodynamic farming.

A stark contrast to biodynamic assessment of soil fertility is represented by what is known as "precision agriculture," also known as satellite farming with the goal of optimizing financial returns on inputs, "based on observing, measuring and responding to inter- and intra-field variability in crops, while preserving resources."[12]

This distinction was echoed by Dr. Walter Goldstein (Research Director at Michael Fields Agricultural Institute in Wisconsin) at a workshop on S&S Homestead Farm in 2006. He discussed alternative ways of assessing soil quality he had presented at various agricultural conferences. He urged farmers to consider not only objective measurements of empirical data with a goal of economic gain, but also contextual observations and intuitive perceptions aimed at the health of the whole farm organism. His recommendations were solidly rooted in Goethe's way of knowing.

Conclusion

It is important to remember that the workshop dealt only with the so-called physical colors in the prismatic spectrum and in atmospheric phenomena. Goethe also investigated other aspects of experiencing color. He studied the physiological responses in the human eye, the chemical qualities of color in plants and other physical objects, as well the Spiritual, moral, and artistic implications of color and color theory.

Goethe's reflection on color reminds the biodynamic farmer that if we remove the human being from the act of knowing, we are left without a frame of reference in how to use knowledge for any given end. This is an urgent concern given the pressing questions in our own time: how to live sustainably in a world of diminished resources, and how to make sense out of what to many appears an increasingly meaningless existence. In following Goethe's way of knowing the world, we can regain a sense of purpose and focus as farmers, as members of our communities, and as human beings embedded in nature.

My daughter, who participated in the workshop, said it well in her written summary of the findings of the workshop, which she presented in a final discussion around the dinner table:

"Johann Wolfgang von Goethe, the German polymath genius of the late 18th and early 19th centuries, addressed the question of how we know and transmit anything by telling us that the act of knowing occurs in three different ways: First, through sensory perception, in which we orient ourselves to something through sight, sound, touch, taste, and smell; secondly, conceptually and linguistically, by applying scientific measurements (and corresponding labels), and lastly, intuitively through non-verbal, non-rational, inspiration.

Goethe said that we only understand something deeply by becoming full and active participants with it. This profound understanding is achieved directly through doing (as in the daily work on the farm), or alternatively through the medium of artistic expression. The artist's vision, embodied in poetry, painting, dance, play, sculpture or drawing, wraps the sensory input, the conceptual framework (observable measurements represented by concepts), and the intuitive, wordless understanding together

into an integrated expression, drawing the observer into the experience as part of the thing itself. Goethe says that in the moment of interaction with the art form (as in the practical work on the biodynamic farm), the observer actually becomes the theory: a full and active participant in the process of knowing.

Goethe's ideas were elegantly articulated, and extraordinarily forward thinking for his time, and they have more recently become accepted as best practice for some forms of education, such as the Waldorf system, but also in new discoveries in the psychology of learning and development. On the biodynamic farm, the Goethean perspective is fundamental not only to experience and knowledge, but to the daily rhythm of working with plants, animals, soil and all the elements in the larger context of the cosmic whole. "[13]

Notes

[1] Based on an introductory workshop on Biodynamics held at S&S Homestead Farm, August, 2008. Photographs by Henning Sehmsdorf.

[2] Goethe, Johann Wolfgang von 1808. From the introduction to "Essay on the Theory of Color." Translation Sall, Pehr 2013. "Goethe's Theory of Color, Part 1 — How It All Started." https://www.youtube.com/watch?v=QnfVlENcHbU. Retrieved August 7, 2021.

[3] Newton, Isaac 1704. *Opicks: Or, A Treatise of the Reflections, Refractions, Inflexions and Colours of Light.* London, UK.

[4] Goethe, Johann Wolfgang von 1963. *Materialien zur Geschichte der Farbenlehre* (Sources for the Study of the Theory of Color), Part I. München, Germany, 18-24.

[5] Schmidt, Lone 1993. *Farven og lyset : studier i Goethes farvelære* (Light and Colors: Studies in Goethe's Theory of Color). Ålborg, Denmark. Figures 1-8 by Lone Schmidt.

[6] For further discussion, see Sehmsdorf, Henning. "Bee Colony Collapse and Biodynamic Strategies to Restore Bee Health," below.

[7] Dictionary. com 2021. "Prism." https://www.lexico.com/en/definition/prism. Retrieved August 10, 2021.

[8] Goethe, Johann Wolfgang von 1963. *Zur Farbenlehre: Didaktischer Teil* (Theory of Color: Didactic Part), München, Germany 48ff. See also Lehrs, Ernst 1958. *Man or Matter.* London, UK 317.

[9] Op.cit, 54.

[10] Wittgenstein, Ludwig 1950. *Bemerkungen über die Farben* (Remarks on Colour), Oxford, UK. https://www.openculture.com/2013/09/goethes-theory-of-colors-and-kandinsky.html. Retrieved July 8, 2022.

[11] Goethe, Johann Wolfgang von 1963. *Maximen und Reflexionen* (Maxims and Reflections), München, Germany 68. Translation mine.

[12] N. a. "Precision Agriculture." https://en.wikipedia.org/wiki/Precision_agriculture. Retrieved August 12, 2021.

[13] Sehmsdorf, Käthe 2008. "Workshop on Biodynamics," S&S Homestead Farm (unpublished manuscript).

Emergy (Embodied Energy) Analysis of S&S Homestead Farm

Andrew C. Haden, 2002; abbreviated, Henning Sehmsdorf, 2021[1]

Summary

This paper uses emergy analysis to assess the ecological sustainability of a small family farm on Lopez Island, Washington. Emergy is defined as the available energy used directly or indirectly for a service or product, usually quantified in solar energy equivalents.[2] By employing a framework that emanates from ecosystem science, the success of the farm in adhering to the goal of ecological sustainability was *quantitatively* assessed. Emergy-based indices and ratios were calculated to estimate the sustainability of different management areas on the farm, as well as for the farm as a whole, based on: a) the emergy yield of the production process studied, b) the load the production process places on the local environment, and c) the overall thermodynamic efficiency of the production process.

The analysis indicates that the various management areas on the farm show widely different levels of emergy yield, environmental load, and energy transformation efficiency, and thus ecological sustainability. In general, it was found that those management areas that relied to a greater extent on locally available renewable emergy flows, and less on purchased inputs, human labor and services, showed higher sustainability than those reliant on purchased and non-renewable emergy flows. The farm enjoys relatively high levels of overall ecological sustainability as presently organized. However, altering some management practices, so that the agro-ecosystem relies to a greater extent on the self-organizing ability of ecosystems and less on inputs from human labor and the market, would further improve the sustainability of some management areas. Finally, the analysis brought to light how the economic and ecological costs of production need to be weighed within the context of holistic goals guiding farm operation, if definitions and estimates of sustainability are to have relevance for agricultural practitioners.

Background

To "farm in nature's image" is a primary goal guiding many current efforts to develop ecologically sustainable agricultural systems.[3] Advocacy for this concept has been inspired by the realization that modern agricultural production systems are dependent upon large quantities of increasingly scarce non-renewable resources to maintain their high yields. There is

246

evidence that many highly mechanized systems of food production can degrade soil, water and genetic resources to such a degree that when access to non-renewable resources becomes limited, the prospects of attaining even the modest yields of pre-industrial agriculture may be dim. Fortunately, recognition of the fact that conventional agriculture deviates from ecological principles has inspired a new generation of scientists and agricultural practitioners who are working to reintegrate the principles of ecology into agriculture.[4]

While the goal of farming to mimic natural systems is incorporated in alternative agriculture movements, measuring the sustainability of such systems remains an important task. The analysis presented here endeavors to meet this task by assessing the ecological sustainability of S&S Homestead, a farm that holds the goal of farming in nature's image as a fundamental organizing principle. By using the theoretical framework of ecosystem science and employing emergy analysis, the success of the farm in adhering to the goals of sustainability and ecomimicry are quantitatively assessed.

Emergy Analysis

Emergy analysis evolved from the field of "eco-energetics" and derives from the study of energy flows that eco- and economic systems develop during self-organization. Self-organized ecological, social, and economic systems exhibit designs and patterns to optimize energy use. The dynamics and performance of such systems are best measured and compared on an objective basis using energy metrics. Emergy analysis is a trans-disciplinary science, and a synthesis of systems theory, ecology and energy analysis.

Emergy Theory of Value

The emergy theory of value states that the more work done, or energy dissipated, to produce something, the greater is its value. Emergy analyses offer a way to objectively assess value in both eco- and economic systems on a common basis. Emergy values are most often quantified and expressed as solar energy equivalents, and the unit used to express emergy values is the solar em-joule (sej). By tracking all resource inputs back to the amount of solar equivalent energy required to make those inputs, emergy analysis accounts for all entropy losses required to make a given product, and thereby allows for qualitatively different resources to be considered on a common basis. In contrast to economic valuation, which assigns value according to market utility and uses willingness-to-pay as its sole measure, emergy offers an opposite view of value. The more energy, time and materials invested in a product, the greater its value.

Energy Systems Language

At the core of an emergy analysis of a given production system is mass and energy flow. A diagram is drawn using the symbols of systems ecology to graphically represent ecological/ energy components, economic sectors, and resource users, and the circulation of money through the system. Originally, energy systems language was developed as a non-quantitative way to visualize energy-constrained mathematical relationships.

Energy Hierarchy & Energy Quality

Odum uses the term "energy hierarchy" to indicate that in all systems a greater amount of energy must be dissipated to produce something containing less energy, but is of higher "quality." This suggests that there is a natural order to how energies of differing qualities can be grouped. The idea stems from the observation that "ecosystems, earth systems, astronomical systems and possibly all systems are organized in hierarchies because this design maximizes useful energy processing."[5] A corollary to this statement is the theory that in open systems away from thermodynamic equilibrium, such as eco- and economic systems, energy hierarchies develop through self-organization for optimum energy use.

Energy quality refers to the observation that energies of different kinds vary in their ability to do useful work. This principle is often illustrated by the example of coal and electricity, where four joules of coal energy must be transformed to supply one joule of electric power. Because of this necessary transformation, electricity occupies a higher position in the energy transformation hierarchy than coal, and is said to be of higher quality. Moreover, human labor is generally of very high "transformity," due to the fact that human beings require large emergy flows and support territories to enable their work.

Transformity

When the energy used to make a product is divided by the energy remaining in the product, one derives the transformity of that product, expressed as the ratio of solar em-joules per Joule (sej/J). Transformities provide an energy quality factor. They account for the convergence of biosphere processes required to produce something, expressed in energy units. Thus, transformity can be used as energy scaling ratio to indicate energy quality and hierarchical position.

At the same time, transformity is an indicator of past environmental contributions that have combined to create a resource. In theory, it is an indicator of the potential effect on a system resulting from the use of that resource.[6] In contrast to other forms of energy analysis which look only at the flows of heat equivalent energy to a process, emergy analysis — through the use of transformities — is able to depict the effect of system inputs with respect to the time, space and energy needed to form those inputs.

Thus transformities articulate the forces driving the self-organizing processes in a given system better than energy analysis alone. Table 1 is a list of common transformities.

When data is not available to calculate all the transformities for the resources converging to form a given product, static averages are used.[7] There is no single established transformity for most products or services. A high transformity input may contribute less energy to a process than a low transformity input, but the overall emergy contribution of the two sources may be similar when adjusted for energy quality using transformities. For example, in hay production at S&S Homestead, gasoline and sunlight contributed roughly equivalent emergy, 5.06 E+14 sej and 5.28 E+14 sej respectively, but the energy contributed by sunlight was 66,000 times greater than gasoline, measured in joules and without adjusting for gasoline's concentration or energy quality with a transformity value.

Table 1

Solar transformity (sej/J)	
Sun	1
Wind, kinetic energy	1,496
Rain, chemical energy	18,199
Earth cycle, geological uplift	34,377
Coal	40,000
Natural gas	48,000
Crude oil	54,000
Top soil organic matter	74,000
Animal feed, concentrates	79,951
Electricity (average)	173,681
Fisheries production	1,200,000
Nitrogen, ammonium fertilizer	1,860,000
Phosphate, mined	10,100,000
Pesticides	19,700,000
Mechanical equipment	75,000,000
Genetic information, single tree species	726,000,000,000
Genetic information, human DNA	14,700,000,000,000,000

Emergy signatures

The energy and resource flows that interact to produce an agricultural product can be thought of as the "emergy signature" of that product. Driving forces — the resources that feed, organize and constrain a system — are a key consideration when assessing the sustainability of agricultural production systems. Within an emergy signature, some flows stand out as primary driving forces. These are key flows and represent the energetic limits of a system. The emergy signature is a convenient way of conceptualizing the energy and resource flows around which an agricultural system has self-organized, and allows the energetic context of a given system to be readily surveyed.[8]

The Maximum Empower Principle

In emergy analysis, the self-organizion of systems to maximize empower is known as the Maximum Empower Principle. MEP describes the thermodynamic law governing self-organization in all systems: "Systems that self-organize to develop the most useful work with inflowing emergy sources, by reinforcing productive processes and overcoming limitations through system organization, will prevail in competition with others."[9] Odum has offered MEP as the fourth law of thermodynamics, positing that it is operating on all systems at all spatial and temporal scales simultaneously.

Materials and Methods

The methodology used to perform the analysis of food production at S&S Homestead Farm followed the format given by Odum: a) The system boundary was defined spatially as the area of land utilized for production, both for the farm as a whole and for the individual subsystems (management areas). The temporal dimension of the study was one calendar year; b) All major energy sources and material resources flowing into, and stored within, the farm system were identified and diagramed using the energy systems language, and the quantities were recorded and converted into energy units

(Joules), mass units (grams) or monetary units (U.S. Dollars); c) The various resource flows were either measured directly, or estimated from production records, financial records and locally available data (e.g. weather data). To derive the emergy values of the resource flows, the quantities were tabulated and multiplied by appropriate transformities chosen from the literature; d) The food items generated by the farm system were converted into energy units (Joules) using standard conversion factors from the USDA Nutrient Data Laboratory (USDA, 2002).

Item, unit	Data (units/yr)	Transformity* (sej/unit)	EMERGY (E13 sej/yr)	EmDollars
Sun, J	7.62E+14	1.00a	76.18	$55.60
Wind, J	3.54E+09	1.50E+03a	0.53	$3.87
Rain, J	9.81E+11	1.82E+04a	1,785.42	$13,032.26

Table 2: Sample Emergy Evaluations

The results of the analyses are given in both diagrammatic and tabular forms. Table 2 is a sample emergy evaluation table. Column 1 of the table gives the name of the item, and the units of raw data for that item — usually joules, grams or dollars — are recorded in column 2. The energy, material or currency flow for each item is then multiplied by its respective transformity, which is given in column 3. The product of the raw data and the transformity equals the total emergy contribution of that component to the system listed in column 4. The majority of the transformities used in this study were gathered from previously published analyses. In

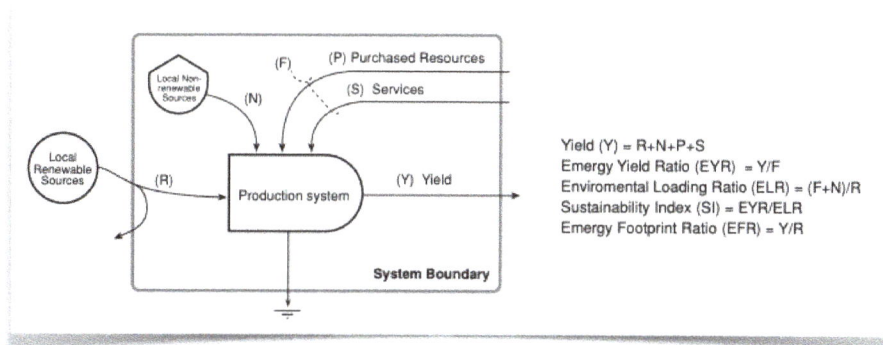

Yield (Y) = R+N+P+S
Emergy Yield Ratio (EYR) = Y/F
Enviromental Loading Ratio (ELR) = (F+N)/R
Sustainability Index (SI) = EYR/ELR
Emergy Footprint Ratio (EFR) = Y/R

Figure 1: Nomenclature of Resource Delineation

column 5 the macro-economic value of the emergy flow is given as Em-Dollars. Em-Dollars are the emergy value of a given flow divided by the emergy/GDP ratio for the year of the study. Em-Dollars are a non-market based indicator of the economic value of inputs to a production process, both from free environmental and purchased resources. In order to assess the contribution of a production system to the long-term sustainability of society, a distinction must be made between the flows supporting a

data for the farm system and adjusting for their energy qualities with transformities, a number of emergy-based ratios and indices were calculated. These aggregated indicators assist in the interpretation of the results of the analysis.

By quantifying the emergy flowing to S&S Homestead Farm, an understanding of the local and external resource base required to operate the farm system was obtained. The overview analysis is intended to give the reader a general picture of the farm system, to show how the various management areas are organized, and to illustrate the connections between the various components that comprise the farm as a whole. Because all buildings on the farm perform multiple duties (excluding the greenhouse), they were counted in the overview analysis, not in analyses of the separate management areas. Therefore, the purchased inputs and services flowing into the farm as a whole will be somewhat greater than the sum of the emergy flows for the management areas. Figure 2 is an energy systems overview diagram of S&S Homestead , showing all the system components that were included in the analysis. Table 3 provides a summary emergy evaluation of S&S Homestead.

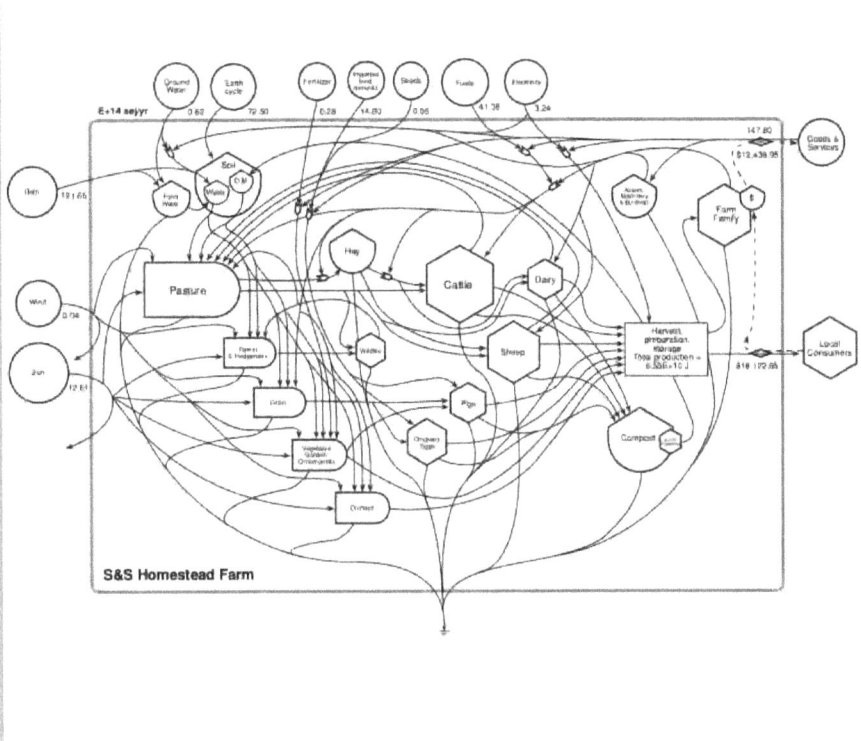

Figure 2: Energy Systems Overview Diagram of S&S Homestead

production process, and whether they are of a renewable or non-renewable character, and whether they are indigenous to the system or must be purchased. Figure 1 is a diagram indicating how resources are delineated and accounted for in this study using the nomenclature developed by Odum,[10] Ulgiati and Brown.[11]

Emergy-Based Indices and Ratios

After tabulating the material and energy flow

For the sake of brevity, the details of the data analysis for all but beef and vegetable production have been dropped from the present essay. They can be viewed in the full version of the paper referenced in footnote 1 below.

Note	Item, unit	Data (units/yr)	Transformity* (sej/unit)	EMERGY (E14 sej/yr)	EmDollars (1993 USD)
RENEWABLE RESOURCES (R)					
1	Sun, J	5.96E+14	1 [a]	5.96	$435.27
2	Wind, J	1.25E+09	1.50E+03 [a]	0.02	$1.37
3	Rain, evapotranspiration, J	3.16E+11	1.82E+04 [a]	57.51	$4,197.46
4	Rain, geopotential, J	4.41E+08	2.79E+04 [a]	0.12	$8.98
5	Earth Cycle, J	1.18E+11	2.90E+04 [a]	34.27	$2,501.66
6	Hay, J (renewable portion, 60%)	1.60E+11	2.27E+04 [a]	39.72	$2,643.82
	Sum of renewable inputs (rain) + hay			93.73	$6,841.28
NONRENEWABLE STORAGES (N)					
7	Net topsoil loss, J	1.29E+09	7.38E+04 [a]	0.95	$69.44
	Sum of free inputs			94.68	$6,910.71
PURCHASED INPUTS (P)					
8	Hay, J (non-renew. portion, 40%)	1.06E+11	2.27E+04 [a]	24.15	$1,762.54
9	Diesel, J	7.66E+09	6.60E+04 [a]	5.06	$369.15
10	Lubricants, J	2.55E+08	6.60E+04 [a]	0.17	$12.30
11	Gasoline, J	3.87E+09	6.60E+04 [a]	2.55	$186.44
12	Electricity, J	7.93E+08	1.60E+05 [b]	1.27	$92.60
13	Mechanical equipment, g	4.29E+04	4.10E+09 [b]	1.76	$128.34
14	Wood posts (fencing), J	1.08E+09	3.49E+04 [i]	0.38	$27.56
15	Iron posts (fencing), g	6.84E+04	3.20E+09 [d]	2.19	$159.86
16	Ironwood posts (fencing), g	3.99E+03	3.90E+08 [a]	0.02	$1.14
17	Barb wire, galvanized steel (fencing), g	3.10E+04	3.20E+09 [d]	0.99	$72.51
18	Electric wire, galvanized steel (fencing), g	2.49E+04	3.20E+09 [d]	0.23	$16.77
19	Insulators, ceramic (fencing), g	2.19E+02	1.00E+09 [a]	0.00	$0.16
20	Plastic (fencing), g	1.16E+03	3.80E+08 [d]	0.00	$0.32
21	Mineral salt blocks, g	9.00E+04	1.00E+09 [a]	0.90	$65.69
SERVICES and LABOR (S)					
22	Labor, J	1.15E+09	2.56E+06 [g]	29.44	$2,148.97
23	Services, USD	1.29E+-03	1.37E+12 [c]	17.68	$1,290.15
	Sum of purchased inputs			86.78	$6,334.50
PRODUCTION, J					
24	**Beef, J**	2.37E+10			

Table 3: Summary Emergy Evaluation of S&S Homestead Farm

The 50-acre farm operates to provide food to members of the surrounding community, to the farmer and his family, and to farm interns. All products sold off the farm are direct-marketed through a CSA (vegetables and fruit), or by contract (meat and dairy). All labor is provided by the farm family, with assistance from two to four interns during the summer. While the interns provide labor during the summer, the analysis presented is of a typical year, and the labor requirements are estimated based on the person-hours needed to perform a given task. Therefore, a single transformity was calculated for the labor inputs, based on the emergy in the services the farmer receives from the economy. The farm strives for self-sufficiency and this is reflected in the farm's diversity.

In emergy terms, the hay fields and pasture are the power base of the farm system, because they are the primary means of solar energy and storage used to provide for the farm animals throughout the winter. The fields are harvested for hay one time per year, usually in late June and early July. The hay is cut, raked, baled and stacked in the field for curing before being stored in the barn for winter feed for cattle and sheep. Labor is provided by the farmer and farm interns, and it requires the labor of approximately six people working eight hours a day for three days to bring the hay in. One thousand bales are usually stored, corresponding to approximately 22.7 tons of hay, with an energy content of 3.55 E+11 Joules.

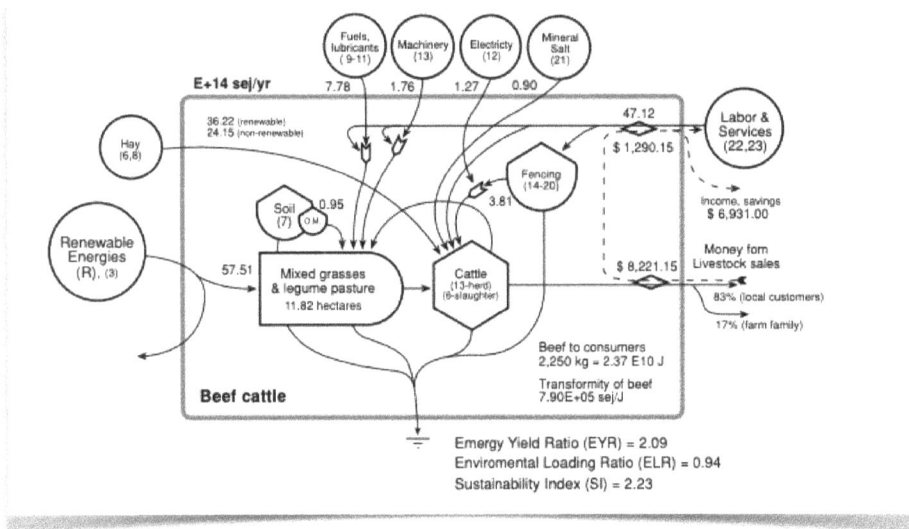

Figure 3: Emergy Systems of Beef Production

Meat Production

The beef cattle are a cross of Simmental, Hereford, Angus and Scottish Highland breeds. The cattle graze on grass pasture throughout the spring and summer months, and are fed hay throughout the late fall and winter seasons. The breeds were selected for their ability to transform grass into high quality protein, as the

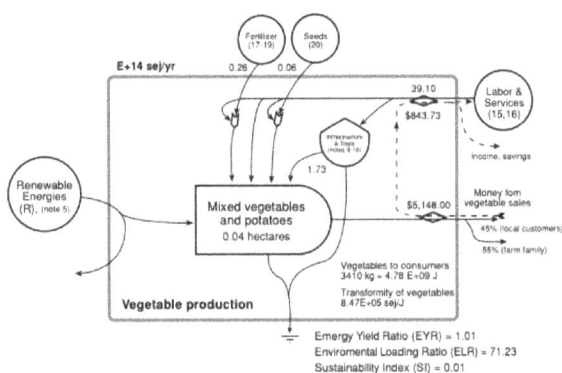

Figure 4: Emergy System of Vegetable Production

cattle are never fed grain. A system of intensive rotational grazing is employed, where the cattle are kept in one and a quarter-acre paddocks for several days before being rotated to fresh

pasture. In this way, the cattle are on fresh pasture every few days and do not return to a paddock until after at least six to eight weeks. This keeps parasite loads very low. In addition, this pulsing of the system may work to maximize the total productivity of the pastures, based on the concept that pulsing maximizes power.[12] Figure 3 is an emergy systems diagram of beef production and Table 3 is the corresponding emergy evaluation.

The sheep are a cross of Romney and Suffolk breeds and are primarily raised for meat, although their skins and fleece are retained for use on the farm, and for sale. The sheep generally graze the margins of the farm, are rotated primarily on the forest edge, and are not on any one pasture for more than one to two weeks. The estimated total land used for sheep was four and a half acres.

The pigs at S&S Homestead Farm are purchased as weaners and pastured in fifty-foot square pens where they are fed farm-grown barley, purchased soy meal, fruit and vegetable culls, as well as whey and other dairy by-products from the farm. The pigs are kept for six months and then slaughtered and sold to local customers, with one pig kept by the farm family.

Vegetable Production

The vegetables produced on the farm are cultivated using the bio-intensive method, based on the techniques developed by John Jeavons.[13] The beds are double-dug and raised four to six inches above soil level to facilitate drainage and to improve aeration and root penetration. All labor is done by hand and purchased fertility is

252

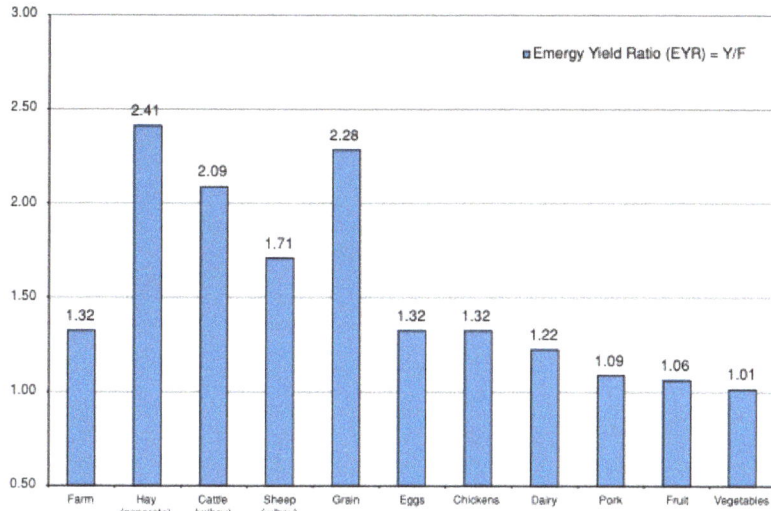

Figure 5: Emergy Yield Ratios for S&S Homestead Farm

kept to a minimum. The analysis shows that the small area used by the garden and the high degree of labor involved lower the sustainability of the garden in emergy terms. However, other factors influence the garden's design, such as securing a source of fresh food year-round, and maintaining high crop diversity.

The garden is not intended, nor designed, to yield net emergy. Figure 4 is an energy systems diagram of vegetable production.

Other Farm Production (Barley, Milk, Eggs, and Fruit)

Barley is the grain crop produced on the farm and is grown for feed for chickens and pigs, and to provide straw for animal stalls during winter. The grain was planted by hand and tilled-in using a rotovator. Harvesting was hired out to a local farmer, registered as a purchased service. The farm has one milk cow that is milked twice a day. The cow is free to forage on approximately three and a half acres of mixed grass and legume pasture and is fed hay during fall and winter. The primary emergy inputs come from human labor, because the milking is done by hand. Egg production is based on twenty-five laying hens that are allowed to range freely around the farm. The estimated range area is two and a quarter acres. The fruits and berries grown on the farm are produced for home consumption only. The orchard contains apple, plum and cherry trees as well as blueberries, strawberries and currants. The grass in the orchard is mowed a few times during the summer months and this accounts for the fuel and machinery

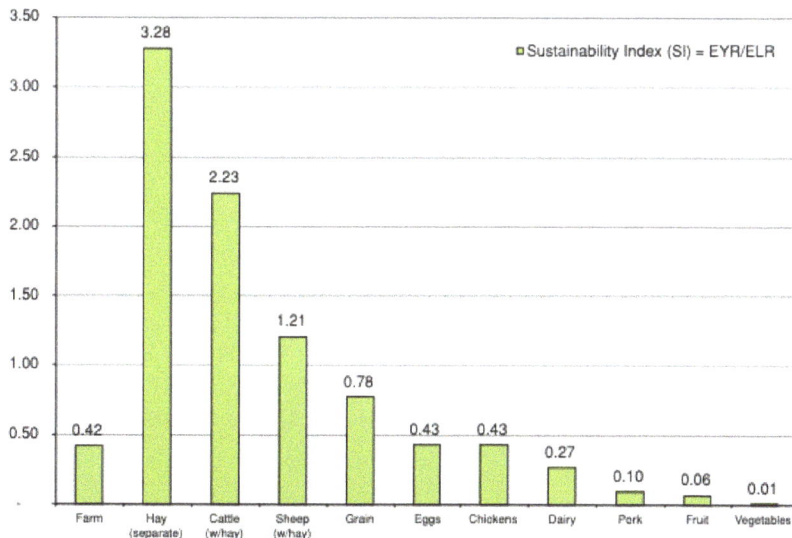

Figure 6: Sustainability Index for S&S Homestead Farm

253

used in fruit production. Otherwise, all other labor is done by hand.

Emergy-Based Ratios and Indices

In order to compare the results of the emergy evaluations of the farm and its subsystems, a number of emergy-based indices were calculated. The indicators reveal that those management areas that are highly dependent on purchased inputs and human labor, and are spatially less extensive relative to the amount of feedback emergy, receive the lowest sustainability ratings when compared to the areas of the farm that are spatially extensive and receive a larger portion of their total emergy requirements from the local environment.

The Emergy Yield Ratio (EYR) is the ratio of the emergy of the output (Y), divided by the emergy of those inputs (F) to the processes that are fed back to the system from outside. Figure 5 is a graph of the Emergy Yield Ratio of the farm and its subsystems or management areas. It indicates that those subsystem management areas of the farm that have a large renewable component in comparison to the economic and labor inputs they require, have the highest EYR.

The EYR indicates those subsystems that are driving, or powering, a given system. In the case of S&S Homestead Farm, it is clear that the grass-fed livestock systems, and the hay and grain fields, are the systems that provide the most yield to the farm organism. This should be taken into consideration when management questions arise regarding these areas. If the maximum power principle is correct, then it would behoove the stewards of the farm to make sure that these large and important subsystems receive feedback — in the form of nutrients and land care — that is commensurate with the relative importance of these systems to overall health of the farm as both an ecological and economic entity.

Environmental Load Ratios (ELR)

The Environmental Load Ratio (ELR) is the ratio of purchased (F) and indigenous non-renewable emergy (N) to free environmental emergy (R). It is an indicator of the amount of stress that a production process places on the local environment, measured against a backdrop of natural undisturbed ecosystems. If a given agricultural system requires large amounts of purchased emergy and draws down the nonrenewable storages that form the productive base of the system (such as soil organic matter), then the system will register a high ELR. In the case of S&S Homestead Farm, none of the management areas are diminishing the soil resources to any great extent, so any high ELR values that are registered are based on high purchased (F) emergy inputs in comparison to the renewable inputs (R). The only exception is the grain which registered comparatively higher uses of nonrenewable storages or (N) values.

The Sustainability Index (SI)

The Sustainability Index (SI) equals EYR/ELR and is an aggregate measure of yield and sustainability that assumes that the objective function for sustainability is to obtain the highest yield ratio at the lowest environmental load.[14] On S&S Homestead Farm, the SI indicated that those land areas that were the smallest relative to the amount of labor and purchased services required for the respective enterprises, registered the lowest SI figures. Specifically pork, fruit and vegetable production, which are all intensively managed enterprises of the farm, register low SI ratings. In the overall farm analysis, the farm infrastructure is considered in its entirety and thus lowers the farm's combined sustainability index below what might be expected from the sum of the subsystem analyses. Figure 6 is a graph showing the SI ratings for S&S Homestead Farm and its enterprise subsystems.

Efficiency of Agricultural Production (Transformity)

When the energy used during production is divided by the energy remaining in the product, one derives the transformity of that product, expressed as the ratio of solar em-joules per

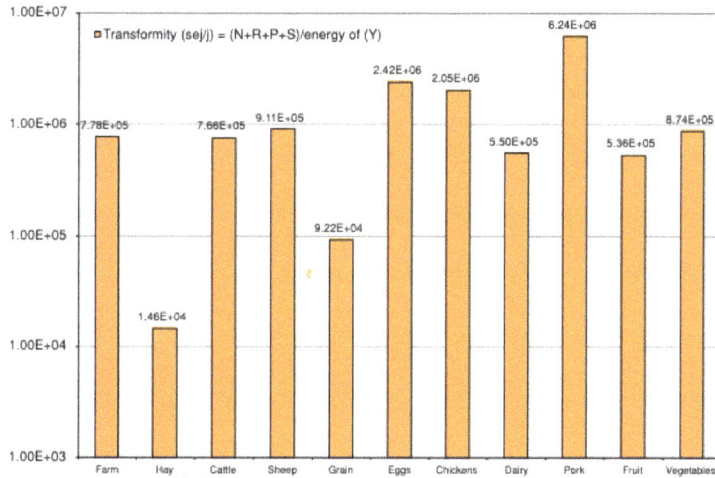

Figure 7: Transformities for S&S Homestead Farm and its subsystems

Joule (sej/J). On S&S Homestead, the various management areas differ widely with respect to the relative efficiency at which a Joule of a given product can be produced. In nature, it has been observed that animals are usually of one to two orders of magnitude higher transformity than green plants. Since this pattern is a sustainable one, it provides a reference frame from which to decipher sustainable patterns in mixed agricultural systems.

Specifically, it is instructive to note the relationship between hay and cattle production on S&S Homestead, which are coupled systems, in comparison to grain and pork production, also coupled systems. The transformity of the animal feed — hay and grain, respectively — are between one and two orders of magnitude lower than the animals they feed. This only makes sense because a large quantity of relatively low quality

energy (hay and grain) is required to produce smaller quantities of higher quality products (beef and pork). It is also interesting to note that the plant food products humans enjoy register transformities that are essentially equal to meat products. If humans were listed on the graph, we would register one to two orders of magnitude higher transformity than the meat, fruits and vegetables that feed us. Figure 7 is a graph showing the transformities for S&S Homestead Farm and its subsystems.

Ecological (Emergy) Footprint

The Ecological Footprint (EF) is a popular concept and an accessible accounting tool used to quantify the amount of resources consumed by a human population within a given area.[15] With EF accounting, the resources consumed by a process or population are translated into an estimation of the amount of productive land needed to produce the resources in question. An emergy-based ecological footprint can also be calculated using data compiled for emergy analyses. After all resource flows to a system have been accounted for and translated into emergy values, one can calculate an Emergy Footprint Ratio (EFR). This is derived by dividing the total emergy yielded by a system (Y) by the

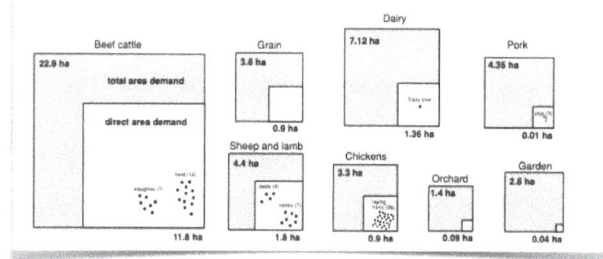

Figure 8: Management Area Ecological Footprints

Reference	Product	Transformity (sej/J)	Emergy flow (sej/yr)			EYR	Emergy-based indices		SI
			R	N	F		ELR	EFR	
(Brandt-Williams, 2001)	Cabbage (1 ha, Florida, 1992)	2.08E+05	1.63E+15	9.50E+14	6.71E+15	1.38	4.70	5.70	0.29
(Brandt-Williams, 2001)	Potatoes (1 ha, Florida, 1992)	1.49E+05	1.49E+15	9.50E+14	1.03E+16	1.24	7.52	8.52	0.16
(Lagerberg & Brown, 1999)	Greenhouse tomatoes (wood heated, Sweden)	5.36E+06	1.00E+17		1.41E+18	1.07	14.10	15.10	0.08
(Brandt-Williams, 2001)	Tomatoes (1 ha, Florida, 1992)	5.97E+05	1.56E+15	-	2.56E+16	1.06	16.40	17.40	0.06
This study	Vegetables (0.05 ha, S&S Homestead)	8.74E+05	5.78E+13	-	4.12E+15	1.01	71.27	72.27	0.01
(Brandt-Williams, 2001)	Dairy cow (1 ha, 1 cow, Florida, 1996)	1.21E+06	4.18E+15	9.50E+14	1.83E+16	1.28	4.61	5.61	0.28
This study	Dairy cow (1.36 ha, 1 cow, S&S Homestead)	5.50E+05	9.28E+14	1.10E+13	4.24E+15	1.22	4.58	5.58	0.27
This study	Beef (11.8 ha, 19 cattle, S&S Homestead)	7.66E+05	9.37E+15	9.50E+13	8.68E+15	2.09	0.94	1.94	2.23
(Brandt-Williams, 2001)	Beef (10 ha, 20 steers, Florida, 1992)	5.64E+05	2.98E+16	1.00E+14	4.23E+16	1.71	1.42	2.42	1.20
This study	Barley (1 ha, S&S Homestead)	9.22E+04	4.87E+14	5.90E+14	8.38E+14	2.28	2.94	3.94	0.78
(Brandt-Williams, 2001)	Oats (1 ha, Florida, 1992)	2.09E+05	1.56E+15	9.50E+14	3.17E+15	1.79	2.64	3.64	0.68
(Rydberg & Jansen, 2002)	Oats (1 ha, Sweden, 1927)	4.47E+04	2.84E+14	-	1.03E+15	1.27	3.64	4.64	0.35
(Ivarsson, J., 2000)	Barley (1 ha, Sunshine Farm, Kansas, 2000)	2.10E+05	3.45E+14	8.53E+14	2.81E+15	1.43	10.63	11.63	0.13
(Andresen, et al., 2000)	Pig production, ecological (0.41 ha, 3 pigs, Sweden)	4.80E+05	1.70E+14	-	1.32E+15	1.13	7.77	8.77	0.15
This study	Pig production (0.34 ha, 3 pigs, S&S Homestead)	6.24E+06	1.66E+14	-	1.84E+15	1.09	11.04	12.04	0.10
(Andresen, et al., 2000)	Pig production, conventional (0.186 ha, 3 pigs, Sweden)	5.80E+05	7.71E+13	-	1.72E+15	1.04	22.31	23.31	0.05
This study	Hay (1 ha, S&S Homestead)	2.49E+04	4.87E+14	8.04E+12	3.50E+14	2.41	0.74	1.74	3.28
(Rydberg & Jansen, 2002)	Hay (1 ha, Sweden, 1927)	1.36E+04	3.86E+14	-	3.83E+14	2.01	0.99	1.99	2.02
(Ivarsson, J, 2000)	Sunshine Farm, Land Institute (Kansas, 2000)	-	7.75E+15	1.73E+16	7.88E+16	1.32	12.40	13.40	0.11
This study	S&S Homstead Farm	7.78E+05	1.22E+16	2.01E+14	3.84E+16	1.32	3.17	4.17	0.42

Table 4: Transformity Comparison

total renewable emergy flows (R) supporting that same system (EFR = (R+N+F)/R). The resulting number indicates how many times larger a production system's support area receiving renewable emergy would have to be for it to meet its emergy requirements locally. Figure 8 depicts this concept graphically by calculating the footprints of the various management areas of S&S Homestead Farm.

Transformity Comparison

Table 4 compares the output and sustainability of S&S Homestead Farm with other systems yielding similar products. It is offered to provide a context for the analysis, and to show how the farm compares with other systems in different locations.

Concluding Discussion

By characterizing a farming system in terms of its energy flow dynamics, including overall energy conversion efficiency and external resource dependency, we can gain an accurate picture of what a particular farming system requires to be maintained, the quality and quantity of its output, and its effects on the local environment. Using energy as measure, food production at S&S Homestead is both efficient and relatively sustainable, given the amount of work from both nature and the human economy required to produce its output. Meat production represents the bulk of the food produced on the farm, and the pasture-based beef and lamb production exhibit both good efficiency (relatively low transformity) and low environmental load in comparison to other systems.

Clearly, the management areas of the farm differ greatly in overall management intensity. When gauging sustainability using the emergy-based ratios, it is important to understand that the EYR and ELR are ratios of local renewable and nonrenewable inputs to feedback from outside, so the number of variables is three, not just two.

This means that a sustainable system is not only characterized by a low requirement of feedback, but also by a large renewable input in comparison with the feedback itself, which can also be large.[16] In the right circumstance, a large purchased input from outside the process can confer sustainability on a production process, so long as the purchased inputs are matched to a large amount of emergy from renewable sources.[17]

S&S Homestead is a good example of a mixed, balanced farm, where some management areas are both spatially extensive and require low labor inputs, while others are more intensive, yielding smaller quantities of specific products that the farmers value for reasons other than their yield to other parts of the system. While emergy analysis is one way to objectively assign non-market values to products, and has proved a useful tool for evaluating S&S Homestead, it cannot capture the intangible properties that are often key components of human value judgements regarding environmental decisions.

In addition to the ubiquitous market valuation of goods and services — based on the neoclassical paradigm of willingness-to-pay as sole measure of value — Spiritual values, cultural mores, ethics and aesthetic preference all inform the management practices of farmers and natural resource stewards and the societies they support.

As a scientific measure of value, emergy analysis seeks only to show, in objective terms, what has gone into making a product and to what extent that product is compatible with its local environment. Emergy analysis, in this case, was used to monitor the performance of S&S Homestead in terms of how closely the farm resembles a natural system and how much the farm system relies on external resources for its operation.

It does not presume to explain the reasoning behind why the farm is organized the way it is. It does show, however, the end result of the reasoning behind the organization of the farm, and that this reasoning is sound, based on its performance in comparison to other systems yielding similar products.

Notes

[1] Abbreviated version of Andrew Haden's farm internship research report, 2002. For the complete paper, which the author submitted to the Department of Rural Development and Agroecology at The Swedish University of Agricultural Sciences in partial fulfillment of the M.S. degree, see https://sshomestead.org/wp-content/uploads/emergy%20analysis.pdf.

[2] Odum, Howard T. 1996. *Environmental Accounting: EMERGY and Environmental Decision Making*. New York, NY.

[3] Soule, Judy. & Jon Piper 1992. *Farming in Nature's Image — An Ecological Approach to Agriculture*. Washington, D.C.

[4] Jackson, Wes 2002. "Natural Systems Agriculture: A Truly Radical Alternative," *Agriculture, Ecosystems and Environment*, 88: 111-117.

[5] Odum, Howard T. 1988. "Self-Organization, Transformity, and Information," *Science* 242: 1132-1139. For a discussion of the philosophical principle of self-organization, see Henning Sehmsdorf, "The Spirituality of the Soil: The Idea of Teleology from Aristotle to Rudolf Steiner," below.

[6] Brown, Mark T. & Sergio Ulgiati 2001. "Emergy Measures of Carrying Capacity to Evaluate Economic Investments," *Population and Environment* 22; 471-501.

[7] Brown & Ulgiati, ibid.

[8] Rydberg, Torbjörn & Jan Jansén 2002. "Comparison of Horse and Tractor Traction Using Emergy Analysis," *Ecological Engineering*, 19(1): 13-28.

[9] Brown, Mark T. & Sergio Ulgiati 1999. "Emergy Evaluation of the Biosphere and Natural Capital," *Ambio* 28, vol 6: 488.

[10] Odum, op.cit

[11] Ulgiati, Sergio & Mark T. Brown 1998. "Monitoring Patterns of Sustainability in Natural and Man-Made Ecosystems," *Ecological Modeling*, 108: 23-36.

[12] Odum, Howard T. 1994. *Ecological and General Systems: An Introduction to Systems Ecology*. Boulder, CO.

[13] Jeavons, John 2017. *How to Grow More Vegetables*, 9th Edition. Berkeley, CA.

[14] Ugliati & Brown, op.cit, 33.

[15] Wackernagel, Mathis. & William Rees 1996. *Our Ecological Footprint: Reducing Human Impact on the Earth*. Philadelphia, PA. Folke, Carl, Åsa Jansson, Jonas Larsson & Robert Costanza 1997. "Ecosystem Appropriation by Cities," *Ambio* 26/3.

[16] Ulgiati & Brown, op.cit.

[17] Ibid.

On-Farm Research: Biodynamic Forage Production

Henning Sehmsdorf, 2007[1]

Summary

The shallow, glacial soils on Lopez Island, WA, carved from virgin forests during the latter half of the nineteenth century, tend to be acidic. Conventionally, farmers adjust soil pH by spreading agricultural lime, with the unintended consequence of releasing substantial levels of CO_2 into the atmosphere. The three-year on-farm research project in collaboration with Washington State University (WSU) researchers in forage production, microbiology and soil science, compared three treatments (Pfeiffer BD Field Spray, Lime, and Control) with four replicates in a randomized design. Data analysis documented for the first time that BD treatments increase soil pH sufficiently to improve nutrient availability, and at the same time increase forage quality as measured by protein levels to a statistically higher level than in plots treated with lime. The economic and ecological savings achieved by substituting BD sprays for lime must be weighed against the labor costs involved in on-farm production and application of BD sprays.

Project Description

Biodynamic farmers consider it an important task "to heal the earth." I occasionally shock people by asserting that at S&S Homestead, we do not farm for profit but for health. A farm, like any enterprise, has to be economically viable; in other words, it has to pay its bills. But that is not the same thing as extracting profit from the land to maximize our participation in the global market. Instead, we farm in response to the potential of the natural ecosystem. This includes species of native and domesticated animals, plants, and living biomass interacting in the local environment.

Our farm is small. We own fifteen acres, and lease another twenty-five from neighbors, and produce beef, pork, lamb, chicken, eggs, dairy products, fruit, and vegetables. We also produce all the animal feed and the farm's sources of fertility. We think of the farm as a self-organizing, self-correcting, self-sufficient, and self-capitalizing organism, what Aristotle might have called an entelechy ($\dot{\varepsilon}\nu\tau\varepsilon\lambda\dot{\varepsilon}\chi\varepsilon\iota\alpha$)[2] of people, animals, plants and soil. After nearly four decades of production, we see emergent qualities resulting from interactions within the complex farm system: for instance, place-specific immunities which make plant disease or pest control, veterinary or medical interventions unnecessary; time- and labor-saving habits of mind and practice arising from years of observation of living processes; and social dynamics strengthening local food security and independence.

The research project I want to describe here plays a role in the unfolding of the farm toward a self-sufficient system in which the amount of energy entering the system is in equilibrium with the amount leaving it, which is another way of describing our goal of ecological food production.[3] The project, which was funded by a small grant from SARE, was started in 2003 and completed in 2006. The goal of the project was to test whether biodynamic preparations can be substituted for lime applications to balance soil pH, thereby increasing the availability of soil nutrients supporting the production of harvestable proteins in grazed and harvested forages.

Ecological Context

Let me begin by placing the farm in the context of the physical ecosystem of which it is a part. S&S Homestead Farm is located on Lopez Island in the San Juan Archipelago. The island lies within the Northwest Pacific climate belt, in the rain shadow of the Olympic Mountain Range, the result of the last major glaciation which occurred about 12,000 years ago. The retreating glacier scraped the uplifting rock masses, leaving behind shallow soils, mostly clay mixed with gravel and erratic boulders that typically strew fields and pastures.[4] The soils on S&S Homestead are clay and gravel loams, rarely more than four to five inches in depth. Since 1970, when we first started farming here, we have improved garden soil depth, texture, aeration, and water holding capacity in berry and vegetable plantations by digging beds up to thirty inches deep, removing rocks and gravel, adding plenty of organic matter, and spraying with farm-produced, fermented herbal (BD) preparations.

Over the years, the combined application of double-digging, composts and biodynamic

sprays has produced dramatically improved garden soil texture and fertility. We have observed a diminished need for soil tillage — we no longer use any rototillers in the gardens — which in turn means vastly reduced weed pressure because dormant seeds are no longer brought to the soil surface, as well as reduced destruction of fungal hyphae by mechanical action, resulting in improved air and water holding capacity.

However, while double-digging works exceptionally well to establish deep soils, and is quite doable in vegetable gardens, it is hardly practical in managing crop fields, pastures, and hayfields. Of course, you could use a so-called tub grinder to scoop up the earth a couple of feet deep, separate rocks and other debris, and lay down a fluffed sheet of soil. Needless to say, this action essentially destroys the soil texture, tears up fungi and macerates earthworms and arthropods. In other words, use of a tub-grinder is a typical industrial, non-ecological process.

There is another reason we don't want to disturb the soils in our pastures and hayfields through mechanical action. The forages we have here were brought originally by Euro-American settlers who came to the islands between the middle and the end of the nineteenth century. They cleared the forests, extirpated large game such as bear, elk and wolves, and brought domestic cattle. They also brought nuisance species, such as the European rabbit, hares, foxes, starlings and, more recently, rats. They also imported a great diversity of grasses, forbs and legumes that naturalized in the islands. Having adjusted to our climate and soils, these forage species have proven remarkably sustainable and resistant to disease, and to excessive winter moisture and summer drought. These forages are not as productive as recently developed hybrids but their longevity is much greater. This eliminates the need for repeated tillage, reseeding, fertilization and irrigation that are economically expensive and ecologically destructive. Since 1970, the only fertilizer we have applied to our pastures and hayfields has come from the droppings of cattle and sheep rotated through the fields after the single annual hay cutting at the height of spring growth, usually in June.

However, we have noted that most of the soils on Lopez Island carved out of forests in the nineteenth century tend to be somewhat acidic (pH of 6.1 on average), which limits the availability of plant nutrients and encourages the growth of intrusive mosses. If the farm could support more cattle than the scale-appropriate rate of two cows per acre, pH levels could probably be modified by applying composted animal manures. Barring that, we are looking to biodynamic preparations to make the difference. Conventionally, most farmers rely on agricultural lime. However, lime is an industrial product prepared from mined sedimentary rock derived from marine invertebrates (calcium carbonate, $CaCO_3$). Because calcium carbonate is insoluble in water, it has to be digested by soil organisms (thereby releasing an estimated 50% of the CO_2 into the atmosphere or the groundwater),[5] before becoming chemically effective in adjusting soil pH. Alternatively, the limestone is heated in special kilns to decompose it into calcium oxide (CAO), which is readily soluble in water, and carbon dioxide which is released during the refinement process. Calcium oxide, also called quicklime, acts swiftly to change soil pH in plowed soils, for example, but it cannot be applied to pastures or hayfields because it would burn the living plants — one wonders, of course, what the application of quicklime does to soil organisms in unplanted fields. Whether in the form of calcium carbonate or calcium oxide, however, the lime has to be ground, bagged, and shipped through various commercial channels to the end user, the farmer, who applies it to the field with special machinery. In other words, while relatively inexpensive to purchase, the indirect energy costs of agricultural lime are substantial and appear low only because the U.S. government subsidizes energy consumption.

Furthermore, the CO_2 released into the atmosphere when applying calcium carbonate to the soil, or during the process of refining calcium carbonate into calcium oxide, is a greenhouse gas that contributes to global warming. In 2001, the estimated emissions of CO_2 from agricultural lime in the U.S. was 4.4-6.6 Tg.[6]

Research Team and Tasks

The on-farm project to test the effectiveness of biodynamic preparations in comparison to lime involved farm collaboration with the following WSU research and extension faculty:

~ Dr. Steve Fransen, forage specialist, who provided a predetermined schedule for grazing and mowing the field, and for taking random forage samples to be evaluated in his lab, and for measuring forage height and weights. All of these tasks were to be carried out by the farmer;

~ Andy Bary, M.S., soil scientist, Small Farms Program, who took baseline pH tests of the field soil, and determined the rate and timing of applying agricultural lime on the designated plots;

~ Dr. Lynne Carpenter-Boggs, microbiologist, and Coordinator for the BioAg Program in the Center for Sustaining Agriculture and Natural Resources (CSANR);

~ Jennifer Reeve, Ph.D. candidate in microbiology, who, together with the farmer, selected and laid out the experimental field, and carried out periodic soil tests at a depth of 0-3" and 3-6" to measure pH, rates of hydrogenation, phosphate and carbon mineralization, and microbial biomass;

~ The team also included Hugh Courtney, M.S., Director of the Josephine Porter Institute of Applied Biodynamics, who selected and supplied the biodynamic preparations and determined their rate and frequency of application;

~ Last, but not least, the team included the resident farmer, who coordinated the three-year project, and managed the designated field, following established farm practices for grazing and mowing. I applied the lime and the Pfeiffer BD Field Spray, took forage samples, and measured forage weights and heights, as instructed. I also kept a central field record integrating all research data.

Project Procedures and Dates

In autumn of 2003, a designated level field located at the center of the farm and measuring 225x144 feet was divided into twelve randomly selected plots to be treated with lime or biodynamic preparations, or left untreated for control. Unlike the garden soils, this field had never been treated with biodynamic preparations before. The entire field was fenced, grazed by sheep on a rotational basis over three years (2004-6), and intermittently mowed.

In December, 2003, Pfeiffer BD Field Spray — a proprietary compound preparation which includes BD 500 (horn manure), BD 502 (yarrow), BD 503 (chamomile), BD 504 (nettle), BD 505 (oak bark), BD 506 (dandelion), and BD 507 (valerian) — was applied to the selected plots at the rate of 2 oz/acre (0.185 oz/18'x225' plot). BD 508 (horsetail) was applied at the rate of 2 oz/10 acres (0.0185 oz/plot). The same selection and rate of Pfeiffer BD Field Spray was applied in November, 2004, July 2005, and May 2006. One unit of BD 501 (horn silica) was applied in July 2004.

Soil tests measuring hydrogenation, phosphate and carbon mineralization, basal respiration, and microbial biomass were taken in March 2004, May 2005, and July 2006.

In May 2004, three root tubes were installed in each of the twelve plots to gauge seasonal root development and sloughing. One tube from each plot was lifted in November 2004 and sent to the lab for evaluation. However, ensuing deterioration of the markers made it impossible to locate the remaining tubes.

After initial pH tests were taken to establish a liming rate of 2000 lbs/acre, 225 lbs of lime were applied to each of the designated plots in November, 2004.

Forage samples were collected by harvesting all the forage inside a randomly thrown two-foot wooden square in April 2004, May 2005, and August 2006. The samples were weighed to establish forage quantity, and lab-tested for crude protein, total dry matter, calcium, phosphorus, potassium, magnesium, ash, and other indicators. Relative forage height was measured in May and August 2006, using a rising plate meter.

Project Data Analysis

Below (next page) are tables 1-3 showing the statistical analysis of the data collected over the course of the project. The soil and forage samples collected over the course of the project were evaluated in WSU laboratories by Dr. Fransen, and by Dr. Carpenter-Boggs with the assistance of Jennifer Reeve,[7] as follows:

"This study compared three treatments with four replicates in a completely randomized design. This design gives it statistical and scientific validity. It means: if/when the data show a treatment effect, that effect has happened in a reliable manner in most plots, not just one instance. Several questions can be addressed from this work. All results are site-specific because this work was not replicated at other sites.

1. Do the BD preparations or lime affect soil pH?
Yes, and yes. Initial pH prior to treatments averaged 6.15 which is slightly acidic. Ideal soil pH is approximately 6.5-7.0. At a low pH many plant nutrients are less available (figure 1), some microbial activities are reduced, and many plants have reduced productivity. Control (untreated) plots had a pH of 6.2 at the end of the study, showing no significant change in the native soil pH over time (figure 2).

As expected and normally observed, liming the soil did increase soil pH to 6.6. What has never been documented before is that there was also an increase in soil pH with the BD preps, to 6.4. Statistically, the untreated plots had the lowest pH, BD plots had higher pH than untreated plots, and limed plots had higher pH than both the untreated and BD plots.

Parameter	BD	Con-trol	Lime
pH	6.4b	6.2c	6.6a
Dehydrogenase (μ TPF/g soil)	11.2a	11.4a	11.1a
Phosphatase (μ p-nitrophenol/g soil)	517a	494a	518a
Readily Mineralizable Carbon (μg C/g soil)	102a	98a	118b
Microbial Respiration (μg C/g soil)	7.3a	5.0a	5.5a
Microbial Biomass (μg C/g soil)	926a	836a	771a
Cmic/Cmin	9.5a	10.2a	7.3b
QCO2	0.008a	0.006a	0.008a

Table 1. Means (n = 24) for soil analyses (depth 0-3 inches)

Parameter	Biodyna	Control	Lime
pH	6.4a	6.3a	6.5a
Dehydrogenase (μ TPF/g soil)	9.0a	8.9a	9.3a

Parameter			
Phosphatase (μ p-nitrophenol/g soil)	287a	386a	298a
Readily Mineralizable Carbon (μg C/g soil)	48a	52a	54a
Microbial Respiration (μg C/g soil)	3.6a	3.9a	4.4a
Microbial Biomass (μg C/g soil)	505a	568a	562a
Cmic/Cmin	11.2a	11.7a	11.9a
QCO2	0.007a	0.007a	0.007a

Table 2. Means (n = 24) for soil analyses (depth 3-6 inches)

Parameter	Biodyn	Contro	Lime
Forage	244a	219a	238a
Crude Protein	7.9a	7.6a	7.4b
ADF	41.6a	40.6a	41.9a
NDF	67.8a	66.3a	67.6a
TDN	55.0a	56.2a	54.8a
REF	77.8a	81.1a	77.9a
Ca	0.44a	0.44a	0.45a
P	0.15a	0.15a	0.15a
K	2.1a	2.1a	2.0a
Mg	0.18a	0.19a	0.19a

| Ash | 8.7a | 7.9a | 8.3a |

Table 3. Means (n = 24) for forage analyses

However, the pH difference between limed and BD plots was small, and in both instances the raised pH level was sufficient to improve the availability of essential nutrients. This increase in pH in the BD plots was consistent, not just in one plot. It cannot be explained by drift or effects from the neighboring plot, since the control plots were also randomly distributed in the field, sometimes neighboring the limed plots, and did not significantly change in pH. This is the first time it has been shown that BD treatments increase soil pH sufficiently to improve nutrient availability.

2. Do the BD preparations or lime affect forage yield?
No, and no. In this study neither BD preparations nor lime treatment significantly

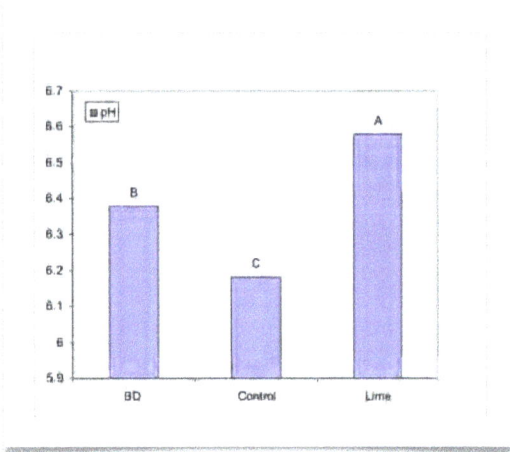

Figure 2. Soil pH as measured in 12 field plots over two years treated with biodynamic preparations, lime, and nothing.

affected forage yield as compared to the untreated plots.

Control plots had just as great a yield as treated plots. This calls into question the need and benefit of liming the slightly acidic soils on Lopez Island if yield increase is the goal.

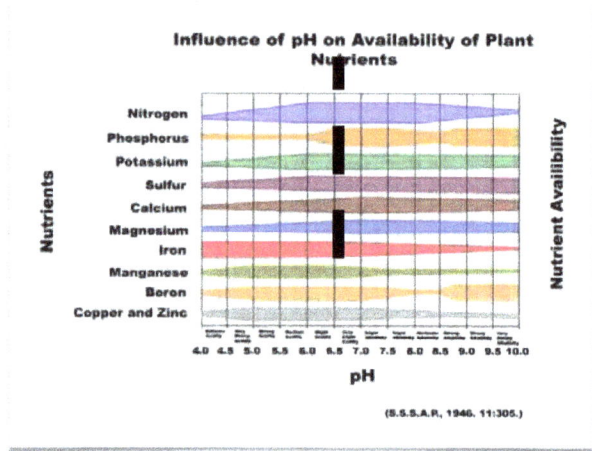

Figure 1. Soil pH of 6.1 is slightly acidic. Raising pH to 6.4-6 increases availability of N, P, Ca, Mg, and Mo.

3. Do the BD preparations or lime affect forage quality?
Yes, and yes. Both BD and limed plots had different forage quality than the untreated plots. However, the direction of change was different between treatments. BD plots improved in forage quality, shown by higher forage protein, while limed plots declined in forage quality, shown by lower forage protein (figure 3). Neither change was very large, but they were statistically significant. This is the first time that it had been shown that BD treatments increased forage quality as measured by protein levels to a statistically higher level than in plots treated with lime.

4. Do the BD preparations or lime affect soil microbial activity?
No, and No. The microbial biomass (the total mass of microorganisms in a gram of soil established by various enzyme assays) was the same among all treatments. However, the proportion of microbial biomass carbon to

total mineralizable carbon was significantly lower in the limed treatment (figure 4). It is possible the lime stimulated humus decomposition (an effect of lime noted by Peter Proctor, *Grasp the Nettle*, 1997, p. 124 as leading to eventual soil degradation), but it is also possible that the extra carbon came from the decomposition of the lime itself, a purely physical process."

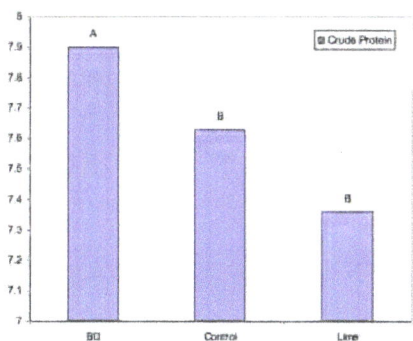

FIGURE 3. Crude protein content of forage measured in May 2005 (sample b) in 12 field plots treated with BD preparations, lime, and nothing.

Concluding Evaluation

The goal of S&S Homestead Farm is to achieve ecological sustainability and resource self-sufficiency, while remaining economically viable. These project results indicate that the farm can maximize forage quality by raising soil pH in fields sufficiently either through lime applications or by BD sprays, and thereby

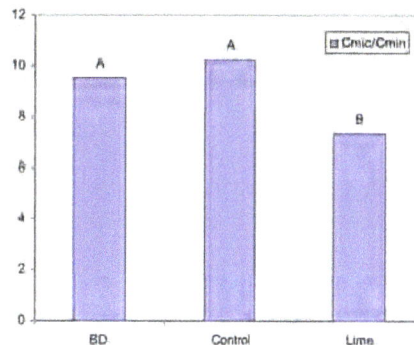

FIGURE 4. Microbial biomass carbon per unit soil mineralizable carbon measured in twelve field plots over two years treated with BD preparations, lime, and nothing.

increase mineral availability. The latter strategy appears preferable, because it reduces the need for purchased inputs that carry both economic and ecological costs to the farm and larger environment, and because it increases farm self-sufficiency. The data further suggest that the forage quality as measured by levels of crude protein, is higher in the BD-treated plots as compared to the lime-treated or control plots.

However, it is not to be overlooked that production and application of the BD sprays is labor intensive and therefore carries its own economic cost. This cost, involving growing, fermenting, processing and applying the sprays, remains to be calculated before a final evaluation of this approach to achieving ecologically and economically viable and scale-appropriate forage production can be made.

Notes

[1] Versions of this preliminary evaluation of the research project were presented at the Biodynamic National Conference held at Rudolf Steiner College, Fair Oaks, CA., 2007, and at the Eco-Farm Conference, Asimolar, CA, 2007. Published in *Biodynamics*, Summer 2008, no. 265. For a more complete analysis in the context of the whole-farm life-cycle, see Reeve, Jennifer, Lynn Carpenter-Boggs & Henning Sehmsdorf 2011. "Sustainable Agriculture: A Case Study of a Small Lopez Island Farm," below.

[2] Coined by Aristotle (384–322 B.C.) from ἐντελής (*enteĺés*, "complete, full, accomplished") + ἔχειν (*ékhein*, "have, hold"). See *Metaphysics* 1050a, 21-23.

[3] See Haden, Andrew 2002. "Emergy Analysis of Food Production at S&S Homestead Farm," below.

[4] Lewis, Mark & Fred Sharpe 1987. *Birding in the San Juan Islands*, Seattle , 9-11.

[5] McBride, A.C. and T.O. West, "Estimating Net CO2 Emissions From Agricultural Lime Applied to Soils in the U.S," *American Geophysical Union*, Fall Meeting 2005, abstract #B41B-0191.

[6] Ibid.

[7] Carpenter-Boggs, Lynn and Jennifer Reeve 2006. SARE Project Report. http://wsare.usu.edu/pro/?sub=fund04#wa. Retrieved July 15, 2021.

Sustainable Agriculture, a Case Study of a Small Lopez Island Farm[1]

J. R. Reeve, L. Carpenter-Boggs, H. Sehmsdorf, 2012

ABSTRACT

Over-reliance on fossil-fuel based inputs, and transport of inputs and products is seen by many as a threat to long-term agricultural and food system sustainability. Many organic, biodynamic, and low-input farmers limit off-farm inputs, attempting instead to farm within the carrying capacity of their land or local environment. These farmers often accept lower farm productivity because they see reduced reliance on non-renewable inputs as more sustainable. Documentation of low-input agricultural systems through both replicated research trials and case studies is needed in order to better understand perceived and real advantages and tradeoffs. The goal of our study was twofold: 1) to compare liming and biodynamic (BD) preparations in improving pasture on a moderately acidic pasture soil through stimulation of soil microbial activity; 2) to place these findings within the context of a whole farm analysis of economic, plant, and animal health.

Treatments included lime, the Pfeiffer Field Spray plus BD compost preparations, and untreated controls. Soil pH, total C and N, microbial activity, forage biomass, and forage quality were evaluated over two growing seasons. Both lime and the Pfeiffer Field Spray and BD preparations were only moderately effective in raising soil pH, with no effect on soil microbial activity or forage yield. Lime significantly reduced forage crude protein but the practical implications of this are questionable given the overall quality of the forage. While the farm is profitable and economically stable and the animals healthy, the need for future targeted nutrient inputs cannot be ruled out for sustainable long-term production.

INTRODUCTION

Many organic, biodynamic, low-input, and geographically isolated farmers limit or entirely eliminate purchase of off-farm inputs, attempting instead to farm within the carrying capacity of their land or local environment. Wastes are recycled, transportation costs to and from the farm, external inputs, and reliance on

267

fossil fuels are all reduced. Animals and plants are managed to be more adapted to their local environment through cross breeding and open pollination. Food is sold to the local community and fertility is sourced within that same local community. Rotational grazing of animals and diverse crop rotations keep nutrients cycling on-farm and increase nutrient use efficiencies so that external inputs can be reduced to a bare minimum or eliminated altogether. Fertility, it is postulated, then becomes an emergent property of the way the farming system is designed and operated. This approach has been described by Edens and Haynes[2] among others as closed system agriculture.

These farmers often accept lower farm productivity in exchange for improved sustainability and reduced reliance on expensive and non-renewable inputs. The potential risk in such an approach is that internally generated fertility may be insufficient to offset the export of nutrients through off-farm sales. However, learning to optimize a locally based system may be the better choice when faced with the alternative of expensive and non-renewable inputs.

The call to reduce reliance on non-renewable and increasingly expensive agricultural inputs has been voiced by many over the past decades.[3] With rising world populations, this presents a dilemma: how can we produce sufficient food to feed this rapidly growing population while at the same time conserving biodiversity and developing food production systems that are less heavily reliant on resources that will ultimately be depleted? One approach is to increase the efficiency with which land and resources are utilized. On the other hand, there is growing evidence that increased agricultural efficiencies, while enhancing crop yield, do not necessarily lead to greater resource conservation. In fact, increased efficiency can stimulate ever greater resource use as prices of inputs drop.

In developing countries, land fragmentation and deforestation increase as rural populations displaced by agricultural intensification move to more marginal lands.[4] This fact is often missed in traditional assessments of agricultural productivity due to the failure to adequately address social, economic, and political realities. Clearly, a thorough assessment of the whole system, or life-cycle analysis,[5] is needed when addressing these questions.

Lopez Island, WA is a small island community of 3,000 people, in the Puget Sound, a fifty-minute ferry ride from the mainland. Until recently, most of the food and fiber consumed on the island was imported at high cost, via the Washington State ferry service. In recent years, the number of small farms has risen significantly, and many of these farms raise grass-fed livestock, fruits and vegetables for local consumption. Although not all are certified, many of these farms are organic or biodynamic. Synthetic inputs are restricted and most fertilizers and animal feeds are produced on the farm.

The soils on Lopez Island tend to be shallow and low in pH due to evergreen forest vegetation and poor drainage. These less-than-ideal soil conditions can make adequate forage production difficult without relying on external inputs. Typically, farmers apply lime to ameliorate soil pH. However, the cost of lime is high on Lopez Island due to shipping charges from the mainland, so many farms do not apply any lime at all. In addition to standard organic practices, biodynamic farmers make use of specially fermented plant and animal-based products that are applied as field sprays and compost starters (Table 1, below). These preparations are applied in very small quantities, which has made them a controversial aspect of the BD approach.[6] However, they do not act as nutrient sources, but as microbial stimulants which lead to greater availability of nutrients to plants, thereby offsetting the need for external inputs. Newer formulations of the BD preparations, such as the Pfeiffer Field Spray (PFS) also include the specific addition of microorganisms known for their plant-growth promoting properties.[7]

The activity of soil organisms is an integral component of soil formation processes, including the chemical weathering of parent material and nutrient cycling.[8] The potential of microbial stimulants to raise soil pH as a cost-effective substitute for lime therefore warrants investigation.

Peer reviewed evaluations of the efficacy of the BD preparations are few, however, particularly in forage systems. Colmenares and Miguel found that the preparations, sprayed on permanent grassland over three and a half years, increased dry matter content in the absence of any fertilization.[9] Some evidence suggest that the BD preparations influence soil microbial processes and carbon and nitrogen dynamics.[10] On the other hand, other studies have proved inconclusive.[11] There is no published research on PFS, or comparison of BD preparations to lime treatment. Moreover, many researchers have expressed concern that BD and other low-input organic and conventional systems may not be adequately replacing nutrients lost through export of agricultural goods.[12]

The goal of the study was to first evaluate the effects of BD preparations and lime on forage yield and quality. Secondly, S&S Homestead farm's soil nutrients, animal health, and economic stability were surveyed in order to place our findings within the context of whole farm health, environmental, economic, and social sustainability.

MATERIALS AND METHODS
Site history and Plot Layout

The trial was located on S&S Homestead Farm on Lopez Island WA, a twenty hectare diversified small holding raising animals, vegetables and forages for on-farm use and local markets. While not certified, the farm has been managed organically for over thirty-eight years. The climate on the island is temperate maritime with average highs of 21°C and lows of 1°C with a mean annual precipitation of 741 mm, most of which falls between November and February. No additional irrigation is supplied to the pasture. Experimental plots 3.7 m by 61 m were laid out in a permanent pasture in the fall of 2003 and baseline soil samples taken. Pasture species composition was approximately thirty-five per cent legumes, fifty-five percent grasses and ten per cent broadleaf forbs (Table 2, below). The soil type is a Bow Gravelly Silt Loam, zero to three per cent slopes, which had received no inputs other than pastured animal urine and manure in over thirty-eight years. The site was subject to seasonal poor drainage due to a heavy clay layer twenty to thirty cm below the soil surface. Treatments of lime, BD preparations, and an untreated control were applied in a randomized design with four replicates.

Lime, composed of 97% $CaCO_3$, 10 mesh; 38.8% Ca, 20 mesh; and 2% $MgCO3$, ten mesh[13] was applied in a single application in the fall of 2004 at a rate of 2.24 Mg ha⁻¹. Liming rate was calculated based on initial soil pH of 5.5. Biodynamic treatment consisted of one unit of PFS together with one unit of BD 502-507 applied in the fall of 2003 and 2004, and one unit of 501 and *equisetum arvensis* applied in the spring of 2004 and 2005 (see Table 1 for unit ha⁻¹ application rates, below). Pfeiffer Field Spray and BD preparations were purchased from and applied according to directions supplied by the Josephine Porter Institute for Applied Biodynamics.[14]

During the field study period 2003-2006, annual temperature was 1.0, 1.4, 1.1, and 0.8°C warmer and precipitation was 48 mm less, 144 mm more, 46 mm more, and 61 mm more than average at the National Oceanic and Atmospheric Administration (NOAA) Anacortes Station twelve linear miles from the farm.

Soil Sampling and Analysis

Soils were sampled from each plot at 0 to10 cm and 10-20 cm at the start of the trial and in May 2005 and June 2006. All samples were a composite of 10 subsamples taken from the plot area a minimum 1 m from the boundary of each plot to avoid edge effects. Samples were transported on ice to Washington State University, Pullman, WA, passed through a 2 mm sieve, and stored at 4°C until analysis. Soil pH was measured in a 1:1 w:v deionized water after 1 hour. Soil was finely ground and total C and N measured by combustion on a Leco CNS 2000. Readily mineralizable carbon (Cmin), basal microbial respiration rate, and active microbial biomass carbon (Cmic) by substrate-induced respiration (SIR) were measured according to Anderson and Domsch:[15] 10g soil was brought to 26% moisture content and incubated at 24°C for 10 days. Total CO_2 released during the 10 days was considered Cmin. Vials were uncapped, evacuated with a stream of air passed through water and covered with parafilm® for 22 hours (to allow soil CO_2 to equilibrate with the atmosphere without loss of soil moisture), recapped for 2 hours and CO_2 measured again for the basal respiration rate.

Samples were again uncapped, evacuated, and covered with parafilm® for 22 hours, then 0.5 mL of a 30 g L^{-1} aqueous solution of glucose was added to the same soil samples, rested for 1 hour before being recapped for 2 hours to measure SIR. Carbon dioxide was measured in the headspace using a Shimadzu GC model GC-17A, with a thermal conductivity detector and a 168 mm HaySep 100/120 column.

Dehydrogenase enzyme activity was measured using 2.5 g dry weight soil and acid and alkaline phosphatase enzyme activity using 1 g dry weight soil as described by Tabatabai.[16] Both enzyme reactions were measured using a Bio-Tek micro-plate reader model EL311s.

Forage Sampling and Analysis

Forage herbage biomass was clipped close to the ground from two random 2 m² samples from each plot in May and August of 2005 and 2006, dried and weighed. Forage height was measured using a rising plate meter. Each year after forage samples were taken the plots were mowed for hay and then rotationally grazed with sheep at an approximate stocking density of 4,500 kg ha[-1]. Stocking duration varied from two to seven days depending on available forage. A subsample of forage biomass was analyzed for crude protein, acid detergent fiber (ADF), neutral detergent fiber (NDF), total digestible nutrients (TDN), relative feed value (RFV), Ca, P, K, Mg, and Ash according to National Forage Testing Association methods.[17]

Whole Farm Survey and Animal Health Assessment

In the summer of 2009 a whole farm survey was conducted to determine overall soil fertility and health. Composite soil samples were taken from four fields on the farm, including the vegetable garden and the former experimental site. Ten to twenty samples were taken per field at a depth of 0-15 and 15-30cm, thoroughly homogenized and shipped to Soiltest Farm Consultants[18] for analysis. Soil samples were passed through a 2 mm sieve, stored at 4° C until analysis, and then analyzed for the following properties according to recommended soil-testing methods by Gavlak et al.[19]

Nitrate-nitrogen (N) was measured with the chromotropic acid method; ammonium-N was measured with the salicylate method; Olsen phosphorus and potassium were measured; DTPA-Sorpitol extractable sulfur, boron, zinc, manganese, copper and iron were measured; Soil pH and electrical conductivity were measured in a 1:1 w/v water saturated paste; calcium, magnesium and sodium were measured in a NH_4OAc extract; cation exchange capacity was measured using the NH_4 replacement method; SMP soil buffer was measured and total bases calculated by summation of extractable bases.

The health and productivity status of the farm's sheep flock were assessed by body condition scoring and reproductive success in June and September of 2009. Farm economic viability was assessed using farm records of purchases and sales.

Statistical Analysis

Data were analyzed as a randomized design (CRD) with treatment as whole plot and year (2005 and 2006) as subplot. Soil properties were analyzed separately by depth. For forage analysis, data from the May and August sampling dates within year were pooled to obtain average forage quality per year. Baseline data were analyzed separately as a CRD. All statistics were analyzed using the SAS system for Windows version 9.1 ANOVA and LSmeans (SAS Institute, Cary, NC). Data were checked for model assumptions and transformed as necessary. When data were transformed, LSmeans reported are in original units. Differences were considered significant at $P < 0.05$ unless otherwise stated.

Results and Discussion

Baseline data revealed no significant differences among plots at the start of the experiment, with an initial soil pH of 6.0 at 0-10 cm and 6.1 at 10-20 cm (data not shown). This soil pH is marginally acidic; however, previous spot testing conducted by the farmer showed soil pH of 5.5 in some areas.

After treatment, a few differences were seen within the 0-10 cm depth (Table 3, below), but there were no differences among treatments in any of the measured parameters at the 10-20 cm depth (Table 4, below). Soil pH rose significantly after treatment from the baseline pH of 6.0 in both limed and BD plots relative to the control in the top 10 cm (Table 3, below). There was no significant treatment x year interaction indicating the treatment response was similar in both years following treatment application. Soil pH in the limed treatment was greatest at pH 6.6, with only a slight change in the untreated control at pH 6.2 and an intermediate change in the BD treatment of pH 6.4. These differences are small and observed in the top 10 cm of the soil only; nevertheless, the rise in pH in both lime and BD spray treatments in the surface soil would be sufficient to improve the availability of several essential nutrients including nitrogen, phosphorus, calcium, magnesium and molybdenum.[20]

Readily mineralized carbon (Cmin) was greater and the Cmic/Cmin[-1] ratio was lower in the limed treatments (Table 3. below). This was due to a greater CO_2 release in the initial 10 days of soil incubation and a lesser relative stimulation of CO_2 release after the addition of soluble glucose to the limed soil. We saw no differences in microbial activity as measured by dehydrogenase and phosphatase enzyme activities. Increased microbial activity, microbial biomass, and soil respiration in response to liming have been reported in lab experiments, forest systems, annual tillage systems, no-till systems, and grassland systems.[21] Some studies indicate long-term use of lime to have negative impacts on organic matter levels in soils through increased microbial activity and C turnover.[22]

Others suggest soil C levels can remain stable in limed soils if the increased C turnover is replaced through increased plant biomass production as a result of improved nutrient availability.[23] Observations of greater Cmin respiration, Cmic biomass (by SIR) and any CO_2-based measures of soil after liming treatment must be considered with some reservations, however, because the inorganic carbonates of the lime itself can contribute to increased CO_2 flux. Bertrand et al. (2007) used labeled tracers to show increased CO_2 production from limed soils originated from mineral sources not organic matter, and cautions that increased soil respiration alone cannot be used as evidence for projected organic matter loss.[24] Thus the increased Cmin measured from limed soils in this study was likely due to continued lime decomposition, not an increased labile organic C pool, as there was no corresponding increase in microbial enzymatic activity, biomass, or reduction in total soil C by treatment. Further long-term monitoring of C dynamics in limed study sites is needed to resolve this debate.

Neither liming nor BD treatment significantly changed forage yield or height over the two years of this study (Table 5, below). Average forage yields were 1170 kg ha-1, which are somewhat low but could be expected as the site could be considered unimproved grassland; it had not been reseeded or received any external fertility inputs in at least thirty-eight years. Bulk forage quality was relatively low also, as characterized by a high ADF content (40.6-41.9%), low crude protein (7.4-7.9%) and TDN (54.8-56.2%). While these ranges are adequate for non-lactating sheep, lactating sheep as were grazed in this study typically require forage crude protein levels of at 9.5% and TDN around 70%.[25] Forage P levels for lactating sheep should be at least 0.18% which is higher than the 0.15% measured in this study.

Most measures of forage quality were similar among the treatments; however, forage biomass from limed plots had significantly (6% less) reduced crude protein as compared to BD (p <

0.01) and (4% less) control (p< 0.05), (Table 5, below). These statistically significant differences are unlikely to be biologically significant, however, as the crude protein was approximately 17-22% below that typically required for lactating sheep. The relative feed value (RVF) which describes overall forage quality was only 77-81 when good forage typically has an RVF of over 100. Lack of yield response to lime in pastures in the absence or with only minimal fertilizer inputs is not unprecedented.[26] This suggests that in low fertility pastures, lime without additional fertilizers may not always increase plant productivity. Low nutrient content may continue to override pH improvements or beneficial chemical effects may be accompanied by negative biological effects such as changes in mycorrhizal associations.[27] Soil fertility measurements indicate this pasture was low in nitrogen, potassium, phosphorus and sulfur (Tables 8 and 9, below).

To our knowledge this is the first study to show a soil pH effect with use of the PFS and BD preparations. Unlike Colmenares and Miguel (1999), however, we found no increase in forage dry matter content in response to BD treatment. There is considerable variability reported in the literature on BD preparations which could represent differences in quality of preparations, differences in site response, as well as the possibility of statistical anomalies. While our results will need to be confirmed, our data suggest that BD treatment may have the potential to ameliorate soil pH. Moreover BD treatment does not involve the mining and transportation of limited resources and BD field sprays can potentially be produced on farm. From the perspective of the farmer, these findings indicate that lime is not a beneficial input to this field at this time. Using BD preparations to moderate pH or using no treatment saves the expense of liming and maintains low non-renewable resource use, while producing similar or better forage quality than liming.

Sustainability of S&S Homestead

As described above, S&S Homestead has been operating organically for over thirty years and represents an example of a farm operated on the principles of a closed system.[28] The boundaries of the system do not represent the actual farm boundaries but include the local community, in this case Lopez Island. In order to assess the overall sustainability of the farm, an informal life cycle analysis was conducted[29] (Table 6, below). The twenty hectare farm is mostly in pasture and produces beef, pork, lamb, a wide selection of vegetables and fruit for sale and home consumption, and a CSA operated on 1.1 hectares of the farm. With the exception of the CSA, the farm also grows most of its own fertility and animal feed. Compost is made from spoiled straw and hay, manure from two dairy cows, and slaughter by-products, effectively recycling on-farm wastes. This compost is used exclusively in the vegetable and fruit growing operations. Pastures receive no inputs other than urine and manure deposited by grazing animals. The farm also has an extensive gravity-fed rainwater collection system that stores water in underground tanks and a holding pond. Water is pumped from the pond using a solar-powered pump and used for irrigating the vegetable gardens. Pastures are not irrigated and rely solely on rainfall for moisture

The farm is economically stable. As an example, in 2007 gross returns for S&S Homestead Farm were $42,666, 42% from animal products, 25% from fruits and vegetables, 19% from hay and grain, and 10% in services provided to the CSA (Table 7, below). Only 27% ($11,707) of the gross farm revenue came from the off-farm sales of meat and vegetables. The remainder represents services rendered to the CSA and dollars saved through on-farm consumption of meat, eggs and dairy products, fruits, vegetables and animal feed. Additional gross returns of $10,146 were earned by S&S Center for Sustainable Agriculture which hosts on-farm workshops and training days as well as interfacing with a local school in a farm-to-school program. Finally, gross returns of $20,146 were generated by Lopez Community Farm CSA, which is run as a subsidiary enterprise by management trainees for their own profit. The gross returns for all three farm enterprises for 2007 totaled $73,146.

Direct production costs for S&S Homestead Farm and S&S Center for Sustainable Agriculture (including on-farm consumption of products) totaled $44,288 and indirect costs (amortization) $8,083. For Lopez Community Farm CSA, direct costs amounted to $9,562, of which $8,282 were paid to the Lopez Community Farm CSA as profit, while S&S Homestead Farm retained $863 in profit.

The farm exports nutrients by off farm sales of animal and vegetable products. The farm sold 6,941kg of meat products which represented 59% of the total generated in 2007. 181kg of vegetables were sold from the vegetable garden and a larger though unrecorded amount from the CSA. Few inputs are purchased.

Some seed is saved but most is purchased from a local company that tests all varieties for local adaption. Open pollinated varieties are purchased whenever available with the exception of sweet corn.

Replacement animals (three pigs) are purchased yearly and some supplemental grain is also purchased in years suffering poor grain yield. In addition to using compost generated on-farm, the CSA purchased 81 kg ha^{-1} each of blood meal, greensand, rock phosphate, and lime, 61 kg ha^{-1} gypsum, and 1.4 kg ha^{-1} zinc and boron and 27 metric tons of wood chip mulch in 2007.

All farm operations are conducted by the

farmer and his wife and two to five seasonal interns. Interns receive lodging and food and a modest stipend. The CSA is managed by a separate couple who are paid exclusively through revenue generated by the CSA. The farm also operates a successful farm-to-school program. High school students come to the farm twice a week for about an hour or more to engage in various farm activities such as bucking hay, planting row crops, and weeding. Students learn about the particular activity in the context of the whole farm system and the larger ecological setting. Students also learn how to bake bread, make sauerkraut, cheese, and sausage, and learn about nutrition and health.

The farm also runs workshops and conferences though the S&S Center for Sustainable Agriculture. The farm balances its cash flows each year and receives no government subsidies. The farm has no equipment overhead or farm loans. It must be emphasized that it is not the ultimate goal of the farm to make as much profit as possible, but rather to maintain economic sustainability within the carrying capacity of the land.

Soil tests taken from three pastures and one of the vegetable gardens show that with the exception of the vegetable garden, most of the farm is low in nitrogen, potassium, phosphorus and sulfur (Tables 8 and 9, below). The vegetable garden is actually high in nutrients suggesting that nutrients are returned to this area in disproportionate quantities. While no records of soil fertility prior to organic conversion is available, the low levels of P, K and S in the pastures are a concern because this indicates the possibility that nutrients removed through the sale of meat are not being sufficiently replaced. Based on the volume of sales relative to purchases (see above) this seems likely.

Despite the concern over adequate replacement of nutrients from pastures sold off farm, it is important to note that soil fertility and forage quality measured against conventional farming standards is problematic for the assessment of organic farming. Firstly, soil tests that measure labile soil nutrients extracted in solution may not adequately reflect available nutrients in organic systems. For example, organic and inorganic P mineralization is increased under low levels of soluble P.[30]

Secondly, conventional nutrient recommendations on soil tests are linked to maximum yield targets. This is because in conventional systems, the low cost of fertilizer has meant that inputs needed for maximum economic return have closely matched fertility targets for maximum yield, despite the law of diminishing returns operating at high input rates.[31] This may not be the case for many organic systems, (and increasingly for conventional systems) particularly when inputs are expensive or not readily available. Regular soil testing should be an integral part of organic farm nutrient management plans as a tool for assessing long-term trends in soil fertility, but interpretations of standard fertilizer recommendations should take these additional factors into account.

Sheep health was assessed as an example of the diverse livestock on S&S Homestead Farm.[32] The flock was established in 1994 and consists of eight breeding ewes and one ram. The initial flock consisted of purebred Suffolks but due to initial health problems the flock was crossbred over the years with a series of different rams in order to increase genetic diversity. The initial ram was a purebred Romney followed after several seasons by a Polypay ram. The sheep flock became stronger, more self-sufficient, and durable. The Polypay was replaced with a Churro ram but this resulted in lambs that were flighty, rough fleeced and not of desirable finish and frame. The Churro was replaced with a Romney/Cheviot ram, which has sired the flock for the past two years.

In order to assess the weight gain of the lambs and health of the ewes, the flock was evaluated for general health, growth rate, and frame condition[33] in the summer of 2009. Unlike conventional sheep farming systems the lambs are not weaned but remain with their dams until they are almost ready for slaughter. Lambing is typically 140% which is similar to that of conventional flocks. Lambs grew at 0.7 to 1.1 kg per day. This is an acceptable rate of growth for conventional sheep enterprises, given that these lambs are not of large frame size. Moreover, the customer or organic base requests, and is satisfied with, the size of these market lambs, which average from 39-45 kg at slaughter. Singles are slaughtered in the fall of the year they were born, and twins are slaughtered as yearlings the following spring. The customer base continues to be willing to pay a premium for these lambs, due in their words, to consistent superior taste and tenderness.

Another product that comes from this flock is tanned hides. The quality of the hides was observed to be superior regarding intact hides (no holes, tears, scarring).

The sheep have year round access to pasture consisting of a considerable number of paddocks, through which they are rotated on a daily or weekly basis depending on season, weather and pasture growth. In addition to year round access to forage and trace minerals, ewes are given 500g of barley per day during early lactation. Four to six week lambs may be given molasses during unusually cold or wet spring seasons. None of the sheep are vaccinated or treated with any form of anthelmintics. Since the genetic diversity of the flock was increased there have not been any illnesses.

Despite the relatively low quality of the forage by conventional standards, the stocking density together with supplemental grain appears to be sufficient to provide adequate nutrition to the sheep. Both ewes and lambs were observed to

be in top condition based on body scoring through palpation and other measurements.

Regarding the forage quality, it is important to note that, unlike a typical conventional pasture, this pasture contains significant percentages of different legumes and forbes (Table 2, below). Bulk forage analyses may underestimate the forage quality as taken in by sheep since foraging behavior allows them to select the more palatable and more nutritious plants. Burkitt et al. showed that while milk yields were lower, parasite loads of biodynamic and conventional dairy herds were similar despite the biodynamic animals' receiving no anthelmintics.[34] Diverse plant mixtures, some high in plant secondary compounds, may allow animals to self-medicate through foraging behavior.[35] Eating plants that contain tannins, for example, reduces the load of internal parasites, alleviates bloat, enhances protein uptake, and improves immune response.[36] The access to diverse forage may explain their excellent weight gain and health assessment despite the low forage quality when compared to conventional standards.

Conclusions

So is S&S Homestead Farm achieving its goals of economic, environmental and social sustainability? As described above, the farm, while not highly profitable, is certainly economically stable, with a balanced cash flow. In terms of a whole farm system or life cycle analysis, S&S Homestead meets the criteria for economic, environmental and social sustainability.

Nevertheless there are concerns that the farm may not adequately balance soil nutrient budgets, in particular nutrient returns to pastures from the off-farm sale of meat products. This is a concern that applies to many biodynamic and closed systems organic farms and could be addressed through increasing off-farm purchases of targeted nutrients when needed. Relying on inputs such as lime and rock phosphate that have been mined and transported long distances cannot ultimately be thought of as a sustainable long-term solution, however. Another approach would be to try to increase mineral weathering of nutrients through stimulating soil microbial populations.

We tested this hypothesis by evaluating the effects of liming and BD preparations on soil pH, microbial activity, and forage yield and quality in a mixed pasture on S&S Homestead Farm. While lime was most effective at raising soil pH, BD preparations were intermediate in effectiveness when compared to untreated controls. No increase in microbial biomass or activity was observed as a result of any of the treatments. A slight negative impact on forage crude protein following lime treatment was observed, although the magnitude of this effect is unlikely to be biologically significant due to the overall forage quality. Lime increased the measured Cmin and reduced Cmic Cmin^{-1} in the top 10 cm of the soil. No differences were observed in soil C or microbial activity and so we suggest that the greater CO_2 release was due to continued dissolution of the lime.

As environmental, economic, and energy pressures increase the urgency of improving long-term agricultural sustainability, the use of local and on-farm resources to generate intrinsic fertility within their farm system and limit external inputs becomes more important. This study found potential to use biodynamic preparations, which can be produced on-farm, to treat moderate soil acidity. Our whole farm survey also highlights the conflicts that can occur between achieving long-term environmental and social sustainability under often harsh economic realities.

Acknowledgements
This research was supported by Western SARE through a farmer/rancher grant, the Crops and Soils Department at Washington State University and the Utah Agricultural Experiment Station, and approved as journal paper number 8222. Thanks in particular to Dr. Steve Fransen for assistance with proposal development and forage data collection and to Dr. Mike Hackett for conducting the sheep health assessment. Many thanks to Dr. Marc Evans for statistical consulting, to Dr. Stuart Higgins for statistical assistance, and to Margaret Davies for much needed assistance in the laboratory. An additional thanks to Ron Bolton for assistance with numerous computer related problems and questions.

Table 1

The main ingredients and recommended (unit) amounts of the biodynamic preparations used per 0.4 ha of land or in 14 t compost.

Preparation	Main Ingredient	Use	Unit volume (cm^3)	Unit mass (g)
500	Cow (*Bos taurus*) manure	Field spray	35	38
501	Finely ground quartz silica	Field spray	2	1.8
502	Yarrow blossoms (*Achilliea millefoilium* L.)	Compost	15	1.1
503	Chamomile blossoms (*Matricaria recutita* L.)	Compost	15	3.0
504	Stinging nettle shoots (*Urtica dioica* L.)	Compost	15	4.4
505	Oak bark (*Quercus robur* L.)	Compost	15	3.9
506	Dandelion flowers (*Taraxacum officinale* L.)	Compost	15	4.7
507	Valerian flower extract (*Valeriana officinalis* L.)	Compost	2	1.2
Pfeiffer Field Spray	Preparations 500–507 plus cultured bacteria	Field spray	78	57

Table 2

Forage composition of the study site.

Legumes ~35% of total	
Red clover	*Trifolium pretense* L.
White clover	*Trifolium repens* L.
Common vetch	*Vicia sativa* L.
Subterranean clover	*Trifolium subterraneum* L.
Birdsfoot trefoil	*Lotus corniculatus* L.
Grasses ~55% of total	
Tall fescue	*Schedonorus phoenix* (Scop.) Holub
Meadow foxtail	*Alopecurus pratensis* L.
Orchard grass	*Dactylis glomerata* L.
Quackgrass	*Elymus repens* (L.) Gould
Reed canary grass	*Phalaris arundinacea* L.
Smooth brome	*Bromus inermis* Leyss.
Annual ryegrass	*Lolium perenne* L. ssp. *multiflorum* (Lam.) Husnot
Perennial ryegrass	*Lolium perenne* L.
Bent grass	*Agrostis, sp.* L.
Sedge	*Carex sp.* L.
Rush	*Juncus sp.* L.
Forbes ~10%	
Dandelion	*Taraxacum officinale* F.H. Wigg.
Coltsfoot	*Tussilago farfara* L.
English daisy	*Bellis perennis* L.
Oxeye daisy	*Leucanthemum vulgare* Lam.
Comfrey	*Symphytum* sp. L.
Dock/sorrel	*Rumex* sp. L.
Plantain	*Plantago major* L.
Mustard	*Brassica* sp. L.
Yarrow	*Achillea millefolium* L.

Table 3

Means ($n = 24$) for soil analyses (depth 0–10 cm) conducted in years 2005 and 2006.

Parameter	Lime	Biodynamic	Control
Total carbon%	5.04a	5.28a	6.03a
Total nitrogen%	0.488a	0.511a	0.575a
C:N ratio	10.3a	10.4a	10.5a
pH	6.6a	6.4b	6.2c
pH change from baseline	0.5a	0.3a	0.1a
Dehydrogenase (μ TPF g^{-1} soil)	11.1a	11.2a	11.4a
Phosphatase (μ p-nitrophenol g^{-1} soil)	518a	517a	494a
Readily mineralizable carbon or Cmin (μg C g^{-1} soil)	118b	102a	98a
Basal microbial respiration (μg C g^{-1} soil)	5.5a	7.3a	5.0a
Microbial biomass carbon or Cmic (μg C g^{-1} soil)	771a	926a	836a
Cmic/Cmin	7.3b	9.5a	10.2a
QCO$_2$ × 1000	7.5a	6.7a	6.7a

Means with different letters in each row are significant at $P < 0.05$.

Table 4
Means (n = 24) for soil analyses (depth 10–20 cm) conducted in years 2005 and 2006.

Parameter	Lime	Biodynamic	Control
Total carbon (%)	4.48a	4.23a	4.55a
Total nitrogen (%)	0.439a	0.420a	0.451a
C:N ratio	10.2a	10.1a	10.1a
pH	6.5a	6.4a	6.3a
Dehydrogenase (μ TPF g^{-1} soil)	9.3a	9.0a	8.9a
Phosphatase (μ p-nitrophenol g^{-1} soil)	298a	287a	386a
Readily mineralizable carbon or Cmin (μg C g^{-1} soil)	54a	48a	52a
Basal microbial respiration (μg C g^{-1} soil)	4.4a	3.6a	3.9a
Microbial biomass carbon or Cmic (μg C g^{-1} soil)	562a	505a	568a
Cmic/Cmin	11.9a	11.2a	11.7a
QCO$_2$ × 1000	7.0a	7.0a	7.0a

Means with different letters in each row are significant at $P < 0.05$.

Table 5
Means (n = 24) for forage analyses conducted in years 2005 and 2006.

Parameter	Lime	Biodynamic	Control
Plant biomass (dry kg ha^{-1})	1190a	1220a	1090a
Plant height (mm)	678a	711a	681a
Crude protein (%)	7.4b	7.9a	7.6a
Acid digestible fiber (%)	41.9a	41.6a	40.6a
Neutral digestible fiber (%)	67.6a	67.8a	66.3a
Total digestible nutrients (%)	54.8a	55.0a	56.2a
Relative feed value	77.9a	77.8a	81.1a
Ca (%)	0.45a	0.44a	0.44a
P (%)	0.15a	0.15a	0.15a
K (%)	2.0a	2.1a	2.1a
Mg (%)	0.19a	0.18a	0.19a
Ash (%)	8.3a	8.7a	7.9a

Means with different letters in each row are significant at $P < 0.05$.

Table 6
Life Cycle Sustainability Indicators for S&S Homestead Farm.

Life Cycle Stage	Indicators		
	Economic	Social	Environmental
Farm production	– Economically stable with marginal return on investment – Farm income sufficient to capitalize operating and infrastructure costs – Farm start up from farmer savings, self capitalizing since 1994 – No debt – Spouse works off farm – CSA manager works part time at school implementing farm-to-school program – No government subsidies	– Farmers nearing retirement age – Limit on number of dwellings allowed on property makes handover to next generation challenging – Two full time farmers plus summer interns – No farm wages for owners, CSA profits returned to managers – Farmers, CSA managers and interns consume farm produced food – Actively involved in intern training, farm to school program, and other community educational programs	– Forage based system. Potential for soil loss very low – Some net loss of nutrients likely from export, leaching and denitrification – No synthetic inputs – Air pollutants low – Species diversity high – Rain based irrigation – Energy partly provided by solar system. Current fossil fuel use less than 378 L per annum – Very low harvest loss to pests and diseases – Animals are healthy and live year-round on pasture with shelter
Origin of genetic resources	– Some seed saved on farm, most is purchased – Animal breeding for desired characteristics conducted on farm – Breeding stock obtained locally when appropriate breeds available	– Most seed purchased from small organic producers – Some livestock, seeds and seedlings sold as local breeding stock, forming a network of genetic resources	– All seeds naturally pollinated – Hybrid seed avoided – Natural reproduction of animals – Disease resistant seed varieties and animals selected – Shipping of seeds and stock is minimized
Food processing and distribution	– Farm products minimally processed and produced on farm – Interns participate in food processing – Products marketed locally or consumed on farm	– High quality meat and vegetables produced – CSA members collect weekly shares at farm – No data on product safety –presumed safe – Intern satisfaction not evaluated	– Mobile slaughter unit visits farm on site – Local consumers collect meat and vegetables from farm in personal vehicles
Consumption and waste	– Farm family and interns spend little on outside foodstuffs – No money spent on food disposal	– Interns share meals of farm-produced foods – Some farm products consumed at local schools – Excess donated to local food bank	– All farm/food waste composted on farm

Table 7
Farm enterprise budget for 2007.

Farm income	
S&S Homestead Farm	$42,666
S&S Center for Sustainable Agriculture	$10,451
Lopez Community Farm CSA	$20,146
Total returns	$73,263
Farm expenses	
Direct costs	$53,850
Indirect costs	$10,268
Total expenses	$64,118
Profit	$9145

Table 8
Soil nutrient profiles from three pastures and one vegetable garden at 0–15 cm S&S Homestead Farm. Pasture 1 represents the study area for the lime response trial.

Soil property	Garden 0–15 cm	Pasture 1 0–15 cm	Pasture 2 0–15 cm	Pasture 3 0–15 cm
Nitrate-N (mg kg^{-1})	48.9	2.5	1.5	2.0
Ammonium-N (mg kg^{-1})	14.4	5.4	4.0	4.9
Organic matter g kg^{-1})	12	9.8	6.4	5.8
Olsen phosphorus (mg kg^{-1})	132	9.0	19	11
Potassium (mg kg^{-1})	1638	151	100	194
Sulfate-S (mg kg^{-1})	39	13	10	10
Boron (mg kg^{-1})	0.80	0.83	0.43	0.70
Zinc (mg kg^{-1})	12.5	2.6	1.2	1.4
Manganese (mg kg^{-1})	18.2	28.5	14.9	29.6
Copper (mg kg^{-1})	2.5	4.5	2.2	2.3
Iron (mg kg^{-1})	166	237	155	196
pH	6.6	6.1	6.1	6.2
EC saturated past (mmhos cm^{-1})	2.44	0.49	0.34	0.26
Calcium (meq 100 g^{-1})	10.6	13.6	5.8	5.3
Magnesium (mq 100 g^{-1})	4.9	6.5	2.5	2.3
Sodium (meq 100 g^{-1})	0.59	0.54	0.30	0.21
CEC (meq 100 g^{-1})	25.9	34.8	14.6	14.6
Buffer capacity (pH)	6.9	6.3	6.6	6.7
Total bases (meq 100 g^{-1})	20.3	21.1	8.8	8.3

Table 9

Soil nutrient profiles from three pastures and one vegetable garden at 15–30 cm S&S Homestead Farm. Pasture 1 represents the study area for the lime response trial.

Soil property	Garden 15–30 cm	Pasture 1 15–30 cm	Pasture 2 15–30 cm	Pasture 3 15–30 cm
Nitrate-N (mg kg^{-1})	53.7	0.8	1.6	3.4
Ammonium-N (mg kg^{-1})	4.3	3.8	3.3	3.0
Organic matter (g kg^{-1})	8.4	5.6	3.9	3.7
Olsen phosphorus (mg kg^{-1})	93	6.0	23	10
Potassium (mg kg^{-1})	736	84	64	143
Sulfate-S (mg kg^{-1})	25	9	6	7
Boron (mg kg^{-1})	0.70	0.72	0.24	0.20
Zinc (mg kg^{-1})	9.8	1.5	0.6	0.5
Manganese (mg kg^{-1})	8.7	8.5	8.0	7.4
Copper (mg kg^{-1})	2.7	4.6	1.5	1.4
Iron (mg kg^{-1})	190	216	116	87
pH	5.9	6.2	6.1	6.3
EC saturated past (mmhos cm^{-1})	1.87	0.23	0.23	0.26
Calcium (meq 100 g^{-1})	10.3	11.3	4.4	5.4
Magnesium (meq 100 g^{-1})	4.5	5.6	1.8	2.2
Sodium (meq 100 g^{-1})	0.35	0.45	0.24	0.21
CEC (meq 100 g^{-1})	22.2	24.5	10.8	11.5
Buffer capacity (pH)	6.6	6.5	6.7	6.8
Total bases (meq 100 g^{-1})	17.0	17.5	6.6	8.1

Notes

[1] Originally published in *Agricultural Systems*, 104: 572-579. Reprinted with permission by Elsevier.com.

[2] Edens, Thomas C. & Dean L. Haynes, D.L. 1982. "Closed System Agriculture: Resource Constraints, Management Options, and Design Alternatives," *Annual Review of Phytophathology*, 20: 363-395.

[3] Edens & Haynes, op.cit.; Hahlbrock, Klaus 2009. *Feeding the Planet: Environmental Protection Through Sustainable Agriculture*. London, UK.

[4] Perfecto, Ivette, John Vandermeer, Angus Wright, 2009. *Nature's Matrix: Linking Agriculture, Conservation and Food Sovereignty*. London, UK, 199-203.

[5] Heller, Martin C. & Gregory A. Keoleian, 2000. "Life Cycle-Based Sustainability Indicators for Assessment of the U.S. Food System," *Center for Sustainable Systems*, U. of Michigan. Report No. CSS00-04.

[6] Carpenter-Boggs, Lynn, John P. Reganold, and Ann C. Kennedy, 2000. "Biodynamic Preparations: Short-Term Effects on Crops, Soils and Weed Populations," *American Journal of Alternative Agriculture* 15: 110-118; Reeve, J. R., L. Carpenter-Boggs, J.P. Reganold, A.L. York, G. McGourty & L.P. McCloskey 2005. "Soil and Wine Grape Quality in Biodynamically and Organically Managed Vineyards," *American Journal of Enolology & Viticulture* 56: 367-376.

[7] Courtney, Hugh, n.d. Josephine Porter Institute for Applied Biodynamics, personal communication.

[8] Brady, Nyle C. & Ray R. Weil, 2002. "Formation of Soils from Parent Materials," *The Nature and Properties of Soils*. Hoboken, NJ, 54-61; Brady et al., "Soil Acidity," op.cit, 390-391.

[9] Colmenares, R. & J.M. de Miguel 1999. "Improving Permanent Pasture's Growth: An Organic Approach, *Cahiers Options Méditerranéennes* 39: 189-191.

[10] Abele, U. 1987. "Produktqualität und Düngung – mineralisch, organisch, biologisch-dynamisch" (Product Quality and Fertilization: Mineral, Organic, Biodynamic). *Schriftenreihe des Bundesministers für Ernährung, Landwirtschaft und Forste* (Publications of the Federal Ministry for Nutrition, Agriculture, and Forests), Series A, Volume 345; Raupp, J. 2001. "Manure Fertilization for Soil Organic Matter Maintenance and its Effects Upon Crops and the Environment, Evaluated in a Long-Term Trial,"*Sustainable Management of Soil Organic Matter CAB International* (Rees, R.M., B.C. Ball, D.C. Campball & C.A. Watson, eds.),Wallingford, UK, 301-308; Mäder, P., A. Fleissbach, D. Dubois, L. Gunst, P. Fried & U. Niggli 2002. "Soil Fertility and Biodiversity in Organic Farming," *Science* 296: 1694-1697; Reeve, et al., 2005, op.cit.; Goldstein, W. 1986. *Alternative Crops, Rotations and Management Systems for the Palouse* (Dissertation, Washington State University, WA), 152-184; Goldstein, W.A. & H.H. Koepf, 1992. "A Contribution for the Development of Tests for the Biodynamic Preparations," *Elemente Naturwissenschaft* (Elements of Natural Science) 36: 41-56.

[11] Pettersson, B.D., H.J. Reents & E. von Wistinghausen, 1992. "Düngung und Bodeneigenschaften: Ergebnisse eines 32-jährigen Feldversuches in Järna, Schweden"(Fertilization and Soil Characteristics: Results of a 32-Year Field Trial in Järna, Sweden)," *Institut für Biologisch-Dynamische Forschung* (Institute for Biodynamic Research), Darmstadt, Germany; Penfold, C.M., M.S. Miyan, T.G. Reeves & T. Grierson, 1995. "Biological Farming for Sustainable Agricultural Production," *Australian Journal of Experimental Agriculture* 35: 849-856; Carpenter-Boggs, et al. 2000, op.cit; Carpenter-Boggs, L, J.P. Reganold & A.C. Kennedy 2000. "Organic and Biodynamic Management: Effects on Soil Biology," *Soil Science Society of America Journal* 64: 1651-1659.

[12] Penfold, op.cit; Burkitt, L.L., D.R. Small, J.W. McDondal, J.W., Wales & M.L. Jenkin, M.L. 2007. "Comparing Irrigated Biodynamic and Conventionally Managed Dairy Farms: 1. Soil and Pasture Properties, 2. Milk Production and Composition and Animal Health," *Australian Journal of Experimental Agriculture*, 47: 479-494.

[13] Imperial brand, J. A. Jack & Sons. Inc, Seattle, WA.

[14] Woolwine, VA.

[15] Anderson, John, P.E. & K.H. Domsch 1978. "A Physiological Method for the Quantitative Measurement of Microbial Biomass in Soil," *Soil Biology & Biochemistry* 10: 215-221.

[16] Tabatabai, M.A. 1994. "Soil Enzymes," *Methods of Soil Analysis, Part 2: Microbiological and Biochemical Properties* (Weaver et al., eds.). Madison, WI, 775-833.

[17] Undersander, Dan J., D.R. Mertens, & N. Thiex. 1993. *Forage Analyses Procedures. National Forage Testing Association*. Omaha, NE.

[18] Moses Lake, WA.

[19] Gavlak, R., D. Horneck, R.O. Miller & J. Kotuby-Amacher, 2003. *Soil and Plant Reference Methods for the Western Region, WREP 125*.

[20] Brady, Nyle C. & Ray R. Weil, op.cit.

21 Condron, L.M., H. Tiessen, C. Trassarcepeda, J.O. Moir, & J.W.B. Stewart 1993."Effects of Liming on Organic-Matter Decomposition and Phosphorus Extractability in an Acid Humic Ranker Soil from Northwest Spain," *Biology & Fertility of Soils* 15: 279-284; Marschner, B. & A.W. Wilczynski 1991. "The Effect of Liming on Quantity and Chemical-Composition of Soil Organic-Matter in a Pine Forest in Berlin, Germany," *Plant Soil* 137: 229-236; Andersson, S., I. Valeur, & C. Lapierre 1994. "Influence of Lime on Soil Respiration, Leaching of DOC, and C/S Relationships in the More Mumus of a Haplic Podsol," *Environment International* 20: 81-88; Haynes, R.J. & R. Naidu, R. 1998. "Influence of Lime, Fertilizer and Manure Applications on Soil Organic Matter Content and Soil Physical Conditions: A Review," *Nutrient Cycling in Agroecosystems* 51: 123-137; Ekenler, M. & M.A.Tabatabai 2003. "Effects of Liming and Tillage Systems on Microbial Biomass and Glycosidases in Soils," *Biology & Fertility of Soils* 39: 51-61; Fuentes, J.P., D.F. Bezdicek, M. Flury, S. Albrecht & J.L. Smith 2006. "Microbial Activity Affected by Lime in a Long-Term No-Till Soil," *Soil &Tillage Research* 88: 123-131; Rangel-Castro, J.I., J.I. Prosser, C.M. Scrimgeour, P. Smith, N. Ostle, P. Ineson, A. Meharg & K. Killham, 2004. "Carbon Flow in an Upland Grassland: Effect of Liming on the Flux of Recently Photosynthesized Carbon to Rhizosphere Soil, *Global Change Biology* 10: 2100-2108.

22 Kreutzer, K. 1995. "Effects of Forest Liming on Soil Processes," *Plant Soil* 168-169: 447-470; Rangel, op.cit.

23 Kemmitt, S., J. Wright, D. Keith, K.W.T. Goulding & D.L. Jones 2006. "pH Regulation of Carbon and Nitrogen Dynamics in Two Agricultural Soils, *Soil Biology & Biochemistry* 38: 898-911.

24 Bertrand, I., O. Delfosse & B. Mary 2007. "Carbon and Nitrogen Mineralization in Acidic, Limed and Calcareous Agricultural Soils: Apparent and Actual Effects," *Soil Biology & Biochemistry.* 39: 276-288.

25 CAN, 1985.

26 Malhi, S.S., M. Nyborg & J.T. Harapiak 1998. "Effects of Long-Term N Fertilizer-Induced Acidification and Liming on Micro-Nutrients in Soil and Bromegrass Hay," *Soil Tillage Research* 48: 91-101; Malhi, S.S., J.T. Harapiak, K.S. Gill & N. Flore 2002. "Long-Term N Rates and Subsequent Lime Application Effects on Macro-Elements Concentration in Soil and Bromegrass Hay," *Journal of Sustainable Agriculture* 21: 79-97.

27 Wallander, H., K. Amebrant, F. Ostrand & O. Karen 1997. "Uptake of N-15-Labelled Alanine, Ammonium and Nitrate in Pinus sylvestris L. Ectomycorrhiza Growing in Forest Soil Treated With Nitrogen, Sulphur or Lime, *Plant and Soil* 195: 329-338; Hart, M.M. & J.T. Trevors 2005. "Microbe Management: Application of Mycorrhizal Fungi in Sustainable Agriculture," *Frontiers in Ecology and the Environment* 3: 533-539.

28 Edens, op.cit.

29 Heller et al., op.cit.

30 Smeck, N.E. 1985. "Phosphorus Dynamics in Soils and Landscapes," *Geoderma* 36 (3-4): 185-199; Trolove, S.N., M.J. Hedley, J.P. Caradus & A.D. Mackay 1996. "Uptake of Phosphorus from Different Sources by *Logus pedunculatus* and Three Genotypes of *Trifolium Repens:* Forms of Phosphate Utilized and Acidification of the Rhizosphere," *Australian Journal of Soil Research* 34: 1027-1040; Oehl, F., E. Frossard, A. Fliessbach, D. Dubois & A. Oberson 2004. "Basal Organic Phosphorus Mineralization in Soils Under Different Farming Systems," *Soil Biology & Biochemistry* 36: 667-675.

31 Tisdale, S.L., W.R. Nelson, J.D. Beaton & J.L. Havlin 1993. *Soil Fertility and Fertilizers.* 5th ed. Upper Saddle River, NJ.

32 Hackett, Mike 2009. "S&S Homestead Farm Biodynamic Farming Study — Sheep Flock Status." https://sshomestead.org/wp-content/uploads/Sheep%20evaluation.pdf. Retrieved November 10, 2021.

33 National Research Council Board of Agriculture and Natural Resources and Earth and Life Studies (BANRELS) 2008. *Changes in the Sheep Industry in the United States: Making the Transition From Tradition.* Washington, DC.

34 Burkitt, op.cit.

35 Provenza, F.D., J.J. Villalba, L.E. Dziba, S.B. Atwood & R.E. Banner 2003. "Linking Herbivore Experience, Varied Diets, and Plant Biochemical Diversity," *Small Ruminant Research* 49: 257-274.

36 Min, B.R. & S.P. Hart 2003. "Tannins for Suppression of Internal Parasites," *Journal of Animal Science* 81: E102-E109; Waghorn, G.C. 1990. "Beneficial Effects of Low Concentrations of Condensed Tannins in Forages Fed to Ruminants," *Microbial and Plant Opportunities to Improve Lignocellulose Utilization by Ruminants* (Akin, D.E., L.G. Ljungdahl, J.R. Wilson & Harris, eds.). New York, NY, 137; Barry, T.N., D.M. McNeill & W.C. McNabb 2001. "Plant Secondary Compounds: Their Impact on Nutritive Value and Upon Animal Production," *Proceedings XIX International Grass Conference,* Sao Paulo, Brazil, 445-452; Niezen, J.H., W.A.G. Charleston, H.A. Robertson, D. Shelton, G.C. Waghorn & R. Green 2002. "The Effect of Feeding Sulla (*Hedysarum Coronarium*) or Lucerne (*Medicago Sativa*) on Lamb Parasite Burdens and Development of Immunity to Gastrointestinal Nematodes," *Veterinary Parasitology* 105: 229-245.

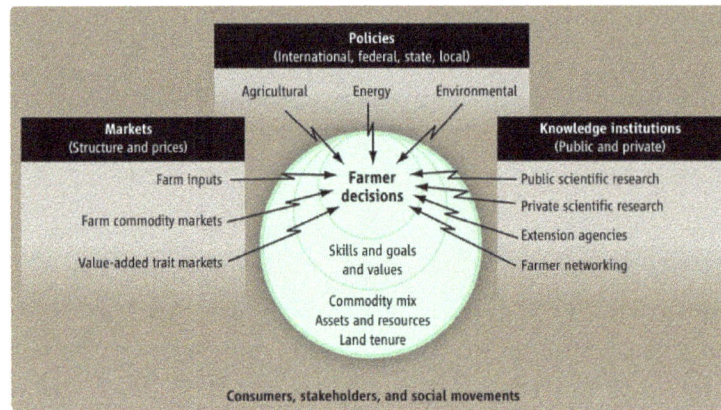

Policies
(International, federal, state, local)

Agricultural   Energy   Environmental

Markets
(Structure and prices)

Knowledge institutions
(Public and private)

Farm inputs

Farm commodity markets

Value-added trait markets

Farmer decisions

Skills and goals
and values

Commodity mix
Assets and resources
Land tenure

Public scientific research

Private scientific research

Extension agencies

Farmer networking

Consumers, stakeholders, and social movements

Transforming U.S. Agriculture[1]

John Reganold, Douglas Jackson-Smith, Sandra Batie, Richard Harwood, Julia Kornegay, Dale Bucks, Cornelia Flora, James Hanson, William Jury, Deanne Meyer, August Schumacher Jr., Henning Sehmsdorf, Carol Shennan, Lori Thrupp, Paul Willis, 2011

Agriculture in the United States and many other countries is at a critical juncture. Public investments and policy reforms will inform landscape management practices to be used by farmers and ranchers for sustaining food and ecosystem security. Although U.S. farms have provided growing supplies of food and other products, they have also been major contributors to global greenhouse gases, biodiversity loss, natural resource degradation, and public health problems. Farm productivity and economic viability are vulnerable to resource scarcities, climate change, and market volatility[2]. Concerns about long-term sustainability have promoted interest in new forms of agriculture that (i) enhance the natural resource base. Achieving sustainable agricultural systems will require transformative changes in markets, policies, and science, (ii) make farming financially viable, and (iii) contribute to the well-being of farmers, farm workers, and rural communities, while still (iv) providing abundant, affordable

> *Achieving sustainable agricultural systems will require transformative changes in markets, policies, and science.*

food, feed, fiber, and fuel. A 2010 report by the U.S. National Research Council (NRC) (see footnote 2) identified numerous examples of innovative farming systems that contribute to multiple sustainability goals but noted they are not widespread. This report joins others[3] critical of aspects of mainstream, conventional farming systems. We argue that the slow expansion of such innovative farming systems in the United States is as much a policy and market problem as a science and technology problem. Incentives for appropriate markets, reform of U.S. farm-related policies, and reorientation of publicly funded agricultural science are needed to hasten implementation of more sustainable agricultural systems.

Incremental, Transformative Approaches

To improve sustainability of U.S. agriculture, the NRC report proposes both incremental and transformative approaches. The former are practices and technologies that address specific production or environmental concerns associated with mainstream, conventional farming systems. Examples include two-year crop rotations, precision agriculture using geospatial technologies that describe field variation, classically bred or genetically engineered crops, and reduced or no tillage. Although incremental approaches offer improvements and should be continued, in aggregate, they are inadequate to address multiple sustainability concerns. In contrast, the transformative approach builds on an understanding of agriculture as a complex socio-ecological system. Transformative change looks to whole-system redesign rather than single technological improvements. Examples of such innovative systems make up a modest, but growing, component of U.S. agriculture and include organic and biodynamic farming, alternative livestock production (e.g., grass-fed), mixed-crop and livestock systems, and perennial grains.[4] Such systems integrate production, environmental, and socio-economic objectives; reflect greater awareness of ecosystem services; and capitalize on synergies between complementary farm enterprises, such as between crop and livestock production. The existence of innovative agricultural systems in the United States suggests that technical obstacles are not the greatest barrier. Rather, change is hindered by market structures, policy incentives, and uneven development and availability of scientific information that guide farmers' decisions (see the figure above).

Market Structures

Most U.S. farmers sell products to a highly consolidated global agri-food industry rewarding primarily the provision of large volumes of low-cost food, feed, fiber, and fuel, often constrained by contract requirements of food processors and retailers. Meanwhile, consumer food consumption habits associated with modern lifestyles have sustained

mainstream farming systems and food markets and have contributed to a national obesity and health crisis. Part of transforming U.S. agriculture is educating more consumers to take responsibility for what they eat and how much they eat.[5] Consumer demand is also growing for more environmental and social accountability from farmers, including considerations of animal welfare, ecosystem services, worker safety and welfare, and resource conservation. In response, "value-added trait" foods and "sustainability brands" have emerged in the marketplace, e.g., U.S. Department of Agriculture Certified Organic and Food Alliance Certified. U.S. and global markets for these value-added trait products have driven the spread of local, organic, and grass-fed livestock systems. Market forces could be accelerated through public-policy incentives.

Policy Incentives

Many international, federal, state, and local agricultural, credit, energy, risk-management, and environmental policies influence farmer decisions (see the figure above). A major policy driver for U.S. agriculture is the Farm Bill, traditionally renewed by the U.S. Congress every four to five years, with the next version expected in 2012. The best-funded provisions of the Farm Bill include financial assistance for low income families to purchase food; commodity subsidies paid to farmers (mostly for corn, cotton, rice, soybeans, and wheat); crop insurance and disaster relief; and conservation programs.[6] Although only roughly a third of U.S. farmers receive commodity or conservation payments under the Farm Bill, it has a major influence on what, where, and how food is produced. Most elements of the Farm Bill were not designed to promote sustainability. Subsidies are commonly criticized for distorting market incentives and making our food system overly dependent on a few grain crops mainly used for animal feed and highly processed food, with deleterious effects on the environment and human health.[7]

Redesigning the bill will be a complex undertaking in light of political and budgetary constraints, as well as knowledge gaps. However, much of the information necessary for Farm Bill redesign is available and not being used.[8] Spending needs to be reduced on programs, such as subsidies, that mask market, social, and environmental risks associated with conventional production systems. Funding needs to be reallocated to encourage markets for sustainability brand products (e.g. by standardizing and defining sustainable product attributes) and to increase support for farming systems that balance all four sustainability goals and are more resilient to resource scarcities and global market variability. With a new version of the Farm Bill due next year, we think the time to start reform is now. In addition, progress in other policy arenas is needed to address conflicting incentives and unintended consequences. Unless we integrate agricultural sustainability into debates over biofuels and other energy policies, climate change, trade agreements, immigration reform, and environmental regulation, we are unlikely to see major changes in policies that created and continue current production systems.

Agricultural Science and Knowledge

The publicly funded agricultural science portfolio could be reoriented toward agricultural sustainability, as this research is less likely to yield marketable inventions for private agribusinesses. The bulk of public and private agricultural science in the United States is narrowly focused on productivity and efficiency, particularly on technologies that fit into existing production systems and lead to private benefits.[9]

A major vehicle for public agricultural research is the National Institute for Food and Agriculture (NIFA). Despite NIFA efforts to solicit proposals addressing sustainability, most NIFA and other federal research grant programs still primarily support incremental research. What is needed is reallocation of public funds to support trans-disciplinary systems research that explores such interlocking issues as farm productivity and resilience at field, farm, and landscape scales.[10] Transition toward transformative agricultural systems currently relies on a smaller, emerging knowledge base developed largely by farmers and nonprofit organizations independent of traditional scientific institutions. Agricultural science and farmers would benefit from an easily accessible information database of farm innovations. Moreover, pilot projects could be funded by reallocation of Farm Bill subsidies to measure multiple sustainability indicators on conventional and innovative farming systems at the landscape or watershed scale.[11]

Final Recommendations

To make difficult choices among competing goals requires public dialogue about what kind of food and agriculture we want, in addition to identifying the roles of markets, policies, and science in delivering them.[12] Successful implementation will require organizations spanning political and institutional boundaries and integrating complex components of agricultural transformation — from research to on-farm implementation, to markets, and to the dinner table.

The Green Lands Blue Waters Initiative[13] to achieve "systemic transformation in the agricultural systems" in the Mississippi River basin is an example of such an effort. This involves community organizers, policy experts, scientists, and farmers from more than a dozen nonprofit organizations, five universities, and multiple government agencies from the Upper Midwest to the Gulf of Mexico.

The goals of agricultural sustainability are not unique to the United States. Although specific market, policy, and science solutions will need to be appropriate to diverse contexts, the importance of viewing sustainability as more than a technical problem applies to developed and less-developed countries. Lessons from experiences in developed countries can help less-developed countries avoid some problems associated with contemporary, industrialized agricultural systems and can reduce exposure to market volatility and climate change risks. Likewise, U.S. farmers can learn from sustainable agricultural practices of less-developed nations.

Notes

[1] Originally published in *Science* 332(6030): 670-1. Reprinted by permission. Written by a sub group of the National Research Council Committee on Twenty-First Century Systems Agriculture 2007-2010 (comprising fifteen experts in crop, soil, horticultural, and water sciences; agricultural economics, rural sociology, agro-ecology and agricultural education; environmental, regulatory, international trade and development policies; and food producers, including one biodynamic farmer — the author of this book), the essay summarizes the major findings of *Toward Sustainable Agricultural Systems in the 21st Century* (The National Academies, Washington, DC, 2010), 570 pp.

[2] Battisti, David S. & Rosamund L. Naylor 2009. "Historical Warnings of Future Food Insecurity With Unprecedented Seasonal Heat," *Science* 323: 240-244.

[3] IAASTD 2009. *International Assessment of Agricultural Science and Technology for Development, Agriculture at a Crossroads: Global Report*, Washington, DC; Rockström, Johan, Will Steffen & [...] Jonathan Foley 2009. "A Safe Operating Space for Humanity," *Nature* 461: 472-475; De Schutter, Olivier 2010. "Final Report: The Transformation Potential of the Right to Food," *United Nations*, Geneva, Switzerland, 1-28; Godfray, H. Charles. J. et al. 2010. "Food Security: The Challenge of Feeding 9 Billion People," *Science* 327: 812-818.

[4] *Toward Sustainable Agricultural Systems in the 21st Century*, 83ff.

[5] U.S. Department of Agriculture and U.S. Department of Health and Human Services 2010. *Dietary Guidelines for Americans*, Washington, DC.

[6] Monke, Jim & Renée Johnson 2010. "Actual Farm Bill Spending & Cost Estimates," *Congressional Research Service Report R41195*, Washington, DC.

[7] Dobbs, Thomas L. & Jules N. Pretty 2004. "Agri-Environmental Stewardship Schemes and Multifunctionality," *Review of Agricultural Economics* 26(2): 220-237; Cox, Craig 2011. "If Not Conservation Districts, What?" *Managing Agricultural Landscapes for Environmental Quality II: Achieving Effective Conservation* (Nowak, Peter & Max Schnepf, eds), Ankeny, IA, 81–94.

[8] Batie, Sandra S. 2009. "Green Payments and the US Farm Bill: Information and Policy Changes," *Frontiers in Ecology and Environment.* 7 (7): 380-388.

[9] National Research Council, op.cit.; Huffman, Wallace E. & Robert E. Evenson 2006 *Science for Agriculture: A Long-Term Perspective*, Ames, IA, 2. ed.

[10] Robertson, G. P. [...] & Diana H. Wall 2008. "Long-term Agricultural Research: A Research, Education, and Extension Imperative," *BioScience* 58 (7): 640–645.

[11] Sachs, Jeffrey [...] & Pedro Sanchez 2010. "Monitoring the World's Agriculture," *Nature*, 466: 558–560; Batie, Sandra S. 2009, op.cit.

[12] Busch, Lawrence 2009. "What Kind of Agriculture? What Might Science Deliver?" *Natures Sciences Sociétés* 17: 241-247.

[13] *Green Lands Blue Waters.* www.greenlandsbluewaters.org. Retrieved November 1, 2021.

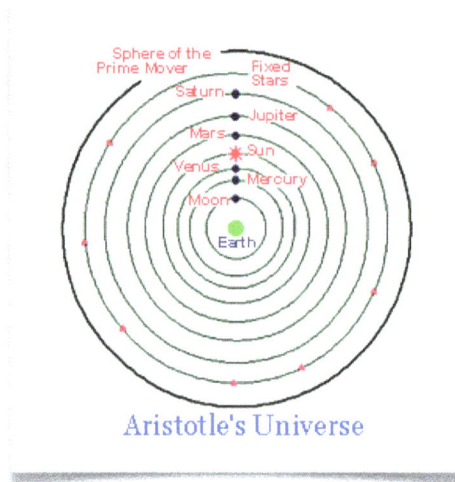

Aristotle's Universe

The Spirituality of the Soil:
The Idea of Teleology from Aristotle to Rudolf Steiner [1]

Henning Sehmsdorf, 2016

I have been farming on Lopez Island for nearly half a century. During thirty of those years I also taught at the university what conventionally is labelled the humanities, but interpreters of Steiner's works would call spiritual science (*Geisteswissenschaften*). The humanities are the disciplines that explore the relation of Spirit and matter in human experience. As an undergraduate student of science, I had puzzled over the difference between the living frog and the pitiful, dead cadaver studied in the laboratory. As the devil says of the scientist, in Goethe's *Faust I*:

> *Wer will was Lebendigs lernen und beschreiben,*
> *Sucht erst den Geist herauszutreiben,*
> *Dann hat er die Teile in der Hand,*
> *Fehlt, leider, nur das geistige Band.*

(To know some living thing and describe it/ He hastens to drive out its Spirit/ Now he holds the parts in his hands/ But, alas, he lacks the Spirit band).

To find what the poet means by the Spirit that binds the material parts of an organism into a living being, I searched for years in philosophy and science, cultural history and art, poetry, folklore, mythology, and religion. Everywhere I found the universal idea of an etheric force that gives life and shape to living beings, plants, animals and humans. For instance, in the Old Testament, the "Breath of God" turns clay into living souls. In the New Testament, the "Creative Logos" is the divine agency by which all things are made. In classical China, *Ch'i* is the natural energy of the whole universe. In Vedic India *Prana* (breath) is the etheric force sustaining life. In Melanesia *Mana* is the impersonal force enlivening nature, including people and animals. The Haida Indians speak of the River of Life manifesting in all creation. In ancient Germanic tribal culture, *Hamingja* is the force identified with the power of the chieftain to give fertility to the soil and to the sea. In pre-industrial European folk belief, the Powers represent life principles imagined as nature beings and elemental spirits. Goethe speaks of the Earth Spirit, Emerson of the Oversoul, as the transcendental force infusing all life.

But I also learned that in order to understand how Spirit and matter are intertwined in life, theoretical and historical knowledge was not enough. What was needed was the intuitive knowledge that comes from participatory experience of elemental nature. In short, I needed to get my hands into the living soil by becoming a farmer, and not just any farmer, but a biodynamic farmer, because in biodynamics the relations between matter and Spirit is made comprehensible both philosophically, and in experiential practice.

Over the years, my older brother, a Biblical scholar and longtime dean of a Lutheran seminary in Stuttgart, Germany, chided me for pagan tendencies whenever I expressed that I had come to experience the presence of Spirit as immanent in Earth, Fire, Air and Water. In husbanding soil, plants and animals biodynamically, I have worked with these four classical elements, mindful of Rudolf Steiner's imperative to heal the earth. Steiner posited the farmer's role as a quintessential element (literally fifth essence) in returning the earth to the elemental balance lost when agriculture first began.

Not long ago, our Lutheran pastor on Lopez Island prepared the congregation for the celebration of the Spirit's presence in the communion bread and wine with the following prayer:

O God, you are Breath: send your Spirit on this meal.
O God, you are Bread: feed us with yourself.
O God, you are Wine: warm our hearts and make us one.
O God, you are Fire: transform us with hope.

After the service I asked the pastor whether her invocation of the divine breath, bread, wine and fire echoed the four natural elements by which the pre-Socratics construed material

reality half a millennium before Christ? Was she thinking of Aristotle's natural philosophy in which these elements became the building stones from which divine consciousness shaped the ordered cosmos out of primordial chaos?[2] The pastor answered, yes, probably so, although she had not thought about it in such specific terms.

Aristotle's concept of the divine as immanent formative force led to his notions of indwelling purpose or potential (*telos*).[3] In his view, an entelechy (*entelechaia*, from *en telei* = purpose + *echein* = becoming complete)[4] is any created thing or living being fully manifesting its intended end. Movement, according Aristotle, is work or evolution that allows a living thing to realize its inner potential or purpose.[5]

This thought carries over into Aristotle's understanding of economics, which term he coined by combining the word for household (*oikos*) with the verb "to take care of" or "steward" (*nemein*). Aristotle conceived of "natural farming" as taking care of the soil, "as a mother would of her children," which he contrasted with "unnatural farming" for money (*chrematistika*, from *chremata*=coins).[6] In other words, for the ancient philosopher, earth was a Spiritual thing, and care of the soil was sacred economics.[7]

Aristotle's unifying vision of Spirit and matter resonated in medieval theology. So did his cosmological treatise (*On the Heavens*), which detailed how the four elements are formed into the perishable bodies of the sublunary sphere ruled by the planetary bodies, whose motions are perfect and eternal.[8]

Aristotle's geocentric model of the universe, based on precise observation of the planetary movements, dominated astronomical thinking for more than 1,800 years, until replaced by Copernicus' heliocentric model in the 1500s. 13th-century theologian, St. Thomas Aquinas, the founder Scholasticism, affirmed Aristotle's concept of motion as inherently teleological evolution,[9] and he made great efforts to reconcile Aristotle's views with the Biblical description of creation. However, Aristotle's pre-Christian vision of divine Spirit as immanent in self-perfecting nature was eventually called into question in the doctrine of the transcendent deity who created material reality from nothing (*ab initio*). When His creation fell by sin, God promised to redeem it by grace. In his disputation against scholastic theology, Martin Luther inveighed that "the whole of Aristotle is to theology as shadow is to light," and that "the entire ethics of Aristotle is the worst enemy of grace."[10]

Furthermore, by the 17th century, Aristotle was rejected by Rationalist philosophers who conceived a new way of knowing the world by focusing their science entirely on material reality to the exclusion of Spirit. For Descartes (1596-1650) "movement" no longer was purposeful development but transport of a measurable material particle from a measurable point A to point B, "nothing more than the action by which any body passes from one place to another." [11]

By contrast, Jewish philosopher Baruch Spinoza (1632-1677) vehemently opposed both the Rationalist and the Judeo-Christian perspectives. His views on the Divine led to his excommunication from the Synagogue and banning of his writings by the Church. For Spinoza, Spirit and matter were the same, and God was immanent in his creation. Right study of nature required focus on *natura naturans* (present active participle of *naturare* = "to produce naturally") by which Spinoza meant the self-causing activity or process of God's presence emerging in nature. This he distinguished from *natura naturata* (perfect passive participle), by which he meant the passive product of causal chains originating in God's creation, i.e. emerged quantities, data, or

material things,[12] i.e. the "facts" studied by reductionist science.

According to 20th-century German philosopher Karl Jaspers, when Spinoza wrote about *deus sive natura* (God or nature), or *deus seu naturam appellamus* (God whom we call nature), he meant Spirit emerging in *natura naturans,* rather than *natura naturata,* the material data of the divinely created world. For Spinoza, God was "dynamic nature in action, growing and changing, not a passive or static thing."[13]

In modern times, Spinoza's views on Spirit and matter are echoed in the writings of scientists and vitalist philosophers as various as Teilhard de Chardin (1881-1955), Hans Driesch (1867-1941), and Werner Heisenberg (1901-1976).

Chardin, a Jesuit priest and paleontologist read the fossil history of life as scientific evidence of teleological evolution towards perfect union with the Godhead. His interpretation of the remains of prehistoric organisms in petrified rocks as proof of the emergence of Spirit in matter[14] cost him excommunication from the Church.

Driesch, a biologist, revived the Aristotelian concept of entelechy on the basis of his studies of sea urchin embryos.. Laboratory experimentation convinced him that embryological development was driven by a non-spatial, intensive, and qualitative life force rather than by spatial, extensive, and quantitative causes.[15] Despite demonstrable proof, his spectacular findings were dismissed by most of the scientific community as religious rather than based on an objective standpoint

Heisenberg, Nobel Prize winner for the Uncertainty Principle in quantum theory, in metaphysical musings on the nature of the universe argued that, according to the logic of the preservation of matter assumed by mechanistic physics, matter and Spirit must be inseparable from each other.[16] He agreed with Albert Einstein that divine Spirit " is somehow involved in the immutable laws of nature."[17]

These are three noteworthy examples of how some modern thinkers echo Aristotle's and Spinoza's thoughts concerning the relation of Spirit and matter in nature, notwithstanding that mainstream scientists today would consider the hypothesis of vitalism "a dead issue in the philosophy of biology,"[18] because the explanation of the origin and vital being of phenomena on the basis of anything other than purely chemical or physical forces lies beyond what can be validated mechanistically.

From a biodynamic perspective, however, Aristotle's and Spinoza's influence was most profound, first on Johann Wolfgang von Goethe (1749-1832), whose scientific writings form the foundation of biodynamic philosophy and agricultural practice, and after him on Rudolf Steiner (1861-1925) and his students. Goethe's approach to the study of nature, on which Rudolf Steiner's concept of agriculture is based, can be characterized as vitalist and phenomenological, in contrast to the reductionist approaches exemplified by the work of Descartes, Newton (1642-1727), Linneaus (1707-1780), and their followers.

In Descartes' view, the physical world is separated from Spirit (God, mind), and the universe a mechanical system governed by natural laws that can be described in mathematical terms. In following Descartes, Newton demonstrated the scientific method of reducing complex phenomena by quantitative measurement substituting mathematically based instruments for the human observer, and by relying on controlled experiments to verify hypotheses. Linnaeus, the founder of the binomial system of nomenclature, based his system of plant classification on counting the number of stamens and pistils in the flower.

Modern science and technology are the direct result of this reductionist method, most strikingly illustrated by the platform tool referred to as nanotechnology, which measures and manipulates matter at the scale of 1 billionth of a meter.

By contrast, Johann Wolfgang von Goethe, natural philosopher, scientist and poet, argued that the life history (ontogenesis) of organisms is not reducible to mechanistic explanation. Life processes must be understood in teleological rather than in causal terms. The *telos* or purpose of life force indwelling an organism is best understood in terms of its holistic "form" or "type." Holistic science must factor in the human observer. Therefore it is necessarily qualitative rather than reducible to quantitative measurement and mathematical abstraction.

Goethe's scientific method called for sustained observation of phenomena in nature through the five senses, but also through intuition and the imagination. Observation intensified through experimental variation would provide the observer with multiple perspectives and gradually lead him to intuit a unifying concept, the "form" or "type" of the organism under study.

Ideally the observer through his imagination would identify with the phenomenon studied, instead of objectifying it through mathematical abstraction. Goethe terms this "a delicate empiricism which intimately identifies itself with the object of study and thus becomes the actual theory."[19] For example, to gain real understanding of the frog dissected in the laboratory, the observer must intuitively become one with the living creature. The goal of Goethe's science is macro-level understanding of the living form rather than of micro-level (molecular or sub-molecular) structures. In other words, Goethe's approach to the study of nature is diametrically opposite to the goals and methodology of nanotechnology.

Goethe demonstrated his methodology in important applied studies to develop a theory of color that corrected and amplified Newton's theory of light.[20] He developed a theory of plant metamorphosis that laid the foundation for the modern study of morphology,[21] in contrast to Linneaus' reductionist model. And he discovered the intermaximillary bone through holistic analysis of human anatomy.[22] Long ignored by natural scientists because of his rejection of mathematical abstraction as an adequate tool to understand vital phenomena in nature, Goethe is today recognized as a founder of ecological systems science.[23]

On the biodynamic farm, Goethean science becomes the model for phenomenological understanding of soil quality, plant and animal life, and of the countless symbiotic relationships that make up the whole farm organism. If nothing else, Goethe teaches the farmer how to see, hear, smell, feel, taste, and intuit the organic and inorganic life forms upon which the success or failure of the farm system depend.

At our farm we teach Goethean science through practical example and application. We base workshops, intern- and apprenticeships, and educational programs, lectures and articles on Goethe's approach to nature.[24]

The Goethean perspective is evident in the work of Alan Chadwick (1909-1980), an early student of Rudolf Steiner and biodynamic gardener extraordinaire. Chadwick's work led to the establishment of the Center for Agroecology and Sustainable Food Systems at the University of California in Santa Cruz. In one of his many talks to his students, Chadwick said that "behind the whole of our living, be-hind all plants and all [material] manifestation is Spirit. Invisible…This is the basic attitude and approach of biodynamics:

Spiritual vision behind everything we do."[25]

Chadwick describes the Spirit behind material manifestation as an invisible presence that can be experienced in biodynamic practice when gardener or farmer work the soil in "reverence and obedience." One of the many times the truth of this statement came to me was when two Fundamentalist preachers approached me in my garden and asked me whether I wanted to "see the kingdom of God," and I answered, spontaneously, that I was standing in it.

Toward the end of his life, Rudolf Steiner gave eight seminal lectures published in English as *Spiritual Foundations for the Renewal of Agriculture,* or simply as The Agriculture Course.[26] In 1924, Steiner responded to requests by farmers in Prussia concerned about the falling fertility of chemically farmed soils and declining animal health. In the Agriculture Course, which laid the foundations of what became later known as biodynamics, Steiner explained that continued applications of chemicals in the form of synthetic fertilizers and pesticides had destroyed the soil micro-biological life.

Fundamentally, farmers had lost the traditional understanding of what life actually is. Soil is not dirt — soil is the dynamic, Spirit-filled basis for life. Soil is a living being. The ways in which humans use soil shape its tilth, health and fertility. Steiner urged farmers to restore humus levels by composts and prescribed biodynamic preparations. More importantly, he called for an holistic agriculture based on the concept of soil, farm, and cosmos as one integrated organism manifesting life force, or Spirit.

But the Agriculture Course is much more than a clarion call to ecological farming. Steiner had studied mathematics, physics and philosophy in Vienna. In 1891, he earned a doctorate at the university in Rostock in Prussia, with a dissertation on Johann G. Fichte's transcendental theory of knowledge. Between 1888 and 1896, he was editor at the Goethe Archives in Weimar and there wrote introductions and commentaries to four volumes of Goethe's scientific writings, as well as two books, in which he developed Goethe's phenomenological approach to the study of nature.[27]

The Agriculture Course can thus be characterized as an exercise in Goethean science, combining observation, qualitative analysis, intuition and imagination. Steiner intends to show how the invisible, Spiritual law of the cosmos manifests in matter, in the life of soil, plants and animals, and in the rhythms embedding the whole farm organism. As did Aristotle, Steiner posits that elemental matter is shaped by etheric, formative and life-giving forces streaming to earth from the cosmos.

The solar system, while objectively heliocentric, is subjectively experienced as geocentric, i.e. the earth is perceived as centered in a rhythm of energy flows. Silica (SiO_2), which science estimates to constitute sixty per cent of the earth's crust (elemental earth), enable plants to have their upright form. Steiner suggests that silica convey formative forces from the outer planets, Mars, Jupiter and Saturn. He holds that lime and related substances which provide plants with volume, fertility and nutritive power, convey astral forces from the inner planets, Moon, Venus, and Mercury.

The sun is the source of all energy in our solar system. In physics, energy is defined as the ability to do work or to produce change (from Greek, *en* = in + *ergon* = work). Solar energy manifests, for instance, in the form of heat, light, sound, electricity and chemical energy. Depending on the material context in which solar energy does its work, it is measured in foot-pounds, joules, ergs, or some other unit of measurement. When solar force acts on a body,

the work performed is the product of the force. But what is the force in itself? Is it the chemical energy in the vitamin D produced by sun shining on human or animal skin, or the electric energy in a photovoltaic panel, or the prismatic colors in the human eye looking at an edge between darkness and light, or the sound of the wind produced by differentially heated air masses? All of these are manifestations of solar energy, but the material embodiment of each phenomenon is radically different from the others, and it tells us nothing about what solar energy is aside from the descriptive scale of observation and measurement used in each instance.

Steiner, as Aristotle, Spinoza, Goethe and others before him, posits that Spirit and matter are the same, seen from different perspectives, one physical, the other metaphysical. Thus the sun is not merely a physical force. The sun is Spirit vested with formative power and indwelling purpose (*telos*). Likewise nitrogen (elemental fire) is a chemical substance that stimulates plant growth in the soil but, in Steiner's view, it is also the carrier of astral force. Oxygen (elemental air) is the carrier of the vital principle in living organisms. Carbon (like silica an aspect of elemental earth) bears the imprint of the cosmic form principle (idea or form). Hydrogen is the force that separates form and life from the physical in dying.

The earth — recently described by Pope Francis as our mother and sister[28] — is envisioned by Steiner as an individual organism, the living diaphragm between the earth's core and the surrounding atmosphere open to Spirit forces flowing from the cosmos. By feeding the earth with a living substance (like compost), organic nitrogen brings astral life into the soil, to support the macro- and microorganisms enabling plant growth and health.

In contrast, chemically farmed soils are poor in micro-organic life because synthetic fertilizers are lifeless, and therefore produce dead soils and nutrient-poor plants. Etheric silica forces from the sun and outer planets shape leaf growth and flower development in a teleological process described by Goethe as plant metamorphosis. Astral lime forces from the inner planets (Moon and Mercury) stimulate the maturation of seeds, and of bunch roots, like legumes, and stem roots, like beets, turnips, or potatoes.

The fermented, herbal preparations Steiner recommended for the restoration of agricultural soils damaged by chemical farming, are based on the same super-sensible, etheric forces at work through elemental carriers. The presence of these forces in the biodynamic preparations cannot be tested in the laboratory through reductive methods, but their effects can be validated through practical application and observation.

To give an example from farm-based research, in 2004-6, we collaborated with Washington State University soil microbiologists and a forage specialist to determine whether pH levels could be raised in our acidic soils to improve the availability of soil nutrients to pasture plants. Conventionally, quantities of lime are applied every few years, but in this blind trial randomly selected plots were treated alternately with lime, biodynamic preparations, or no input. Soil and forage samples taken and tested in university laboratories over three seasons demonstrated that the minute inputs (less than a gram per acre) of biodynamic preparations were equally effective in adjusting pH levels as were massive, and expensive, inputs of lime, but the researchers could not explain the effect of the preparations using conventional scientific methods. The report for this field-project was published in a peer-reviewed science journal.[29] The same kind of farm-produced, fermented soil stimulants have been applied consistently to our fields for several decades, resulting in five-fold increases

in soil organic matter (from three to fifteen per cent), a clear indication that the biodynamic preparations are materially effective, even though the quantities applied are minute, at rates of one teaspoon to one quarter cup per acre.[30]

Another biodynamic practice based on the observation that subtle cosmic energies are correlated with the daily and seasonal rhythms of sun, moon, planets, and stars, was developed by a student of Steiner's. Over fifty years of longitudinal field trials, Maria Thun developed an astronomical calendar for soil preparation, planting and husbanding crops. Steiner argued that the forms of all of organic life, including humans, animals, and plants, evolved in response to Spiritual energies pulsing from the cosmos. In her work as a farmer, Maria Thun documented the growth rhythms of root, leaf, flower, fruit and seed to make visible the impact of etheric and astral forces in the plants' shape, structure, and quality. She correlated these rhythms with the protein, fat, carbohydrate, and salt content of the produce, and with its flavor and stability in storage. The accumulated data resulted in a predictive calendar that relates quality differences in plants to the temporal plane of astronomical rhythms.

Today three such calendars are published annually, two for the Northern hemisphere, and one for the Southern. Besides Maria Thun's prototype in Germany (also published as *North American Maria Thun Biodynamic Almanac*), Sherry Wildfeuer publishes *Stella Natura* in the U.S., and Brian Keats publishes the *Antipodean Astro-Calendar* in Australia. Interestingly, Keats' calendar includes astronomical charts predicting the schedules when ruminants chew their cuds, showing that animals, too, respond to cosmic rhythms in their daily behaviors, an observation I have been able to verify in cow rumination on our farm.

Steiner's concept of nutrition is also based on the notion that what nourishes all living beings are cosmic energies carried in the elements and streaming from the cosmos.

Animals and humans assimilate the elements of air, light, and warmth (elemental fire) directly through their nerve-sense and respiratory systems, but the elements of earth and water indirectly through their digestive systems. By contrast, plants directly assimilate light, air, earth, and water through their leaves and roots. Humans eliminate carbon dioxide through their breath, and fecal and other waste elements through glandular and alimentary excretion. These are recycled as nutrients for plants. Plants give off air and warmth through respiration and soil nutrients through their roots. Plants, animals, and humans nourish each other in a symbiotic relationship, where the waste products of plants feed humans and animals, and vice-versa.

Steiner's concept of human nutrition posits that our true digestive organ is the nerve-sense system located in the brain and throughout the body, with the stomach and intestines in support roles. Through the food we put in our stomachs, we internalize the astral and etheric principles vested in elemental substances derived from plant and animal sources. The energies derived from this material food empower the brain and nerve-sense system to receive non-material principles of form and Spirit by way of breath, feeling, thought, intuition and imagination, and by input from our five senses. If the food we put in our stomachs is nutritionally inferior and without life (as most ultra-processed foods), the nerve-sense system cannot do its work of absorbing etheric energies from the cosmos.

The analogy that has helped me understand Steiner's concept of human nutrition is that of a radio (representing the nerve-sense system), plugged into a power source (representing the food ingested though the alimentary system). If the power source is poor or faulty (as in lifeless food), the radio cannot receive the music streaming in from the ether, and if I turn the radio dial (my five senses) to stations broadcasting trashy contents, the brain will be left without true nourishment for thoughts, will, feeling, intuition and creative imagination.

Steiner predicted that unless agriculture returned to pre-industrial, holistic methods of food production, not only would soil organic life be depleted, but human society would soon suffer massive losses of physical and mental health. A hundred years later, the truth of Steiner's prediction is plain to see.

What, then, are the practical consequences of managing the soil, and the whole farm organism, from a Spiritual perspective? Can you be a biodynamic farmer without recognizing the Spirituality of the soil? Demeter certification of biodynamic farms does not require demonstration of Spirituality. How could it? It is not possible to quantify and certify Spirit. Nevertheless, the sense that matter is embedded in Spirit is of the essence of the biodynamic world view and practice.

Biodynamic farming is inherently practical in that it deliberately and systematically makes use of Spiritual forces to secure the fertility of the soil. It thereby ensures the health of every living thing dependent on the soil: plants, animals, farmers, and consumers, the whole of the farm organism, and the ecological and social community in which it is embedded. More than that, as Steiner put it, biodynamic farming practice promotes the inner development of the farmer by guiding "the Spiritual in the human being to the Spiritual in the universe."[31]

Personal progress toward a unifying and transformative apprehension of cosmos and life as Spiritual is fundamental to Steiner's thinking and to biodynamic practice. It also informs the writings of numerous modern ecologists from Emerson[32] and Thoreau[33] to Rachel Carson[34], and more recently the work of Per Espen Stoknes[35], John Vaillant[36], Fred Kirschenmann[37] and Wendell Berry, who speaks of the "Spirit (of God) astir in the world." [38]

Let me conclude with a particularly moving example of the sense that divine Spirit is immanent in living nature and in the soil, as it occurs in the work of William Bryant Logan, a soil scientist and ecologist.

In his book, *Dirt: The Ecstatic Skin of the Earth*, Logan asks what to make of the Biblical story of the Burning Bush. He refers the reader to the two equations by which all organic life fundamentally exists, one the equation of photosynthesis, the other that of burning. In photosynthesis, the plant rooted in the soil (earth element) makes the food it needs from sunlight (fire element), carbon dioxide (air element), and H2O (water element). In the process of burning (fire), plants, humans, and animals unlock the stored solar energy to turn it into biological energy fueling all physical and mental processes, thought, feeling and motion.

Thus, as Logan puts it, it is a "fundamental fact of nature" that "all that is living burns." The remarkable thing about the Biblical story of the Burning Bush, Logan observes, is that Moses beheld this natural fact directly, and recognized it as the presence of God speaking to him through nature.

God commands Moses to take off his shoes because the soil he is standing on is holy ground.[39] Like the gardener, Alan Chadwick, long after him, Moses beholds the indwelling divine Spirit "in reverence and obedience," and is utterly transformed.

Notes

[1] Paper given at Harvard Divinity School Conference, "The Spirit of Sustainable Agriculture," March 31-April 1, 2016.

[2] See. Wicksteed, Philip H. & Francis Cornford, 1934. *Aristotle: The Physics.* Cambridge, Mass, xvff.

[3] The modern term "teleology" based in Aristotle's thinking was coined by German philosopher, Christian von Wolff in 1728 in *Philosophia rationalis, sive logica* (Rational Philosophy, or Logic), Frankfurt, Germany.

[4] Aristotle, 4th century B.C. Φυσικὴ ἀκρόασις (*Physics*), III, i, 201a, 10-16: "Διῃρημένου δὲ καθ' ἕκαστον γένος τοῦ μὲν ἐντελέχεια, τοῦ δὲ δυνάμει, ἡ τοῦ δυνάμει ὄντος ἐντελέχεια, ᾗ τοιοῦτον, κίνησίς ἐστιν, § 8. οἷον τοῦ μὲν ἀλλοιωτοῦ, ᾗ ἀλλοιωτόν, ἀλλοίωσις, τοῦ δὲ αὐξητοῦ καὶ τοῦ ἀντικειμένου φθιτοῦ (οὐδὲν γὰρ ὄνομα κοινὸν ἐπ' ἀμφοῖν) αὔξησις καὶ φθίσις, τοῦ δὲ γενητοῦ καὶ φθαρτοῦ γένεσις καὶ φθορά, τοῦ δὲ φορητοῦ φορά" (We have now before us the distinctions in the various classes of being between what is fully real and what is potential. Definition: *The fulfillment of what exists potentially [indwelling end, purpose, or power to become actual— H.S.], insofar as it exists potentially, is motion* — namely, of what is alterable *qua* alterable *alteration*: of what can be increased and its opposite what can be decreased (there is no common name), *increase* and *decrease*: of what can come to be and can pass away, *coming to be* and *passing away*: of what can be carried along: *locomotion*). Translation: Richard McKeon 1941. *The Basic Works of Aristotle*, New York, NY, 254.

[5] Ibid.

[6] Aristotle. *Politics*, I, iii, 1258a, 34 -1258b, 8: "μάλιστα δέ, καθάπερ εἴρηται πρότερον, δεῖ φύσει τοῦτο ὑπάρχειν. Φύσεως γάρ ἐστιν ἔργον τροφὴν τῷ γεννηθέντι παρέχειν· παντὶ γάρ, ἐξ οὗ γίνεται, τροφὴ τὸ λειπόμενόν ἐστι. Διὸ κατὰ φύσιν ἐστὶν ἡ χρηματιστικὴ πᾶσιν ἀπὸ τῶν καρπῶν καὶ τῶν ζῴων. § 23. Διπλῆς δ' οὔσης αὐτῆς, ὥσπερ εἴπομεν, καὶ τῆς μὲν καπηλικῆς τῆς δ' οἰκονομικῆς, καὶ ταύτης μὲν ἀναγκαίας καὶ ἐπαινουμένης, τῆς δὲ μεταβλητικῆς ψεγομένης δικαίως (1259a) οὐ γὰρ κατὰ φύσιν ἀλλ' ἀπ' ἀλλήλων ἐστίν, εὐλογώτατα μισεῖται ἡ ὀβολοστατικὴ διὰ τὸ ἀπ' αὐτοῦ τοῦ νομίσματος εἶναι τὴν κτῆσιν καὶ οὐκ ἐφ' ὅπερ ἐπορίσθη. Μεταβολῆς γὰρ ἐγένετο χάριν, ὁ δὲ τόκος αὐτὸ ποιεῖ πλέον (ὅθεν καὶ τοὔνομα τοῦτ' εἴληφεν· ὅμοια γὰρ τὰ τικτόμενα τοῖς γεννῶσιν αὐτὰ ἐστιν, ὁ δὲ τόκος γίνεται νόμισμα ἐκ νομίσματος)· ὥστε καὶ μάλιστα παρὰ φύσιν οὗτος τῶν χρηματισμῶν ἐστιν" (But, strictly speaking, as I have already said, the means of life must be provided beforehand by nature; for the business of nature is to furnish food to that which is born, and the food of the offspring is always what remains over of that from which it is produced. Wherefore the art of getting wealth out of fruits and animals is always natural. — There are two sorts of wealth-getting, as I have said; one of them is part of household [stewardship], the other is retail trade: the former necessary and honorable, while that which consists in exchange is justly censured; for it is unnatural, and a mode by which men gain from one another. The most hated sort, and with the greatest reason, is usury, which makes gain out of money itself, and not from the natural object of it. For money was intended to be used in exchange, but not to increase at interest. And this term interest, which means the birth of money from money, is applied to the breeding of money because the offspring resembles the parent. Wherefore of all modes of getting wealth it is the most unnatural). Translation: Richard McKeon 1941, op.cit., 1140-1141.

[7] For an applied discussion of Aristotelian economics in the context of biodynamics, see Sehmsdorf, Henning 2013, "The Economics of the Small-Scale, Self-Sufficient Farm," *Stella Natura;* see also Sehmsdorf, Henning 2021, "Farming for Health: The Economics of Stewardship," below.

[8] Aristotle. Περὶ οὐρανοῦ (On the Heavens), in: Richard McKeon, op.cit., 395-466.

[9] See Saint Thomas Aquinas, 1963. *Commentary on Aristotle's Physics* (1258-1260). Venice, Italy, 136-137. At an ecumenical forum in 2007, Catholic theologian, Otto-Hermann Pesch, commented on Thomas Aquinas' reading of Aristotle as follows ("Gnade und Rechtfertigung am Vorabend der Reformation und bei Luther,"(Grace and Justification on the Eve of the Reformation and in Luther), Heidelberg, Germany, 13: "Thomas hat…das ganze Instrumentarium der Philosophie des alten Griechen Aristoteles zur Verfügung. Er geht davon aus, dass Gott allen Geschöpfen nicht nur die Fähigkeit, die "Potenz", wie das Aristoteles nennt, zum artgemäßen Tätigsein gegeben hat, sondern darüber hinaus auch die angeborene oder durch Übung zu erwerbende Tätigkeitsvorprägungen, sogenannte *habitus* ("Gehaben"), durch die sie in bestimmter Richtung sicher, spontan, leicht und lustvoll tätig sind (*sponte, faciliter et delectabiliter*) [Thomas has the entire range of philosophical instruments of the ancient Greek, Aristotle, at his disposal. He presupposes that God has endowed all created beings not only with the capacity, the "potential," as Aristotle calls it, for species-appropriate activity, but beyond that with innate or acquired tendencies, so-called habitus ("behaviors"), by which they become active in a particular direction (sponte, faciliter et delectabiliter)]. Translation mine.

[10] Luther, Martin, 1517. *Disputatio pro declaratione virtutis indulgentiarum* (Disputation for the Carification of Indulgences: Ninety-Seven Theses: Disputation Against Scholastic Theology.) Wittenberg, Germany, theses 51 and 42.

[11] Descartes, Renée, 1644 *Les Principes de la Philosophie* (Principles of Philosophy). Paris, France, II, 24: "Le mouvement donc, n'est autre chose que l'action par laquelle un corps pass d'un lieu en un autre." Translation mine.

12 Spinoza, Baruch, 1677. *Ethica: ordine geometrico demonstrata* (Ethics, Demonstrated in Geometrical Order), Amsterdam, The Netherlands: I, Propositio (Thesis) XXIX: "Nam ex antecedentibus jam constare existimo, nempe, quod per Naturam naturantem nobis intelligendum est id, quod in se est, & per se concipitur, sive talia substantiae attributa, quae aeternam, & infinitam essentiam exprimunt, hoc est Deus, quatenus, ut causa libera, consideratur" — (Because, as I believe, it follows from what has been said before, that by *Natura naturans* we are to understand that which exists in itself and is conceived through itself, or such attributes of the substance which expresses eternal and infinite being, i.e. God, inasmuch as he is regarded as free cause) — "Per naturatam autem intelligo id omne, quod ex necessitate Dei naturae, sive uniuscujusque Dei attributorum sequitur, hoc est, omnes Dei attributorum modos, quatenuus considerantur, ut res, quae in Deo sunt, & quae sine Deo nec esse, nec concipi possunt" — (By *Natura naturata*, however, we understand all that which follows by necessity from God's nature, i.e. all modifications of God's attributes, in as much as they can be regarded as material things, which are in God and without God can neither exist nor be conceptualized). Translation mine.

13 Spinoza, op.cit, IV, Introduction. See Jaspers, Karl, 1974 *Spinoza*. New York, NY & London, UK, 14 and 95.

14 Chardin, Teilhard de 1955. *The Phenomenon of Man*. New York, NY, 308.

15 Driesch, Hans, 1891. "Entwicklungsmechanische Studien: I. Die Werthe der beiden ersten Furchungszellen in der Enchiodermenentwicklung. Experimentelle Erzeugung von Theil- und Doppelbildungen. II. Über die Beziehungen des Lichtes zur ersten Etappe der thierischen Formbildung" (The Potency of the First Two Cleavage Cells in Enchioderm Development. Experimental Production of Partial and Double Formations), *Zeitschrift für wissenschaftliche Zoologie* (Journal for Scientific Zoology), no. 53: 160–84.

16 Heisenberg, Werner, 1955. *Das Naturbild der heutigen Physik* (The Image of Nature in Today's Physics). Hamburg, Germany, 108-109: "Ist also der Geist eine Eigenschaft oder Wirkung des Stoffes im Gehirn, so muß diese Eigenschaft gemäß dem Gesetz von der Erhaltung des Stoffes den von der Mechanistik vorausgesetzten Atomen unter allen Umständen zukommen, und der Stein, der Tisch, die Cigarre sind beseelt, ebenso wie der Baum, das Thier und der Mensch. In der That drängt sich dieser Gedanke, wenn man die Voraussetzung zugiebt, so unwiderstehlich auf, daß man in der neueren philosophischen Literatur ihn entweder als richtig oder doch wenigstens als angemessen empfiehlt, oder aber zu seiner Vermeidung einen entschlossenen und unüberbrückbaren Dualismus zwischen Geist und Materie aufstellt" (If Spirit is a property or effect of matter in the brain, then this property must accrue to atoms according to the law of the preservation of matter assumed by mechanistic physics, and the stone, the table, the cigar must be endowed with Spirit, as much as trees, animals and humans. Indeed, given the stated assumptions, this thought becomes so irresistible that current philosophers recommend it as correct or at least reasonable, or in order to avoid it, take refuge in a resolute and insurmountable dualism of Spirit and matter). Translation mine.

17 Heisenberg, Werner 1907. *Science and Religion*. https://www.edge.org/conversation/werner_heisenberg-science-and-religion. Retrieved September 12, 2022.

18 Nagel, Ernest 1961. *The Structure of Science*. New York, NY, 429ff.

19 Goethe, Johann Wolfgang von, 1963. *Maximen und Reflexionen* (Maxims and Reflections). München, Germany, 68: "Es gibt eine zarte Empirie, die sich mit dem Gegenstand innigst identisch macht und dadurch zur eigentlichen Theorie wird." Translation mine. Much like Goethe, 20th century philosopher and scientist, Michael Polanyi, argues that "it is not by looking at things, but by dwelling in them that we understand their joint meaning" (1983. *The Tacit Dimension*, Gloucester, MA, 18).

20 Goethe 1963 (1808). *Zur Farbenlehre* (The Color Theory), 3 vols. München, Germany.

21 Goethe 1963 (1790). *Versuch die Metamorphose der Pflanzen zu erklären* (An Essay to Explain the Metamorphosis of Plants). München, Germany.

22 Goethe 1962 (1784, 1820, 1831). "Über den Zwischenkiefer des Menschen und der Tiere" (Concerning the Intermaximillary Bone in Humans and Animals). München, Germany.

23 Bortoft, Henri, 1996. *The Wholeness of Nature: Goethe's Way of Science*. Edinburgh, UK, passim.

24 See www.sshomestead.org/research/publications.

25 Chadwick, Alan, 2013. *Reverence, Obedience and the Invisible in the Garden*. Asheville, NC, 45-51.

26 Steiner, Rudolf 1924. *Geisteswissenschaftliche Grundlagen zum Gedeihen der Landwirtschaft*. Dornach, Switzerland.

27 Steiner, Rudolf 1886. *Grundlinien einer Erkenntnistheorie der Goetheschen Weltanschauung mit besonderer Rücksicht auf Schiller* (Outline of an Epistemology of Goethe's World View With Particular Consideration of Schiller). Berlin & Stuttgart, Germany. Ibid 1891. *Wahrheit und Wissenschaft : Vorspiel einer Philosophie der Freiheit* (Truth and Science: Prologomena to a Philosophy of Freedom), Rostock, Germany.

28 Pope Francis, 2015. *Laudato Si' mi' Signore* (Praise Be to You, My Lord). Rome, Italy, 5.

[29] Reeve, Jennifer R., Lynn Carpenter-Boggs & Henning Sehmsdorf 2011. "Sustainable Agriculture, a Case Study of a Small Lopez Island Farm," *Agricultural Systems*, 104: 572-579. Reprinted below.

[30] For details on how the preparations are produced and applied, see www.sshomestead.org/offerings/biodynamic.

[31] Steiner, Rudolf 1993. *Spiritual Foundations for the Renewal of Agriculture*. Junction City, OR, ix.

[32] Emerson, Ralph Waldo 1934. "On Self-Reliance," *Essays*. New York, NY, 32: "Prayer is the contemplation of the facts of life from the highest point of view. It is the soliloquy of a beholding and jubilant soul. It is the Spirit of God pronouncing his works good. But prayer as a means to effect a private end is meanness and theft. It supposes dualism and not unity in nature and consciousness. As soon as the man is one with God, he will not beg. He will then see prayer in all action. The prayer of the farmer kneeling in his field to weed it, the prayer of the rower kneeling with the stroke of his oar, are true prayers heard throughout nature, though not for cheap ends."

[33] Thoreau, Henry. D. 1849. *A Week on the Concord and the Merrimack Rivers*. Boston, and Walden (1845), Boston, Mass, are infused with the understanding that the divine is immanent in nature.

[34] Rachel Carson 1962, *Silent Spring*. New York, NY, 203, 298. Carson couches her analysis of the impacts of man-made radiation and chemicals on human and ecological health in mythopoeic terms when she refers to adenosine triphosphate (ATP), which by means of the mitochondria furnishes energy to muscle and nerve cells, as "the universal currency of energy," and in her conclusion urges that we take into account the "life forces" and cautiously seek "to guide them into channels favorable to ourselves."

[35] Stoknes, Per Espen, 2015. *What We Are Thinking When We Are Not Thinking About Climate Change*. White River Junction, VT, 210: "We have forgotten that the air is a sacred, intelligent creative being."

[36] Vaillant, John 2005. *The Golden Spruce: A True Story of Myth, Madness, and Greed*. New York, NY & London, UK, 147: "The Haida refer to the Yakoun (river) as the River of Life, and just as the islands seem to represent the life force in concentrated form, the golden spruce represented the concentrated essence of the Yakoun."

[37] Kirschenmann, Frederick L. 2010. *Cultivating an Ecological Conscience: Essays from a Farmer Philosopher*, Lexington, KY, 17: "[I see God] in every thistle in our fields and every calf humping another calf in our pasture ... In my theology, the divine always meets us in the flesh — *all* flesh — *all* relationships, not just our relationship with humans or relationships we like. This seems to me to be at the heart of the concept of incarnation."

[38] Berry, Wendell, n.d. "The Satisfactions of the Mad Farmer," *The Mad Farmer Poems*, Berkeley, CA, 9-11, last stanza: "What I know of spirit is astir/ in the world. The god I have always expected/ to appear at the wood's edge, beckoning/ I have always expected to be/ a great relisher of this world, it's good/ grown immortal in his mind."

[39] *Exodus* I, 5. Quoted in Logan, William Bryant 2007. *Dirt: The Ecstatic Skin of the Earth*. London, UK, 93.

Abele, U. 1987. "Produktqualität und Düngung – mineralisch, organisch, biologisch-dynamisch" (Product Quality and Fertilization: Mineral, Organic, Biodynamic). *Schriftenreihe des Bundesministers für Ernährung, Landwirtschaft und Forste* (Publications of the Federal Ministry for Nutrition, Agriculture, and Forests), Series A, Volume 345.

Adhikary, Sujit 2012. "Vermicompost, the story of organic gold: A review," *Agricultural Sciences* 03 (07): 905-917.

Anderson, John, P.E. & K.H. Domsch 1978. "A Physiological Method for the Quantitative Measurement of Microbial Biomass in Soil," *Soil Biology & Biochemistry* 10: 215-221.

Andersson, S., I. Valeur, & C. Lapierre 1994. "Influence of Lime on Soil Respiration, Leaching of DOC, and C/S Relationships in the Mor Mumus of a Haplic Podsol," *Environment International* 20: 81-88.

Angst, Gerrit et al. 2019. "Earthworms Act as Biochemical Reactors to Convert Labile Plant Compounds Into Stabilized Soil Microbial Necromass," *Communications Biology* 2: 441.

Aquinas, Thomas Saint 1963. *Commentary on Aristotle's Physics* (1258-1260).Venice, Italy.

Aristotle, 4th century B.C. τὰ μετὰ τὰ φυσικά (Metaphysics), Book VIII.

Aristotle. 4th century B.C. Περὶ οὐρανοῦ (On the Heavens).

Aristotle,4th century B.C. Φυσικὴ ἀκρόασις (Physics), III.

Aristotle, 4th century B.C., Πολιτικά (Politics), Book I.

Arnarson, Atli 2019. "A1 vs. A2 Milk — Does It Matter?" https://www.healthline.com/nutrition/a1-vs-a2-milk#definition.

Asgarpanah, Jinous 2012. "Phytochemistry and Pharmacological Properties of *Equisetum arvense*," *Journal of Medicinal Plant Research* 6: 21.

Ashbrook, Frank G. 1973. *Butchering, Processing and Preservation of Meat: A Manual for the Home and Farm*, New York.

@presbyformed 2016. "Gothic Cathedrals & Medieval Symbolism." https://presbyformed.com/2016/09/07/gothic-cathedrals-medieval-symbolism/.

Bairacli, Juliette de 1984. *The Complete Herbal Handbook for Farm and Stable.* London,UK.

Balch, James & Phyllis 1997. *Prescription for Nutritional Healing*, 2nd ed. New York, NY.

Barkley, Melanie 2017. "Prevent Parasites Through Grazing Management," *Penn State Extension*, University Park, PA.

Barnett, Tanya M. 2003. "Gratia Plena," *Earth Ministry Newsletter*, Seattle, WA.

Barry, T.N., D.M. McNeill & W.C. McNabb 2001. "Plant Secondary Compounds: Their Impact on Nutritive Value and Upon Animal Production," *Proceedings XIX International Grass Conference,* Sao Paulo, Brazil, 445-452.

Batie, Sandra S. 2009. "Green Payments and the US Farm Bill: Information and Policy Changes," *Frontiers in Ecology and Environment.* 7 (7): 380-388.

Batra, Suzanne W. T. 1968. "Behavior of Some Social and Solitary Halictine Bees Within Their Nests: A Comparative Study," *Journal of the Kansas Entomological Society,* 41 (1): 120–133.

Battisti, David S. & Rosamund L. Naylor 2009. "Historical Warnings of Future Food Insecurity With Unprecedented Seasonal Heat," *Science* 323: 240-244.

Bent, Stephen, et al. 2006. "Valerian for Sleep: A Systematic Review and Meta-Analysis," *American Journal of Medicine,* 119 (12): 1005-12.

Bergström, Anders et al. 2020. "Origins and genetic legacy of prehistoric dogs", *Science.* 370: 557–564.

Berry, Wendell, n.d. "The Satisfactions of the Mad Farmer," *The Mad Farmer Poems*, Berkeley, CA.

Bertrand, I., O. Delfosse & B. Mary 2007. "Carbon and Nitrogen Mineralization in Acidic, Limed and Calcareous Agricultural Soils: Apparent and Actual Effects," *Soil Biology & Biochemistry*. 39: 276-288.

Bess, Vicki 2000. "Understanding Compost Tea," Biocycle (October).

Bethel, John P. (General Editor) 1956. *Webster's New Collegiate Dictionary*. Springfield, MA.

Blamire, John 1998. "Classification" *Science at a Distance* http://www.brooklyn.cuny.edu/bc/ahp/CLAS/CLAS.Linn.html.

Blank, Stephen 1998. *The End of Agriculture in the American Portfolio*, Westport, CT.

Blank, Stephen 1999. "The End of the American Farm?," *The Futurist*, 22-27.

Blunden, Andy 2010. *An Interdisciplinary Theory of Activity*. Boston, Mass.

Bockemühl, Jochen 1995. "Morphic Movements in the Vegetative Leaves of Higher Plants" *The Metamorphosis of Plants: Essays by Jochen Bockemühl and Andreas Suchantke*, Cape Town, 21-46.

Bollier, David 2011. "Commons — Short and Sweet." http://bollier.org/cp\commons-short-and-sweet.

Bortoft, Henri, 1996 *The Wholeness of Nature: Goethe's Way of Science*. Edinburgh, UK.

Brady, Nyle C. & Ray R. Weil, 2002. "Formation of Soils from Parent Materials," *The Nature and Properties of Soils*. Hoboken, NJ, 54-61.

Brady, Nyle 2002. "Soil Acidity," *The Nature and Properties of Soils*. Hoboken, NJ, 390-391.

Brown, Mark T. & Sergio Ulgiati 1999. "Emergy Evaluation of the Biosphere and Natural Capital," *Ambio* 28, vol 6: 488.

Brown, Mark T. & Sergio Ulgiati 2001. "Emergy Measures of Carrying Capacity to Evaluate Economic Investments," *Population and Environment* 22; 471-501.

Burlacu, Ema et al. 2020. "A Comprehensive Review of Phytochemistry and Biological Activities of Quercus Species," *Forests*, 11: 0904.

Burns, David & Sheri 2021. "Requeening a Beehive."https://www.honeybeesonline.com/requeening-a-bee-hive/.

Burrows, Sara 2018. "Vegetables Lose Half Their Nutritional Value by the Time They Get to the Store: Another Reason to Grow Your Own," *Return to Now*. https://returntonow.net/.

Busch, Lawrence 2009. "What Kind of Agriculture? What Might Science Deliver?" *Natures Sciences Sociétés* 17: 241-247.

Rachel Carson 1962, *Silent Spring*. New York, NY.

Chadwick, Alan, 2013. *Reverence, Obedience and the Invisible in the Garden*. Asheville, NC.

Campbell, Stu 1998. *Let it Rot*. Pownal, VT.

CAN, 1985.

Carlson, Laurie W. 2001. *Cattle: An Informal Social History*. Chicago, Il.

Carpenter-Boggs, Lynne 1997. *Effects of Biodynamic Preparations on Compost, Crop, and Soil Quality* (Ph.D. Dissertation), Pullman, WA.

Carpenter-Boggs, Lynn, John P. Reganold, and Ann C. Kennedy, 2000. "Biodynamic Preparations: Short-Term Effects on Crops, Soils and Weed Populations," *American Journal of Alternative Agriculture* 15: 110-118.

Carpenter-Boggs, L, J.P. Reganold & A.C. Kennedy 2000. "Organic and Biodynamic Management: Effects on Soil Biology," *Soil Science Society of America Journal* 64: 1651-1659.

Carpenter-Boggs, Lynn and Jennifer Reeve 2006. SARE Project Report. http://wsare.usu.edu/pro/?sub=fund04#wa.

Carroll, William C. 1994 "'The Nursery of Beggary': Enclosure, Vagrancy, and Sedition in the Tudor-Stuart Period," *Enclosure Acts: Sexuality, Property, and Culture in Early Modern England*, eds. Richard Burt and John Michael Archer, Ithaca & London, 34–47.

Ceccanti, Constanza et al. 2020. "Comparison of Three Domestications and Wild-Harvested Plants for Nutraceutical Properties and Sensory Profiles in Five Wild Edible Herbs: Is Domestication Possible?" *Foods* 9(*): 1065.

Chardin, Teilhard de 1955. *The Phenomenon of Man*. New York, NY.

Charles, Dan 2015. "The Ancient City Where People Decided To Eat Chickens." https://www.npr.org/sections/thesalt 2015/07/20/424707879/the-ancient-city-where-people-decided-to-eat-chickens.

Chevallier, Andrew 2016. *Encyclopedia of Herbal Medicine: 550 Herbs and Remedies for Common Ailments*. London, UK.

Cogger, Craig 2015. "Soil Management for Small Farms," *WSU Extension Food & Farm Connection*, EB1895.

Cole, Jason 2001. "Micas," U. of Waterloo Earth Sciences Museum. https://uwaterloo.ca/earth-sciences-museum/resources/detailed-rocks-and-minerals-articles/micas.

Coleman, Eliot 1995. *Four Season Harvest*. White River Junction, VT.

Coleridge, Samuel T. 1884. *The Complete Works of Samuel T. Coleridge*. Shed, W.G.T. (ed.). New York, NY.

Colmenares, R. & J.M. de Miguel 1999. "Improving Permanent Pasture's Growth: An Organic Approach," *Cahiers Options Méditerranéennes* 39: 189-191.

Condron, L.M., H. Tiessen, C. Trassarcepeda, J.O. Moir, & J.W.B. Stewart 1993."Effects of Liming on Organic-Matter Decomposition and Phosphorus Extractability in an Acid Humic Ranker Soil from Northwest Spain," *Biology & Fertility of Soils* 15: 279-284.

Cox, Craig 2011. "If Not Conservation Districts, What?" *Managing Agricultural Landscapes for Environmental Quality II: Achieving Effective Conservation* (Nowak, Peter & Max Schnepf, eds), Ankeny, IA.

Crespi, Bernard et al. (1995). "The Definition of Eusociality," *Behavioral Ecology*, 6: 109–115.

Culpeper, Nikolas 2019. *Culpeper's Complete Herbal: Over 400 Herbs and Their Uses*. London,UK.

Davis, Ellen F. 2014. *Scripture, Culture, and Agriculture: An Agrarian Reading of the Bible*, Cambridge, UK.

Davis, William 2013. *Wheat Belly 30-Minute (or Less!) Cookbook: 200 Quick and Simple Recipes*. Kutztown, PA.

Descartes, Renée, 1644 *Les Principes de la Philosophie* (Principles of Philosophy). Paris, France.

De Schutter, Olivier 2010. "Final Report: The Transformation Potential of the Right to Food," *United Nations*, Geneva, Switzerland.

Dictionary. com 2021. "Prism." https://www.lexico.com/en/definition/prism.

Diver, Steve 1999. "Biodynamic Farming and Compost Preparation," *Attra* (February).

Dobbs, Thomas L. & Jules N. Pretty 2004. "Agri-Environmental Stewardship Schemes and Multifunctionality," *Review of Agricultural Economics* 26(2):220-237.

Drewnowski, Adam 2007. "Disparities in Obesity Rates: Analysis by ZIP Code Area," *Social Science & Medicine*, Dec; 65(12): 2458-63.

Driesch, Hans, 1891. "Entwicklungsmechanische Studien: I. Die Werthe der beiden ersten Furchungszellen in der Enchiodermenentwicklung. Experimentelle Erzeugung von Theil- und Doppelbildungen. II. Über die Beziehungen des Lichtes zur ersten Etappe der thierischen Formbildung" (The Potency of the First Two Cleavage Cells in Enchioderm Development. Experimental Production of Partial and Double Formations), *Zeitschrift für wissenschaftliche Zoologie* (Journal for Scientific Zoology), no. 53: 160–84.

Edens, Thomas C. & Dean L. Haynes, D.L. 1982. "Closed System Agriculture: Resource Constraints, Management Options, and Design Alternatives," *Annual Review of Phytophathology*, 20: 363-395.

Edwards, S. 1998. "Vermicompost," *Biocycle*. July: 63-66.

Ekarius, Carol 1999. *Small-Scale Livestock Farming: A Grass-Based Approach for Health, Sustainability, and Profit*. North Adams, Mass.

Ekenler, M. & M.A.Tabatabai 2003. "Effects of Liming and Tillage Systems on Microbial Biomass and Glycosidases in Soils," *Biology & Fertility of Soils* 39: 51-61.

Elmadfa, Ibrahim et al. 2002/2003. *Die Grosse GU Nährwert-Kalorien-Tabelle* (The Comprehensive GU Nutrition & Calorie Tables), Munich, Germany.

Emerson, Ralph Waldo 1934. "On Self-Reliance," *Essays*. New York, NY, 18-38.

Evandrou, Maria and Jane Talking 2005. "Demographic Changes in Europe: Implications for Future Family Support for Older People," in *Aging Without Children: European and Asian Perspectives on Elderly Access to Support Networks*, eds. Kreager, Philip and E. Schröder-Butterfill. Oxford, UK.

Ewais, E.A. et al. 2019. "Phytochemical Contents of White and Pink Flowers of Marshmallow (Althaea officinalis L) Plants and their Androgenesis Potential on Anther Culture in Response to Chemical Elicitors," *Bioscience Research* 16(2): 1276-1289.

Fallon, Sally 2001. *Nourishing Traditions: The Cookbook that Challenges Politically Correct Nutrition*. Washington, D.C.

Fineman, Mia 2005. "The Most Famous Farm Couple in the World: Why American Gothic still fascinates." https://slate.com/culture/2005/06/the-most-famous-farm-couple-in-the-world.html.

Folke, Carl, Åsa Jansson, Jonas Larsson & Robert Costanza 1997. "Ecosystem Appropriation by Cities," *Ambio* 26/3.

Freeman, Jacqueline 2014. *The Song of Increase: Returning to Our Sacred Partnership with Honeybees*. Battleground, WA.

Fuentes, J.P., D.F. Bezdicek, M. Flury, S. Albrecht & J.L. Smith 2006. "Microbial Activity Affected by Lime in a Long-Term No-Till Soil," *Soil &Tillage Research* 88: 123-131.

Gao, Si 2016. "Locally produced wood biochar increases nutrient retention in agricultural soils of the San Juan Islands, WA," *Research Works Archive*.

Gavlak, R., D. Horneck, R.O. Miller & J. Kotuby-Amacher, 2003. *Soil and Plant Reference Methods for the Western Region (WREP 125)*. Corvallis, Or.

Gillam, Carey. 2015. "U.S. Honeybee Losses Soar Over Last Year, USDA Finds," *Reuters (*May 14).

Giuffra, E. et al. 2000. "The Origin of the Domestic Pig: Independent Domestication and Subsequent Introgression," *Genetics* 154 (4): 1785-1791.

Godfray, H. Charles. J. et al. 2010. "Food Security: The Challenge of Feeding 9 Billion People," *Science* 327: 812-818.

Goethe, Johann Wolfgang von. "On the Theory of Color." Transl. Pehr Sall 2013. "Goethe's Theory of Color, Part 1 — How It All Started." https://www.youtube.com/watch?v=QnfVlENcHbU.

Goethe, Johann Wolfgang von 1962. *Faust : Eine Tragödie, I* (Faust: A Tragedy), München, Germany.

Goethe, Johann Wolfgang von 1963. *Schriften zur Botanik und Wissenschaftslehre* (Writings on Botany and the Theory of Science), München, Germany.

Goethe 1963. *Zur Farbenlehre* (The Color Theory), 3 vols. München, Germany.

Goethe, Johann Wolfgang von 1963. *Materialien zur Geschichte der Farbenlehre* (Sources for the Study of the Theory of Color). München, Germany.

Goethe 1963. *Versuch die Metamorphose der Pflanzen zu erklären* (An Essay to Explain the Metamorphosis of Plants). München, Germany.

303

Goethe 1963. *Versuch die Metamorphose der Pflanzen zu erklären* (An Essay to Explain the Metamorphosis of Plants). München, Germany.

Goethe, Johann Wolfgang von 1963. *Maximen und Reflexionen* (Maxims and Reflections), München, Germany.

Goldstein, W. 1986. *Alternative Crops, Rotations and Management Systems for the Palouse* (Ph.D. dissertation), Pullman, WA).

Goldstein, W.A. & H.H. Koepf, 1992. "A Contribution for the Development of Tests for the Biodynamic Preparations," *Elemente Naturwissenschaft* (Elements of Natural Science) 36: 41-56.

Goldstein, Walter, et al. 2019. "Biodynamic Preparations, Greater Root Growth and Health, Stress Resistance, and Soil Organic Matter Increases Are Linked," *Open Agriculture* 4: 187–202.

Graville, Iris et al. 2008. *Hands at Work - Portraits and Profiles of People Who Work with Their Hands,* Flemington, NJ.

Graville, Iris et al. 2016 . *Bounty — Lopez Island Farmers, Food, and Community,* Lopez Island.

Greenbuilders Registry. 2001. https://sbregistry.greenbuilder.com/search.straw?RID=12.

Green Lands Blue Waters. www.greenlandsbluewaters.org.

Grotzke, Heinz 1999. "Growing Cucumbers," *Biodynamics Journal*, no. 222: 13-17.

Hackett, Mike 2009. "S&S Homestead Farm Biodynamic Farming Study — Sheep Flock Status." https://sshomestead.org/wp-content/uploads/Sheep%20evaluation.pdf. Retrieved November 10, 2021.

Haden, Andrew C. "Emergy Analysis of Food Production at S&S Homestead Farm." https://sshomestead.org/wp-content/uploads/emergy%20analysis.pdf.

Hahlbrock, Klaus 2009. *Feeding the Planet: Environmental Protection Through Sustainable Agriculture.* London, UK.

Hart, M.M. & J.T. Trevors 2005. "Microbe Management: Application of Mycorrhizal Fungi in Sustainable Agriculture," *Frontiers in Ecology and the Environ* 3: 533-539.

Hauck, Günther 2002. *Toward Saving the Honeybee.* San Francisco, CA.

Haynes, R.J. & R. Naidu, R. 1998. "Influence of Lime, Fertilizer and Manure Applications on Soil Organic Matter Content and Soil Physical Conditions: A Review," *Nutrient Cycling in Agroecosystems* 51: 123-137.

Heckman, Joseph R. et al. 2004. "Plant Nutrients in Municipal Leaves," Fact Sheet 824 (Rutgers U.), New Brunswick, NJ.

Heisenberg, Werner 1907. *Science and Religion.* https://www.edge.org/conversation/werner_heisenberg-science-and-religion.

Heisenberg, Werner, 1955. *Das Naturbild der heutigen Physik* (The Image of Nature in Today's Physics). Hamburg, Germany.

Heller, Martin C. & Gregory A. Keoleian, 2000. "Life Cycle-Based Sustainability Indicators for Assessment of the U.S. Food System," *Center for Sustainable Systems,* U. of Michigan. Report No. CSS00-04.

Hellerer, Ulrike & K.S. Jarayaman 2000. "Greens Persuade Europe to Revoke Patent on Neem Tree…," *Nature,* vol 45: 266-267.

Hemleben, Johannes 1975. *Rudolf Steiner: A Documentary Biography.* Rye, UK.

Hindes, Daniel 2009. "Rudolf Steiner and Anthroposophy." http://www.rudolfsteinerweb.com/Rudolf_Steiner_and_Anthropsophy.php.

Høk, Johanna 2002. "Nutrient Recycling and Composting on S&S Homestead Farm" (unpublished internship report).

Høk, Johanna 2002. "Medicinal Plants." https://sshomestead.org/wp-content/uploads/Herb%20project.pdf.

Howell, Edward 1994. *Food Enzymes. Health and Longevity,* 2nd ed., Twin Lakes, WI.

Howell, Edward 1995. *Enzyme Nutrition.* New York, NY.

Huffman, Wallace E. & Robert E. Evenson 2006 *Science for Agriculture: A Long-Term Perspective*, Ames, IA, 2. ed.

Huntington, Brian 2000. "Soil Health and Fertility" (unpublished internship report)

Hylton, William H. 1974. *The Rodale Herb Book.* Emmaus, PA.

IAASTD 2009. *International Assessment of Agricultural Science and Technology for Development, Agriculture at a Crossroads: Global Report.* Washington, DC.

Ikerd, John 2008. *Small Farms Are Real Farms: Sustaining People Through Agriculture.* Austin, TX.

Ingham, Elaine, et al. 2001. *Compost Tea Manual.* Corvallis, OR.

Iowa State U. 2009. "What is Organic Agriculture?" http://extension.agron.iastate.edu/organicag/whatis.html.

Isom, Cathy 2019. "Predators that Prey on Bees and Your Harvest." https://agnetwest.com/predators-prey-bees-harvest/.

Jackson, Wes 2002. "Natural Systems Agriculture: a Truly Radical Alternative," *Agriculture Ecosystems and Environment*, 88: 111-117.

Jantsch, Erich. *The Self-Organizing Universe: Scientific and Human Implications of the Emerging Paradigm of Evolution.* New York, NY.

Jaspers, Karl, 1974 *Spinoza.* New York, NY & London, UK.

Jeavons, John 1995. How to Grow More Vegetables…Berkeley, CA.

Jefferson, Thomas 1785. "Letter to John Jay."

Jenkins, Joseph 1999. *The Humanure Handbook.* Grove City, PA.

Johnson, John 2020. "What to Know About Valerian Root." https://www.medicalnewstoday.com/articles/valerian-root.

Johnson, John 2021. "Bee Pollen: What to Know," *Medical News Today.* https://www.medicalnewstoday.com/articles/bee-pollen.

Josephine Porter Institute 2021. https://jpibiodynamics.org.

Jung, C. G. 1958. *Psychology & Religion: West and East.* Princeton, NJ.

Kambartel, Friedrich & Jürgen Mittelstrass 1978. "Zum Normativen Fundament der Wissenschaft," *Syntese,* 37/3: 471-477.

Kant, Immanuel. https://en.wikipedia.org/wiki/Immanuel_Kant#Theory_of_perception.

Karp, Robert 2007. "Toward an Associative Economy in the Sustainable Food and Farming Movement," *New Spirit Ventures.* https://www.biodynamics.com/content/toward-associative-economy-sustainable-food-and-farming-movement-robert-karp.

Kemmitt, S., J. Wright, D. Keith, K.W.T. Goulding & D.L. Jones 2006. "pH Regulation of Carbon and Nitrogen Dynamics in Two Agricultural Soils, *Soil Biology & Biochemistry* 38: 898-911.

King County 2005. "Building a Sustainable Community Food System in Seattle and King County: Concept for Developing a Local Food Policy Council."https://archive.northwestpublichealth.org/web-specials/resources-obesity.

Kingsolver, Barbara 2003. "A Good Farmer," *The Nation* (November 3).

Kirschenmann, Fred 1997. "On Becoming Lovers of Soil," in: Madden, J. Patrick (editor) 1997. *For All Generations: Making World Agriculture More Sustainable.* Glendale, WA.

Kirschenmann, Frederick L. 2010. *Cultivating an Ecological Conscience: Essays from a Farmer Philosopher*, Lexington, KY.

König, Karl 1982. *Earth & Man*. N. p.

Kreutzer, K. 1995. "Effects of Forest Liming on Soil Processes," *Plant Soil* 168-169: 447-470.

Kühne, Petra 2003. "Milk as Part of Our Food," *Biodynamics* 245: 35-37.

McKeon, Richard (trans.) 1941. *The Basic Works of Aristotle*, New York, NY.

Lair, Cynthia 1996. *Feeding the Whole Family*. Seattle.WA.

Lamb, Gary 2010. *Associative Economics: Spiritual Activity for the Common Good*. Ghent, NY.

Lang, Adrian 2020. "What is Vervain?" https://www.healthline.com/nutrition/vervain-verbena.

LaPado-Breglia, Christine G.K. 2011. "Smitten with Bees," *Chico News & Review Archives*. https://www.newsreview.com/chico/content/smitten-with-bees/2978078/.

Lewis, Mark & Fred Sharpe 1987. *Birding in the San Juan Islands*. Seattle, WA.

Life Garden Program. https://www.lopezislandschool.org/cms/one.aspx?pageId=500997.

Logan, William Bryant 2007. *Dirt: The Ecstatic Skin of the Earth*. London, UK.

Logsdon, Gene 1995. *The Contrary Farmer*, White River Junction, VT.

Loon, Dirk 1976. *The Family Cow*. Montgomery, IL.

Lovejoy, Ann 2001. "Compost Tea," *Seattle Post Intelligencer* (March 4).

Lovel, Hugh 2000. *A Biodynamic Farm: For Growing Wholesome Food*. Austin, TX.

Luther, Martin, 1517. *Disputatio pro declaratione virtutis indulgentiarum* (Disputation for the Carification of Indulgences: Ninety-Seven Theses: Disputation Against Scholastic Theology.) Wittenberg, Germany.

Madden, J. Patrick (editor) 1997. *For All Generations: Making World Agriculture More Sustainable*. Glendale, WA.

Mäder, P., A. Fleissbach, D. Dubois, L. Gunst, P. Fried & U. Niggli 2002. "Soil Fertility and Biodiversity in Organic Farming," *Science* 296: 1694-1697.

Malhi, S.S., M. Nyborg & J.T. Harapiak 1998. "Effects of Long-Term N Fertilizer-Induced Acidification and Liming on Micro-Nutrients in Soil and Bromegrass Hay," *Soil Tillage Research* 48: 91-101.

Malhi, S.S., J.T. Harapiak, K.S. Gill & N. Flore 2002. "Long-Term N Rates and Subsequent Lime Application Effects on Macro-Elements Concentration in Soil and Bromegrass Hay," *Journal of Sustainable Agriculture* 21: 79-97.

Marschner, B. & A.W. Wilczynski 1991. "The Effect of Liming on Quantity and Chemical-Composition of Soil Organic-Matter in a Pine Forest in Berlin, Germany," *Plant Soil* 137: 229-236.

McBride, A.C. and T.O. West, "Estimating Net CO2 Emissions From Agricultural Lime Applied to Soils in the U.S," *American Geophysical Union*, Fall Meeting 2005, abstract #B41B-0191.

McKibben, Bill 2012. "Global Warming's Terrifying New Math," *Rolling Stone*. https://www.rollingstone.com/politics/politics-news/global-warmings-terrifying-new-math-188550/.

Metropulos, Megan. 2017. "The Benefits and Risks of A2 Milk." https://www.medicalnewstoday.com/articles/318577.

Merrill, A.L. & Watt, B.K. 1973. "Energy Value of Foods: Basis and Derivation." *Agriculture Handbook* No. 74, ARS, USDA, Washington DC.

Merrill, Richard and John McKeon 2002. "Compost Tea: A Brave New World," *Organic Farming and Research Foundation Information Bulletin* No. 9 (Winter).

Miles, Carol A. 2003 "Growing the Dry Bean Market," *Agrochemical and Environmental News*, August, Issue No. 208.

Miles, Carol A. et al. 2005. "Dry Bean Variety Trial Comparison WSU Vancouver REU & Moses Lake". Sustainable Seed Systems. http://SustainableSeedSystems.wsu.edu/nicheMarket/02DBVarietyReport.pdf.

Min, B.R. & S.P. Hart 2003. "Tannins for Suppression of Internal Parasites," *Journal of Animal Science* 81: E102-E109.

More, Thomas 2006. *Utopia*. San Diego, CA.

Monke, Jim & Renée Johnson 2010. "Actual Farm Bill Spending & Cost Estimates," *Congressional Research Service Report R41195*, Washington, DC.

N.a. 2009. "Permaculture Reflections." www.permaculturereflections.com/2009/02/species-of-month-comfrey.html.

N.a. 2011."Lotus Eaters help define Meaning of Humanity."

N.a. 2016. "Lopez Island Farm Trust (LIFT)." *Lopez Community Land Trust*. https://www.lopezclt.org/lopez-island-farm-trust-lift/.

N.a. 2018 "Biochar Is a Valuable Soil Amendment," *International Biochar Initiative*. https://biochar-international.org/biochar/.

N.a. 2019. "Nutrient Management for Recycled Orchards," *U. of California Agriculture and Resources*. https://orchardrecycling.ucdavis.edu/nitrogen-management.

N.a. 2020. "*Føderåd*." The Norwegian Tax Administration. https://www.skatteetaten.no/en/rettskilder/type/handboker/skatte-abc/2020/føderad/.

N.a. 2021. "Wonder Bread's 100th Anniversary." https://www.wonderbread.com/about-us.

N.a. 2021. "*Altenteil*." https://de.wikipedia.org/wiki/Altenteil#Rechtsgrundlagen.

N.a. 2021 "Yarrow." https://www.rxlist.com/yarrow/supplements.htm.

N.a. 2021. "Typical Broiler Body Weights & Feed Regiments." https://www.pinterest.com/pin/484277766154093950/.

N.a. 2021. "Oak Bark." https://www.rxlist.com/oak_bark/supplements.htm.

N.a. 2021. "The Benefit of Bees," *The New Agriculturist*. http://www.new-ag.info/00-5/focuson/focuson8.html.

N. a. 2021. "Precision Agriculture." https://en.wikipedia.org/wiki/Precision_agriculture.

N.A. 2021. "Lacticaseibacillus rhamnosus." https://en.wikipedia.org/wiki/Lacticaseibacillus_rhamnosus.

N.a. 2021. 'American Gothic." https://en.wikipedia.org/wiki/American_Gothic.

N.a. 2021. "*Føderåd* or *kår*," *Norway-Heritage*. http://www.norwayheritage.com/snitz/topic.asp?TOPIC_ID=5109.

N.a. 2021. "*Auszugshaus*." https://de.wikipedia.org/wiki/Auszugshaus.

N.a. 2021. "Γεροντομοίρι." https://www.slang.gr/definition/28189-gerontomoiri.

N.a. 2022. "Commons." https://en.wikipedia.org/wiki/Commons.

Nagel, Ernest 1961. *The Structure of Science*. New York, NY.

National Research Council Board of Agriculture and Natural Resources and Earth and Life Studies (BANRELS) 2008. *Changes in the Sheep Industry in the United States: Making the Transition From Tradition*, Washington, DC.

National Research Council of the National Academy of Science 2007. *Status of Pollinators in North America*, Washington, D.C.

National Research Council of the National Academies 2010. *Toward Sustainable Agricultural Systems in the 21st Century*, Washington, D.C., 21.

Newton, Isaac 1704. *Opticks: Or, A Treatise of the Reflections, Refractions, Inflexions and Colours of Light.* London, UK

Niezen, J.H., W.A.G. Charleston, H.A. Robertson, D. Shelton, G.C. Waghorn & R. Green 2002. "The Effect of Feeding Sulla (*Hedysarum Coronarium*) or Lucerne (m*edicago sativa*) on Lamb Parasite Burdens and Development of Immunity to Gastrointestinal Nematodes," *Veterinary Parasitology* 105: 229-245.

Noraini, Jaafar, 2014. "Biochar as a Soil Amendment and Habitat for Microorganisms." https://research-repository.uwa.edu.au/en/publications/biochar-as-a-soil-amendment-and-habitat-for-microorganisms.

Odum, Howard T. 1988. "Self-Organization, Transformity, and Information," *Science* 242: 1132-1139.

Odum, Howard T. 1994. *Ecological and General Systems: An Introduction to Systems Ecology.* Boulder, CO.

Odum, Howard T. 1996. *Environmental Accounting: EMERGY and Environmental Decision Making.* New York, NY.

Oehl, F., E. Frossard, A. Fliessbach, D. Dubois & A. Oberson 2004. "Basal Organic Phosphorus Mineralization in Soils Under Different Farming Systems," *Soil Biology & Biochemistry* 36: 667-675.

Opalco Newsroom 2012. "More Local Power: S&S Homestead Farm on Lopez Island" *Energy Efficiency & Conservation, Membership Programs.* (https://www.opalco.com/4873/2012/09/).

Organic Consumer Organization 2005. "Biopirates Lose Patent on Seeds of India's Sacred Tree, the Neem."

Ovadje P., et al. 2016."Dandelion Root Extract Affects Colorectal Cancer Proliferation and Survival Through the Activation of Multiple Death Signaling Pathways," *Oncotarget.* 7(45):73080-100.

Paine, Thomas 2007. *Common Sense for the Modern Era.* (Begler, Elsie and R. F. King, eds.) San Diego, CA.

Palmer, Brian 2015. "Would a World Without Bees Be aWorld Without Us?" (https://www.nrdc.org/onearth/would-world-without-bees-be-world-without-us).

Patočka, Jiří, et al. 2010. "Biomedically Relevant Chemical Constituents of *valeriana officinalis*," *Journal of Applied Biodmedicine, 8: 11-18.*

Pennisi, Elizabeth 2018. "Quaillike Creatures Were the Only Birds to Survive the Dinosaur-Killing Asteroid Impact". *Science.* doi:10.1126/science.aau2802.

Perfecto, Ivette, John Vandermeer, Angus Wright, 2009. *Nature's Matrix: Linking Agriculture, Conservation and Food Sovereignty.* London, UK.

Perlman, Deanna 2014 "Wool Processing at S&S Homestead: Grow, Create, Teach, Learn." https://sshomestead.org/research2/.

Pesch, Otto-Hermann 2007. "Gnade und Rechtfertigung am Vorabend der Reformation und bei Luther,"(Grace and Justification on the Eve of the Reformation and in Luther), *Ökumenisches Forum*, Heidelberg, Germany.

Pettersson, B.D., H.J. Reents & E. von Wistinghausen, 1992. "Düngung und Bodeneigenschaften: Ergebnisse eines 32-jährigen Feldversuches in Järna, Schweden"(Fertilization and Soil Characteristics: Results of a 32-Year Field Trial in Järna, Sweden)," *Institut für Biologisch-Dynamische Forschung* (Institute for Biodynamic Research), Darmstadt, Germany.

Penfold, C.M., M.S. Miyan, T.G. Reeves & T. Grierson, 1995. "Biological Farming for Sustainable Agricultural Production," *Australian Journal of Experimental Agriculture* 35: 849-856.

Pfeiffer, Ehrenfried 1958. "Preface." *Agriculture Course: The Birth of the Biodynamic Method.* Dornach, Switzerland, 5-16.

Piccaglia, Roberta et al. 1993. "Antibacterial and Antioxidant Properties of Mediterranean Aromatic Plants," *Industrial Crops and Products,* 2 (1): 47-50.

Parnes, Robert 1990. *Fertile Soil,* Davis, CA.

Parnes, Robert 2013. "Soil Fertility: A Guide to Organic and Inorganic Soil Amendments." https://soilandhealth.org/book/soil-fertility-a-guide-to-organic-and-inorganic-soil-amendments/.

Penfold, op.cit; Burkitt, L.L., D.R. Small, J.W. McDondal, J.W., Wales & M.L. Jenkin, M.L. 2007. "Comparing Irrigated Biodynamic and Conventionally Managed Dairy Farms: 1. Soil and Pasture Properties, 2. Milk Production and Composition and Animal Health," *Australian Journal of Experimental Agriculture*, 47: 479-494.

Pollan, Michael 2006. *The Omnivore's Dilemma: A Natural History of Four Meals*. New York.

Pollan, Michael 2013. Cooked: A Natural History of Transformation. New York.

Pope Francis, 2015. *Laudato Si' mi' Signore* (Praise Be to You, My Lord). Rome, Italy.

Poppelbaum, Hermann 1992. "Horns & Antlers," *Star & Furrow* 78: 11-17.

Provenza, F.D., J.J. Villalba, L.E. Dziba, S.B. Atwood & R.E. Banner 2003. "Linking Herbivore Experience, Varied Diets, and Plant Biochemical Diversity," *Small Ruminant Research* 49: 257-274.

Raupp, J. 2001. "Manure Fertilization for Soil Organic Matter Maintenance and its Effects Upon Crops and the Environment, Evaluated in a Long-Term Trial," *Sustainable Management of Soil Organic Matter CAB International* (Rees, R.M., B.C. Ball, D.C. Campball & C.A. Watson, eds.),Wallingford, UK, 301-308.

Reed, William 2001. "FDA Bans Sale of Comfrey Herb." https://www.foodnavigator.com/Article/2001/07/09/FDA-bans-sale-of-comfrey-herb.

Reeve, J. R., L. Carpenter-Boggs, J.P. Reganold, A.L. York, G. McGourty & L.P. McCloskey 2005. "Soil and Wine Grape Quality in Biodynamically and Organically Managed Vineyards," *American Journal of Enology & Viticulture* 56: 367-376.

Reeve, Jennifer R., Lynne Carpenter-Boggs, & Henning Sehmsdorf 2011. "Sustainable Agriculture: A Case Study of a Small Lopez Island Farm," *Agricultural Systems* 104(7): 572-579.

Reese, Frank et al. 2021. "Definition of a Heritage Turkey." https://livestockconservancy.org/heritage-turkey-definition/.

Reganold, John 1989. "Farming's Organic Future," *New Scientist* (June 10).

Reich, Lee 2016. "The Jury is Still Out on Compost Tea: Compost Tea is Currently Hot in the Gardening World, But Will it Also Move Beyond Fad Status?" *Fine Gardening* (Issue 107).

Ritter, Georg 1971. "Horns & Antlers," *Star & Furrow* 47: 26-29.

Robertson, G. P. [...] & Diana H. Wall 2008. "Long-term Agricultural Research: A Research, Education, and Extension Imperative," *BioScience* 58 (7): 640–645.

Rockström, Johan, Will Steffen & [...] Jonathan Foley 2009. "A Safe Operating Space for Humanity," *Nature* 461: 472-475.

Roehl, Evelyn 1996. *Whole Food Facts: The Complete Reference Guide*. Fairfield, CT.

Riotte, Louise 1998. *Roses Love Garlic: Companion Planting and Other Secrets of Flowers*. Pownal, VT.

Robinson, Jo 2000. *Why Grassfed Is Best!: The Surprising Benefits of Grass-fed Meats, Eggs, and Dairy Products*. Vashon Island, WA.

Rohr, Richard 2019. *The Universal Christ: How a Forgotten Reality Can Change Everything We See, Hope For, and Believe*. New York, NY.

Rangel-Castro, J.I., J.I. Prosser, C.M. Scrimgeour, P. Smith, N. Ostle, P. Ineson, A. Meharg & K. Killham, 2004. "Carbon Flow in an Upland Grassland: Effect of Liming on the Flux of Recently Photosynthesized Carbon to Rhizosphere Soil, *Global Change Biology* 10: 2100-2108.

Rural Roots 2003. *Northwest Direct Farmer Case Studies*. Moscow, Idaho.

Russel, James, et al. 2018. Pasture Management Guide for Livestock Producers. Ames, Iowa.

Rydberg, Torbjörn & Jan Jansén 2002. "Comparison of Horse and Tractor Traction Using Emergy Analysis," *Ecological Engineering*, 19(1): 13-28.

Sachs, Jeffrey [...] & Pedro Sanchez 2010. "Monitoring the World's Agriculture,"*Nature*, 466: 558–560.

Saeidnia, S., et al 2011. "A Review of Phytochemistry and Medicinal Properties of the Genus *Achillea*," *DARU Journal of Pharmaceutical Science*, 19 (3): 173-186.

Salamon, Sonya 1992. *Prairie Patrimony: Family, Farming, and Community in the Midwest*. Chapel Hill, N.C.

Sandars, Nancy K. 1985. *Prehistoric Art in Europe*. New York.

Sather, David 2018-2019. "LIFE on Lopez," *Washington Principal*, 28-31. https://cdn5-ss19.sharpschool.com/UserFiles/Servers/Server_176833/File/Unique/Garden/life_on_lopezfall18.pdf.

Schad, Wolfgang 1977. *Man & Mammals: Toward a Biology of Form*. Garden City, NY.

Schmid, Ron 2003. *The Untold Story of Milk*. Brandywine, MD.

Schmidt, Lone 1993. *Farven og lyset : studier i Goethes farvelære* (Light and Colors: Studies in Goethe's Theory of Color). Ålborg, Denmark.

Schweitzer, Albert 1960. *The Philosophy of Civilization*. New York, NY.

Scilipoti, Jan 2001. "Strawbale Construction on S&S Homestead Farm," *The Last Straw: The International Journal of Straw Bale and Natural Building*, No. 34 (Summer), 33-34.

Sehmsdorf, Henning 2002-3. "Kids Can Make a Difference." See Simpson.

Sehmsdorf, Henning 2004. "Class on a Lopez Island Farm" *Western Sustainable Agriculture Research & Education*. https://sshomestead.org/wp-content/uploads/FarmSchool_txt.pdf.

Sehmsdorf, Henning 2004. "Solar-Powered Micro-Irrigation on S&S Center forSustainable Agriculture and Homestead Farm." https://sshomestead.org/wp-content/uploads/Solar%20Micro-Irrigation.pdf.

Sehmsdorf, Henning 2008. "On-Farm Research: Biodynamic Forage Production," *Biodynamics*, no.265.

Sehmsdorf, Henning 2011. "Sustainable Agriculture: A Case Study of a Small Lopez Island Farm." See Reeve et al.

Sehmsdorf, Henning 2013. "The Economics of the Small-Scale, Self-Sufficient Farm," *Stella Natura*. Kimbertown, PA.

Sehmsdorf, Henning 2016. "Economic Valuation of S&S Homestead Farm." (https://docs.google.com/spreadsheets/d/1Na6dxEs-3AfhDFISvr7Jnoh5eRsGdoy5Qi5r0TfglC0/pubhtml.)

Sehmsdorf, Henning 2016. "S&S Homestead Farm Production." https://docs.google.com/spreadsheets/d/1Na6dxEs-3AfhDFISvr7Jnoh5eRsGdoy5Qi5r0TfglC0/pubhtml.

Sehmsdorf, Henning 2016. "Future Farm Project," https://sshomestead.org/future-plans/.

Sehmsdorf, Henning 2020. *Myth & Tradition in Norwegian Literature & Folklife*. Lopez Island, WA.

Sehmsdorf, Henning 2021. "Retiring on the Commons," *Biodynamics* , no. 301.

Sehmsdorf, Henning 2021. "Small-Scale, Self-Sufficient Farming for Health,"*Lilipoh* (Summer): 36-57.

Sehmsdorf, Henning. 2021. *Eating Locally...* See Simpson.

Sehmsdorf, Henning 2023. *Trauma & Blessings: Biography of a Prussian Immigrant*. Lopez Island (forthcoming).

Shaffer, Bob 2010. "Curing Compost: An Antidote for Thermal Processing." *Acres USA*, Vol. 40, No. 11.

Sheldrake, Rupert et. al. 2001. *Chaos, Creativity and Cosmic Consciousness*. Rochester, VT.

Shenker, Israel 1969. "E. B. White: Notes and Comment by Author," *New York Times* (July 11).

Singh, R. et al."Phytochemical Composition and Functional Properties of Dandelion," *Acta Horticulturea, 1287: 24.*

Shiva, Vandana 2000. *Stolen Harvest: The Hijacking of the Global Food Supply.* Cambridge, MA

Simpson, Elizabeth and Henning Sehmsdorf 2002-3. "Nutrition – Body and Soul," (keynote at the First National Conference on Farm-to-Cafeteria), Seattle. Published in *Community Food Security News* (Spring 2003) https://foodsecurity.org/CFSCSpring2003, and *Kids Can Make a Difference* (Spring 2003, vol. 8, no. 2.) https://kidscanmakeadifference.org/.

Simpson, Elizabeth & Henning Sehmsdorf 2021. *Eating Locally and Seasonally: A Community Food Book for Lopez Island (and All Those Who Want to Eat Well)*, Lopez Island, WA.

Smeck, N.E. 1985. "Phosphorus Dynamics in Soils and Landscapes," *Geoderma* 36 (3-4): 185-199.

Smith, Casey 2017. "Cats Domesticated Themselves, Ancient DNA Shows," *National Geographic.* https://www.nationalgeographic.com/science/article/domesticated-cats-dna-genetics-pets-science.

Snežana Jarić et.al. 2015 "Review of Ethnobotanical, Phytochemical, and Pharmacological Study of Thymus serpyllum L," *Evidence-Based Complementary and Alternative Medicine*, volume 2015, article ID 101978. https://doi.org/10.1155/2015/101978.

Soule, Judy. & Jon Piper 1992. *Farming in Nature's Image — An Ecological Approach to Agriculture.* Washington, D.C.

Spinoza, Baruch, 1677. *Ethica: ordine geometrico demonstrata* (Ethics, Demonstrated in Geometrical Order). Amsterdam, The Netherlands.

Steiner, Rudolf 1886. *Grundlinien einer Erkenntnistheorie der Goetheschen Weltanschauung mit besonderer Rücksicht auf Schiller* (Outline of an Epistemology of Goethe's World View With Particular Consideration of Schiller). Berlin & Stuttgart, Germany.

Steiner, Rudolf 1891. *Wahrheit und Wissenschaft : Vorspiel einer Philosophie der Freiheit* (Truth and Science: Prologomena to a Philosophy of Freedom), Rostock, Germany.

Steiner, Rudolf 1923. *What Is Anthroposophy? Three Lectures.* Dornach, Switzerland.

Steiner, Rudolf 1923-4. *Das Wesen der Bienen* (On the Nature of Bees). Dornach, Switzerland.

Steiner, Rudolf 1928. "Story of My Life," *Rudolf Steiner Archive (Steiner Online Library). https://rsarchive.org/Books/GA028/English/APC1928/GA028_index.html.*

Steiner, Rudolf 1924. *Geisteswissenschaftliche Grundlagen zum Gedeihen der Landwirtschaft* (Spiritual-Scientific Foundations. for a Thriving Agriculture). Dornach, Switzerland.

Steiner, Rudolf 1958. *Agriculture Course: The Birth of the Biodynamic Method.* Dornach, Switzerland.

Steiner, Rudolf 1993. Spiritual Foundations for the Renewal of Agriculture. Junction City, OR.

Steiner, Rudolf 1993. *Economics: The World as One Economy.* Chartham, UK.

Steiner, Rudolf 2000. *Nature's Open Secrets : Introductions to Goethe's Scientific Writings*, London, UK.

Steiner, Rudolf 2009. *Nutrition: Food, Health and Spiritual Development.* New York, NY.

Stoknes, Per Espen, 2015. *What We Are Thinking When We Are Not Thinking About Climate Change.* White River Junction, VT.

Storl, Wolf D. 2013. *Culture & Horticulture.* Berkeley, CA.

Suchantke, Andreas 1995. "The Leaf: The True Proteus," *The Metamorphosis of Plants: Essays by Jochen Bockemühl and Andreas Suchantke*, Cape Town, South Africa, 7-19.

Tabatabai, M.A. 1994. "Soil Enzymes," *Methods of Soil Analysis, Part 2: Microbiological and Biochemical Properties* (Weaver et al., eds.). Madison, WI, 775-833.

Tautz, Jürgen 2007. *Phänomen Honigbiene*. Heidelberg, Germany. Sandmann, David C., transl. 2008. *The Buzz About Bees: Biology of a Superorganism*, Berlin, Germany.

Taylor, Kate 2017. "These 10 Companies Control Everything You Buy." https://www.independent.co.uk/life-style/companies-control-everything-you-buy-kelloggs-nestle-unilever-a7666731.html.

The Humane Farming Association, www.hfa.org.

Thoreau, Henry David 1845. *Walden*. Boston, Mass.

Thoreau, Henry. D. 1849. *A Week on the Concord and the Merrimack Rivers*. Boston, Mass.

Thun, Maria 1999. *Gardening for Life: The Biodynamic Way*. Stroud, UK.

Tillich, Paul 1953-63. *Systematic Theology*. Chicago, Ill.

Tisdale, S.L., W.R. Nelson, J.D. Beaton & J.L. Havlin 1993. *Soil Fertility and Fertilizers*. 5th ed. Upper Saddle River, NJ.

Trolove, S.N., M.J. Hedley, J.P. Caradus & A.D. Mackay 1996. "Uptake of Phosphorus from Different Sources by *Logus pedunculatus* and Three Genotypes of *Trifolium Repens*: Forms of Phosphate Utilized and Acidification of the Rhizosphere," *Australian Journal of Soil Research* 34: 1027-1040.

Ulgiati, Sergio & Mark T. Brown 1998. "Monitoring Patterns of Sustainability in Natural and Man-Made Ecosystems," *Ecological Modeling*, 108: 23-36.

Undersander, Dan J., D.R. Mertens, & N. Thiex. 1993. *Forage Analyses Procedures. National Forage Testing Association*. Omaha, NE.

U.N. Millennium Ecosystem Assessment 2005. https://www.millenniumassessment.org/en/Index-2.html.

U.S. Department of Agriculture and U.S. Department of Health and Human Services 2010. *Dietary Guidelines for Americans*, Washington, DC.

Vaillant, John 2005. *The Golden Spruce: A True Story of Myth, Madness, and Greed*. New York, NY & London, UK.

Van der Ploeg, Jan Douwe 2008. *The New Peasantries: Struggles for Autonomy and Sustainability in an Era of Empire and Globalization*, London & Sterling, VA.

Wackernagel, Mathis. & William Rees 1996. *Our Ecological Footprint: Reducing Human Impact on the Earth*. Philadelphia, PA.

Waghorn, G.C. 1990. "Beneficial Effects of Low Concentrations of Condensed Tannins in Forages Fed to Ruminants," *Microbial and Plant Opportunities to Improve Lignocellulose Utilization by Ruminants* (Akin, D.E., L.G. Ljungdahl, J.R. Wilson & Harris, eds.), New York, NY.

Wallander, H., K. Amebrant, F. Ostrand & O. Karen 1997. "Uptake of N-15-Labelled Alanine, Ammonium and Nitrate in Pinus sylvestris L. Ectomycorrhiza Growing in Forest Soil Treated With Nitrogen, Sulphur or Lime, *Plant and Soil* 195: 329-338.

Washington State Department of Agriculture 2009. "Skirting-the-Law-With-Cow-Share-Agreements." https://www.foodsafetynews.com/2009/11/skirting-the-law-with-cow-share-agreements/.

Wesner, Erik J. 2015. "Old Age in Amish America." https://amishamerica.com/old-age/.

Wicksteed, Philip H. & Francis Cornford, 1934. *Aristotle: The Physics*. Cambridge, Mass.

Wilhelmi, Christy 2013. "Biodynamic Beekeeping with Michael Thiele," *Gardenerd. https://gardenerd.com/blog/biodynamic-beekeeping-with-michael-thiele/*.

Wines, Michael 2015 "A Sharp Spike in Honeybee Deaths Deepens a Worrisome Trend, "*New York Times* (May 13).

Wistinghausen, Christian von, et al. 2000. *The Biodynamic Spray and Compost Preparations Production Methods, Book I*. Stroud, UK.

Wistinghausen, Christian von, et al. 2000. *The Biodynamic Spray & Compost Preparations Production Methods, Booklet 1;* and *The Biodynamic Spray and Compost Preparations: Directions for Use, Booklet 2.* East Troy, WI.

Wittgenstein, Ludwig 1950. *Bemerkungen über die Farben* (Remarks on Colour), Oxford. https://www.openculture.com/2013/09/goethes-theory-of-colors-and-kandinsky.html.

Wolff, Christian von 1728. *Philosophia rationalis, sive logica* (Rational Philosophy, or Logic), Frankfurt, Germany.

Wong, Cathy 2020. "Health Benefits of Dandelion Root." https://www.verywellhealth.com/the-benefits-of-dandelion-root-89103.

Wordsworth, William 1802. "The Rainbow."

Wu, Qingli. et al. 2018. "Phytochemical Analysis and Anti-Inflammatory Activity of Nepeta cataria Accessions." *Journal of Medicinally Active Plants* 7, (1):19-27.

Zabel, Steph 2017. "Raspberry: Beyond the Fruit." https://www.cambridgenaturals.com/blog/raspberry-beyond-the-fruit#.

Zinniker Farm 2021. https://zinnikerfarm.com/biodynamic-preps.

Index

(The Index references central concepts of biodynamic thinking and practice across the individual essays. Page numbers indicate the same item appearing elsewhere in the volume. References marked "see" point to conceptual contexts in which the item should be considered.)

oikologia (ecology) 3, 131, 140n, 154, 181, 204f, 247, 258n, 283n, 287n, 292, 300n, 302n, 304n, 308n
teleology 162, 183, 217, 235, 237n, 258n, 288, 297n
telos (immanent purpose) 155, 159,161, 204, 289, 291, 293
"the whole is greater than the sum of its parts" 23, 155, 161, 179

ascorbic acid (vitamin C)121, see **Macro-/ Micro-Nutrients/Phytochemical(s)**
astral, see **Steiner**

Astronomical
- calendar 12f, 294, see **Rhythm(s)**
- charts 294
- rhythm 294, **see Rhythm(s)**
- system 247, see **System(s)**
- thinking 289

Aurochs (primordial oxen) 126
Auszugshaus (German: old folks' residence), 218-222, 221n, 307n, see also *Daudihaus*
Aupumla (Old Norse: primordial cow)127

B

bacteria, see **Bread/Compost/Dairy/Food/ Soil**

Balance(d)
acid-alkaline balance 36
bacterial balance 149f
- (in) bees 161, see **Bees**
- (in) "calculation & experience" 76, see CSA
- cash flow 274f, see **Economics**
- (in) compost 95, see **Compost**
- (in) CSA 72, see CSA
cyclical balance 21f, see **Farm/System(s)**
- economy 213, see **Economics**
- ecosystem 204-206, see **System(s)**
elemental balance 289, see **Element(-al)**
- ecological enterprise(s) 138
- environment 208, 131
- farm 22, 72f, 86, 131, 138, 230, 257, 286, see **Economics**
- fertilizer 135, see **Fertility**
- finance 215, 274f, see **Economics**
- food system 111, see **Food/Healing**
homeostasis 160, see **System(s)**
- industrial system 111, see **Agriculture**
- life 86, 230, see **Steiner**

- meal 116, see **Food/Healing/Nutrition**
natural balance 131, 230
- nutrient budget 100-103, 275, see **Fertility**
- plant food 101, see **Fertility**
- production 13, see **Farm**
- rotation 103, see **Soil**
self-ordering balance 230, see self-correcting/-healing/-organizing /**Sustainability**
- soil, 32, 36, 68, 135, 200, 207, 260, 275, see **Soil**
- stasis 208-209
sustainable balance 208, 286, see **Sustainability**
- tillage 78, see **Tillage**
- (of) "mutual dependence" 126, see livestock
- whole 86, 95, see **Whole(-ness)**

Barnett, Tanya 141-144, 144n
basal respiration, see **Soil**
bats 241, see **Pests/Beneficials**

BD Preparations
BD 500 (horn manure)/501 (horn silica)/502 (yarrow), 503 (chamomile)/504 (nettle)/ 505 (oak bark), 506 (dandelion)/ 507 (valerian), 508 (horsetail) 262, see **Herbs**
- application(s) 25, 27, 58, 65, 67f, 69n, 76f, 96, 97n, 99, 101, 105, 135f, 140n, 198, 201, 207, 227, 260-262, 264f, 267-269, 272, 276, 282n, 294, 299n, 301, 304, 313
barrel compost 77
- compost 88f, 101, 201, 202n, see **Compost**
- economic value 214
- function in the farm organism 55
- infuse vital, life-forces 25, 58, 207, 198
- observational validation 294
- plants 59, 65-68, see **Herbs**
- production 12f, 59, 65-69, 69n, 77, 88f, 135f, 140n, 299n, 312f
- provide farm fertility/nutrient cycling 77, 135f
- research 96, 97n, 99-101, 105n, 202, 259-266, 267-287, 294, 301, 304
- soil, crop, compost amendments 12
- soil management 13, 77
- support nutrient transfer 27
- sustain whole farm organism 55f, 77
Steiner on BD preparations 227, 292, 294, see **Steiner**

Bee(s) (*apis mellifera*) 125, 153-162
- benefits 139n, 140n, 306n, 307n, 308n
biodynamic bee-keeping 130f, 140n, 312n
- biology 140n
bumble bee 130
- colonies 155

E

habitable earth 218
"healing the earth" 121, 208, 227, 259, 289, see **Healing/Steiner**
- health 24
- life force 25, see **Energy**
life one earth 154
- ministry 2, 144n, 300
"Sister Earth" 293, see Pope Francis
- mystery 142
- plaster 166-169
preparations buried in earth 66, see **BD Preparations**
sacred earth 9, 13, see sacred
- sciences/studies 283n, 302, 307
Earthway seed drill 82
- species 87
spherical shape of earth 115
- stewards 13, 206, see stewardship
-systems 247, see **System(s)**
tub grinding of earth (soil) 26
- wide animal extinction 129, 131, 154
- worms 27, 67, 100, 105n, 135, 206f, 241, 260, 300, see **Pests/Beneficials**
- Spirit (*Erdgeist*) 288f, see **Goethe**

Ecological(-ly)
- approach 258n, 311
- balance 205, see **Balance**
- based system(s) 11f, see **Balance**
- beekeeping, 131, see **Bees**
- benefits 64, 156, 217n, see **Bees**
- community 295
- components 247, see **Element(-al)**
- conscience 299n, 306, see Kirschenmann
- context 259, 274
- cost(s) 246, 265
- crises & disasters 5, 13, 131, 205, see **Climate**
- decision(s) 20
destructive ecology 260, see **Climate/Steiner**
- economics 134, see **Economics**
- engineering 258n, 310
- enterprise 137, see **Economics**
- entity 254, see **Organism**
- farm(ing) 138, 292 see **Farm(s)**
- food production 12, 260
- Food Production Class 1, 4f, 32, 41, 179, 210
-footprint 255, 258n, 312
- forage production 265, see **Forage**
- health 103, 126, 154, 299n, see **Healing**
- livestock raising, 13, 105n, 125-140, see livestock
- modeling 258n, 312
non-ecological breeding 139, see livestock

non-ecological process 259
- opportunities 138
- principles 216, 247
- resources 179, see **Resources**
- responsibility 12, 138, 212
- role 160
- savings 259
- small-farm financing 200, see **Economics**
- stewardship 1, 205, 208, 213, see **Aristotle/Economics/Steiner**
- survival 212, see **Climate/Steiner**
- sustainability 14, 24 , 31, 246f, see **Sustainability**
- system(s) 246f, 258n, 285, 292, 308, see **System(s)**
- theology 2, see world view
- vs. industrial livestock raising 125-140

Economic(s) 187-220
agricultural economics, 4, 28, 31f, 200, 205, 209-212, 287n, 302n, see **Agricultural**
Aristotelian economics, 205, 289, 287n, see **Aristotle**
associative economics 11, 13, 15n, 30, 196, 211f, 217n, 305f, see Lamb/**Steiner**
biodynamic economics 2, 21, 191, 203
community economics 43
competitive (market-based)economics 211
conventional (transactional) economics 210
- (of) culture and health 33, 140n, see **Healing**
dairy economics 150f, see **Dairy**
ecological economics 2
- (of) financial health 204
- (of) happiness 210, see **Aristotle**
holistic economics 26f
- (of) money 29f, 42, 72, 81, 86, 132, 134, 171, 188, 191, 196, 198-201, 203, 205, 209f, 212, 247, 280, 289, 297n, see **Aristotle**
- (of) non-profits 198, 211, 217n, 250, 286
opportunity cost 210
- (of) pollination, 160, see **Bees**
- (of) price 27f, 49, 52, 72f, 74f, 85f, 134, 138, 150, 190, 197, 201, 209, 211-213, 268, see Lamb/**Steiner**
- (of) profit 12, 27-31, 51, 85, 103, 107, 122, 131, 151, 154, 161, 179, 188, 190f, 195-199, 205, 208, 210f, 216, 217n, 220, 250, 267, 273f, 279f, 303
- of self-sufficiency, 23-34, 190, 199, see self-sufficiency
- of small-scale farming, 197f, 202f, 310, see **Farm**
- of soil biology 227
- of stewardship 205, 208
"pastoral economics" 210, see Logsdon

322

humus, see **Soil**

Huxley College for the Environment 1, 5, 162

hybrid(-ization) 107, 121, 128, 279, see **Seeds**

hydrogen, see **Macro-/Micro-Nutrients**

(de)hydrogenation (ated) 36, 261f, 263, 270f, 277f

hydroponics 65

I

idealism 10, 121, 212

Ikerd, John 210, 217n, 305

imagination 25, 161, 232f, 236, 291, 293, 295, see **Goethe**

immigration reform 286

imitation (image) of nature 133, 220, 298n, 304, see **Aristotle/Nature**

immunity(-ies) 13, 23, 30, 50f, 131, 161, 188, 197, 226, 259, 283, 308

incentives 175, 207, 284f, 286

incremental 284, 286

incubate 270

indicator 96, 99f, 248, 250, 254, 262, 278, 282n, 286, 304

(pre) industrial(-ization) 2, 19f, 25f, 29, 37, 47-54, 52, 55, 69n, 103, 106-108, 111, 120f, 131, 137f, 140n, 142, 154, 161, 178f, 192, 205, 208, 210, 219, 247, 260f, 287, 288, 295, 308

industry and consumer associations 212, see **Economics/Steiner**

indwelling, see **Aristotle**

infrastructure 2, 24, 29, 85, 103, 149-151, 192, 195, 200, 203, 209-216, 213, 220, 254, 279

Ingham, Elaine 87, 93, 97n, 305

inorganic 97, 102, 272, 274, 292, 308

insecticides 52

insects 41, 58, 64, 100, 130, 154f, 188, 226, 229, see **Pests/Beneficials**

inspiration, 201, 244, see **Goethe**

internalizing nature 211, see imitation of nature

intern(ship) 1-5, 21, 28, 31f, 69n, 84, 87f, 97, 104n, 134, 140n, 149, 151, 165, 171, 174, 180, 191, 194-200, 208, 210f, 216, 228, 233, 251, 258n, 274, 279f, 292, 295, 305, see apprentice(ship)

intestines 66f, 126, 136, 145f, 295, see **BD Preparations**

intuition (intuitively), 10, 30, 58, 161, 243f, 247, 291, 293, 295, see **Goethe**

iron see minerals/**Macro-/Micro-Nutrients**

irrigation, see **Water**

J

Jackson, Wes 97n, 258n, 305

Jaspers, Karl 290, 298n, 305

Jay, John 32, 34, 97n, 305

Jeavons, John 69n, 104n, 252, 258, 305

Jefferson, Thomas 32, 34, 129, 220, 305

Jenkins, Joseph 88, 97n, 104n, 105n, 305

Judeo-Christian 26, 206, 290

Jung, Carl G. 14, 15n, 305

jungle fowl (*gallus gallus*), 129, see livestock

K

Kant, Immanuel 10, 15n, 305

Keats, Brian 294

kår (Norwegian: means of livelihood), 221n, 307, 218-220, see also *føderåd*

Kids Can Make a Difference 202n, 311

kiln 207, 261

Kingsolver, Barbara 26, 34n, 305

Kirschenmann, Fred 220, 296, 299n, 305f

König, Karl (Uli) 11, 140, 152, 306

L

labor (-intensive) 28f, 31, 49, 74, 78, 81f, 84-86, 103, 121, 134, 137f, 151, 166, 168, 191f, 198f, 210f, 213-216, 227, 246, 248, 251-254, 257, 259f, 265, see **Economics**

laboratory 131, 243, 249, 276, 288, 290f, 294

ladybugs 111, **see Pests/Beneficials**

Lamb, Gary 217n, 306

landscape 12, 178, 283n, 284, 286, 287n, 302, 311

land trust 1f, 5, 32, 43, 54n, 86, 177, 190, 198, 201, 212, 217n, 220, 221n, 307

Larsen, Rick 180

"Laws of Life" 106

leaf (leaves) 25, 36, 42, 47, 58, 60-64, 66, 68, 83, 101f, 103, 105n, 115, 131, 133, 145f, 156, 158, 206, 229f, 232-236, 236n, 237n, 269, 294, 301, 304, 311, see **Goethe**

leek, see **Herb(s)**

legumes 23, 64, 98, 103f, 133, 137, 206, 260, 269, 275, 277, 294, see **Forage**

livestock, 125-162, see **Bees/Cat/Cattle/Chicken/Dog/Sheep/Turkeys**

life-cycle analysis 267-283, see **Farm**

light 25, 50, 67, 100, 125, 147, 156f, 240-243, 245n, 293, 295, 308, 310, see **Goethe**

Linnaeus (Linné, Carl) 291f

M

335

Tillich, Paul 203, 217n, 312
Title I 116

Tool(s)
accounting 255
compost/cover crops/BD preparations 76
creativity 31
energy analysis 257
farmer/apprentices/interns 203-217, 230
hand/machine tools 79, 82, 84, 103, 116, 126, 166, 168, 170, 211, 213-216
labor 31
mathematical abstraction 292
money/profit 31, 196, 209
platform (nanotechnology) 291
rototiller 12, 78f, 207, 260
soil testing 274

Toxic (-in)
- chemicals/phytochemicals 57, 62, 135, 167, 208
- plants 56, 64f, 67
Washington Toxics Coalition 2

trace minerals, see **Fertility/Macro-/Micro-Nutrients**
Property Rights 107, 219f
transactional compensation 211, see **Economics**

Transform(-ative/-ation)
transformity (energy) 246-258, 308, see **Emergy**
- force 14, 24, 204, 288, 293f, 295
- knowledge 227
material transformation 98, 141, 147, 151, 156, 207, 232f, 241, 271
morphological transformation 225-237, 296, see **Goethe**
social transformation 43, 121, 122n, 123n, 309, see **Steiner**
spiritual transformation 121, 121n, 123, 289, 295, see **Spirit/Steiner**
systemic transformation (agriculture) 2, 5, 284-287, 287n, 302, see **Steiner**

turkeys (*meleagris gallopavo*), 129f, see livestock

U

"ultimate concern" 203
University of Washington (UW) 1, 181, 195, 209
UN Millenium Ecosystem Assessment 217n, 312

UN Panel on Climate Change 13
urine, see **Fertility**
USAID (US Agency for International Development) 108
US Bureau of Labor Statistics (BLS) 191
US Department of Agriculture (USDA) 12f, 24, 27, 42, 51, 64, 105n, 108, 138, 162n, 173, 191f, 194, 198, 202n, 210, 215, 217n, 249, 303, 306
US Department of Labor (DOL) 28, 191f
Utah Agricultural Experiment Station 276
utopia 218f, 221n, 307
ultraviolet 157, 241

V

Vaillant, John 296, 299n, 312
valerian, see **Herbs/BD Preparations**
van der Ploeg, Douwe J. 210f, 217n, 312
vegetable(s), 47-123
vermifuge 58
vetch, 77, 135, 227, see **Crops**
viable 23f, 29, 130, 134, 187-189, 204, 259, 265, 284
viability 2, 13, 28-30, 33, 43, 113n, 151, 196f, 213, 215, 271, 284, see **Economics/Farm/Sustainability**
vital principle 293, see also **Steiner**
vitality 12, 24, 65, 122, see also **Steiner**
vitalism 291
vitamin 26f, 35-40, 47, 55-70, 57f, 116, 125, 130, 135, 146-152, 191, 293, see **Macro-/Micro-Nutrients**
volatile oil, see **Phytochemicals**

W

Waldorf 5, 11, 244
Washington Association of Churches (WAC) 191
Washington Principal 202n, 310
Washington State University (WSU) 1, 34n, 96, 199, 233, 259, 270, 276, 282n, 294
waste(s) 13, 21f, 24, 27, 43, 51, 57, 88, 90f, 95, 98, 100f, 103, 134, 137, 143, 175, 180, 201, 228, 267, 273, 280, 295

Water 80, 174f, 273
- catchment 2, 21, 24, 32, 41, 80, 136, 170, 174f, 178, 203, 206, 214-216, 273
- (in) clay construction 168
cloudy water (*Trübes Mittel*) 242
- (in) compost 88, 96, 100, see **Compost**
- (from) cosmos, 147, see **Dairy**
- conservation 12, 80

X

Y

Z

www.ingramcontent.com/pod-product-compliance
Lightning Source LLC
Chambersburg PA
CBHW052340210326
41597CB00037B/6205